CISM International Centre for Mechanical Sciences

Courses and Lectures

Volume 594

For more than 40 years the book series edited by CISM, "International Centre for Mechanical Sciences: Courses and Lectures", has presented groundbreaking developments in mechanics and computational engineering methods. It covers such fields as solid and fluid mechanics, mechanics of materials, micro- and nanomechanics, biomechanics, and mechatronics. The papers are written by international authorities in the field. The books are at graduate level but may include some introductory material.

More information about this series at http://www.springer.com/series/76

Matthew S. Allen · Daniel Rixen ·
Maarten van der Seijs · Paolo Tiso ·
Thomas Abrahamsson · Randall L. Mayes

Substructuring in Engineering Dynamics

Emerging Numerical and Experimental Techniques

Springer

Matthew S. Allen
Department of Engineering Physics
University of Wisconsin–Madison
Madison, WI, USA

Daniel Rixen
Lehrstuhl für Angewandte Mechanik
Technische Universität München
Munich, Bayern, Germany

Maarten van der Seijs
Vibes Technology
Delft, The Netherlands

Paolo Tiso
D-MAVT
ETH Zürich
Zürich, Switzerland

Thomas Abrahamsson
Chalmers University of Technology
Göteborg, Sweden

Randall L. Mayes
Sandia National Laboratories
Albuquerque, NM, USA

ISSN 0254-1971 ISSN 2309-3706 (electronic)
CISM International Centre for Mechanical Sciences
ISBN 978-3-030-25531-2 ISBN 978-3-030-25532-9 (eBook)
https://doi.org/10.1007/978-3-030-25532-9

This Springer imprint is published by the registered company Springer Nature Switzerland AG
The registered company address is: Gewerbestrasse 11, 6330 Cham, Switzerland

Preface

One fundamental paradigm in engineering is to break a structure into simpler components in order to simplify test and analysis. In the numerical world, this concept is the basis for finite element discretization and is also used in model reduction through substructuring. In experimental dynamics, substructuring approaches such as Transfer Path Analysis (TPA) are commonly used, although the subtleties involved are perhaps not always adequately appreciated. Recently, there has been renewed interest in using measurements alone to create dynamic models for certain components and then assembling them with numerical models to predict the behavior of an assembly. Substructured models are also highly versatile; when only one component is modified, it can be readily assembled with the unchanged parts to predict the global dynamical behavior. Substructuring concepts are critical to engineering practice in many disciplines, and they hold the potential to solve pressing problems in testing and modeling structures where nonlinearities cannot be neglected.

This book originates from lecture notes created for a short course in Udine, Italy in July 2018. We will review a general framework, which can be used to describe a multitude of methods and the fundamental concepts underlying substructuring. The course was aimed at explaining the main concepts as well as specific techniques needed to successfully apply substructuring both numerically (i.e., using finite element models) and experimentally. The course centered around the following topics, which range from classical substructuring methods to topics of current research such as substructuring for nonlinear systems:

1. Introduction and motivation.
2. Primal and dual assembly of structures.
3. Model reduction and substructuring for linear systems including Guyan and Hurty/Craig–Bampton reduction, McNeal, Rubin, Craig–Chang interface reduction methods, and model reduction in the state space.
4. Experimental–analytical substructuring including frequency-based substructuring or impedance coupling, substructure decoupling methods including the transmission simulator method, measurement methods for substructuring, the

virtual point transformation, state-space substructuring, and an overview of finite element model updating (a common alternative to experimental substructuring).
5. Industrial applications and related concepts including Transfer Path Analysis and finite element model updating.
6. Model reduction and substructuring methods for nonlinear systems are highlighted with a focus on geometrically nonlinear structures and nonlinearities due to bolted interfaces.

This text was designed to provide practicing engineers or researchers such as Ph.D. students with a firm grasp of the fundamentals as well as a thorough review of current research in emerging areas. The reader is expected to have a solid foundation in structural dynamics and some exposure to finite element analysis. The material will be of interest to those who primarily perform finite element simulations of dynamic structures, to those who primarily focus on the modal test, and to those who work at the interface between test and analysis.

Udine, Italy
July 2018

Matthew S. Allen
Daniel Rixen
Maarten van der Seijs
Paolo Tiso
Thomas Abrahamsson
Randall L. Mayes

Contents

Chapter 1
Introduction and Motivation

Abstract "Divide and Conquer" is a paradigm that helped Julius Cesar to dominate on the wide Roman Empire. The power of dividing systems, then analyze them as parts before combining them in an assembly, is also an approach often followed in science and engineering. In this introductory chapter, we shortly discuss the main idea behind domain decomposition and substructuring applied to mechanical systems.—*Chapter Author: Daniel Rixen*

1.1 Divide and Conquer Approaches in the History of Engineering Mechanics

The work of engineers consists of understanding systems in order to build new solutions or optimize the existing ones. To analyze systems, it is very common and efficient to decompose them into subcomponents. This reduces the complexity of the overall problem by considering first its parts and can provide very useful insight for optimizing or troubleshooting intricate structures. In this introductory chapter, the general idea of substructuring is discussed and an overview of the topics treated in these lecture notes is given.

One often refers to Schwarz (1890), who proved the existence and uniqueness of the solution of a Poisson problem on a domain that was a combination of a square and a circle. By considering the problem as two separate problems on the simple overlapping square and circular domains for which existence and uniqueness of the solution were already known, and devising an iteration strategy between the two domains with guaranteed convergence, he could conclude his very important proof. The algorithm that he devised as a means for his proof is still used today, underlying some very popular solution strategies on multiprocessor computers (see for instance Toselli and Widlund 2006).

But in essence, approximation techniques such as the Rayleigh–Ritz method (Rayleigh 1896; Ritz 1909; Géradin and Rixen 2015), where the solution space is approximated by a reduced number of admissible functions, can also be considered as divide and conquer methods: the solution space is decomposed into functional subspaces in which an approximate solution can be found. The Finite Element technique,

M. S. Allen et al., *Substructuring in Engineering Dynamics*,
CISM International Centre for Mechanical Sciences 594,
https://doi.org/10.1007/978-3-030-25532-9_1

pioneered by mathematicians like Courant (1943), Courant and Hilbert (1953), can in fact be seen as a clever application of the Rayleigh–Ritz technique on small region of the computational domain, resulting therefore in not only a function decomposition of the problem within the element, but also a spatial decomposition of the problem as the physical domain is broken into elements.

Today when we consider substructuring, most often we envision dividing a finite element model into subcomponents that each contain many elements. This substructuring approach was developed in the 60s, motivated by the need to solve structural problems in aerospace and aeronautics of several tens of nodes (considered to be very large problems at that time, given the fact that computers were just arising). Hence, the idea was to add another level of decomposition and reduction by subdividing the mesh of a finite element model into substructures and to represent the dynamics per substructure in a reduced and approximate way. Some of the substructuring techniques still commonly used today originate from that time where memory and CPU power were very limited.

Later, in the 80s and especially in the 90s, solving very large problems using domain decomposition became a very active research topic. In order to efficiently use the computational power of new multiprocessor machines, it was essential to minimize the communications between processing units and to ensure that each CPU was fed with enough work that could be done independently. Domain Decomposition is the paradigm to achieve this, by letting each CPU solve the local problems of each domain and searching for the global solution by iterating on the interface solutions (Toselli and Widlund 2006). Nowadays, high-performance computing techniques solve engineering problems of close to a billion degrees of freedom on several hundreds of thousands of processors thanks to that paradigm (Toivanen et al. 2018).

The last wave of intensive research in substructuring came at the beginning of the twenty-first century, this time not directly powered by exponential grows of computing power, but by significant progress in techniques to accurately measure the dynamics of components. Experimental substructuring aims to build models of assemblies from the data measured on components, possibly in combination with numerical models. Although the theory of experimental dynamic substructuring is essentially the same as its analytical counterpart, many novel techniques needed to be developed in order to alleviate the adverse effects of measurement errors (even if strongly reduced thanks to novel sensing and acquisition techniques) when building a model based on experimentally characterized components.

1.2 Advantages of Substructuring in Mechanical Engineering

In mechanical engineering practice, substructuring techniques are useful for the following cases:

- When working on a large project (for example, the design of an aircraft), different groups and departments work on different components. In that case, it is essential that each team can concentrate on the modeling of their part, and then have an efficient means for assembling to other components in order to predict the overall dynamics and optimize their design accordingly.

- In industry, it is common that models of subsystems must be shared. A classical example is the coupled load analysis of the assembly of a satellite and a launcher: in that case, the satellite company needs to share its model with the launcher operator in order to perform a global dynamic analysis and guarantee that vibration limits will not be crossed. In that case, exchanging information can be efficient if the different components have been modeled separately in such a way that their complexity (number of degrees of freedom) is low and such that they can be assembled on their interface.
- In many engineering problems, parts of the system to be analyzed are significantly more complex than the rest. In structural dynamics, for instance, parts of the model could be highly nonlinear whereas the remainder can be considered as linear. In that case, it is very advantageous to substructure the problem in such a way that only the parts that are nonlinear are treated with the appropriate techniques, whereas the linear parts are, for instance, condensed once and for all. Such a subdivision of the problem into substructures of different complexities can greatly improve the analysis speed and give better insight into the dynamic behavior.
- When optimizing the design during the development of a new product, or during an improvement of an old design, one often wants to modify only a small number of parts (for instance, change the bushings connecting the power train of a car to the chassis in order to modify the noise and vibration harshness). When the parts that will not be modified are available as substructures and already precomputed (e.g., their dynamics on the interface is known), then recomputing the global behavior when only the components under optimization are changed requires a significantly reduced cost, enabling much faster design cycles.
- In experimental testing, it is not always possible to test the full system; often system-level analysis is needed before all of the components have been manufactured. In other cases, it is impossible to test the full assembly because appropriately exciting a large structure would require forces that are beyond the capabilities of existing shakers. It is then very advantageous to experimentally characterize the structure part by part. This allows troubleshooting problems arising from the local dynamics in those parts, but also to build a full model by assembling the measured parts using experimental substructuring techniques.
- When parts of the system have not yet been built, it is possible to combine the measured dynamics of hardware components with other parts that are modeled only numerically (hybrid substructuring). In some cases, the combination in real time of the transient response of a hardware component (the physical substructure) with the dynamics of a numerical substructure is essential to predict the dynamics of the hardware part in realistic conditions. Such hardware-in-the-loop tests (also called Real-Time Substructuring, Hybrid Testing or Cyber-Physics) allow testing very complex systems by having only a part in the lab, the rest of the system being co-simulated in real time.

- Substructuring methods open new opportunities for efficient analyses and design, including the development of digital twins, or a model that can be used to monitor the health of a system.

These are only a few examples of cases in which substructuring can be advantageous. For these and other reasons, substructuring has attracted a lot of attention over the years and is yet today a very active research field.

References

Courant, R. (1943). Variational methods for the solution of problems of equilibrium and vibrations. *Bulletin of the American Mathematical Society*, *49*(1), 23.

Courant, R., & Hilbert, D. (1953). *Methods of Mathematical Physics* (Vol. 1). New York: Interscience Publishers.

Géradin, M., & Rixen, D. (2015). Mechanical vibrations. *Theory and Application to Structural Dynamics* (3rd ed.). Chichester: Wiley.

Rayleigh, J. (1896). *The Theory of Sound* (2nd ed.). Mineola: Dover.

Ritz, W. (1909). Uber eine neue methode zur lösung gewisser variations probleme der mathematishen physik. *Journal für die Reine und Angewandte Mathematik*, *135*, 1–61.

Schwarz, H. A. (1890). *Gesammelte Mathematische Abhandlungen* (Vol. 2, pp. 133–143). Berlin: Springer; First published in Vierteljahrsschrift der Naturforschenden Gesellschaft in Zürich (1870) (Vol. 15, pp. 272–286).

Toivanen, J., Avery, P., & Farhat, C. (2018). A multilevel feti-dp method and its performance for problems with billions of degrees of freedom. *International Journal for Numerical Methods in Engineering*, *116*(10–11), 661–682.

Toselli, A., & Widlund, O. (2006). *Domain decomposition methods-algorithms and theory* (Vol. 34). Berlin: Springer Science & Business Media.

Chapter 2
Preliminaries: Primal and Dual Assembly of Dynamic Models

Abstract There are several ways to formulate the dynamics of a substructure. The different domains in which the dynamics can be described will be reviewed since the manner in which substructures are characterized will later determine the substructuring methodology that can be applied. In addition to how the substructures are formulated, the way in which the coupling/decoupling problem is expressed will allow us in the subsequent chapters to develop different numerical and experimental techniques. Two conditions must be satisfied on the interface between substructures: a condition on the displacement field (compatibility) and on the interface stresses (force equilibrium). Those conditions can be accounted for following several different formulations, all mathematically equivalent, but each leading to different numerical methods, experimental approaches, and approximation techniques, as will be explained in the following chapters. In this chapter, we outline the basic concepts of the so-called three-field formulation, dual and primal assembly.—*Chapter Author: Daniel Rixen*

2.1 The Dynamics of a Substructure: Domains of Representation

In these lecture notes, we assume that the problem has already[1] been discretized (using, for instance, an appropriate Finite Element or Boundary Element formulation). The discretized problem describing the dynamics in a substructure $\Omega^{(s)}$ can be written in the form

$$\mathbf{M}^{(s)}\ddot{\mathbf{u}}^{(s)} + \mathbf{f}_{nl}^{(s)}\left(\dot{\mathbf{u}}^{(s)}, \mathbf{u}^{(s)}\right) = \mathbf{f}^{(s)} + \mathbf{g}^{(s)}. \tag{2.1}$$

The superscript (s) indicates that the equation is written for a given substructure s. $\mathbf{M}$ denotes the discretized mass matrix,[2] $\mathbf{f}_{nl}$ the discretized nonlinear internal force

[1] A similar classification can be found in a different form in de Klerk et al. (2008).

[2] Here we assume that the mass matrix is constant. This is not the case for models in multibody dynamics for instance Géradin and Cardona (2001), but this will not be considered here.

M. S. Allen et al., *Substructuring in Engineering Dynamics*,
CISM International Centre for Mechanical Sciences 594,
https://doi.org/10.1007/978-3-030-25532-9_2

function, which depends in general on the nodal velocities and displacements $\dot{\mathbf{u}}$ and $\mathbf{u}$. Two types of forces are applied on the substructures: the externally applied forces, denoted $\mathbf{f}$, and the forces due to interactions between substructures, denoted $\mathbf{g}$. The latter are, in fact, internal forces when considering the entire structure, but are considered as applied forces when analyzing the substructure problem.

In case of small displacements, the internal forces $\mathbf{f}_{nl}$ can be linearized and the substructure dynamics are described by

$$\mathbf{M}^{(s)}\ddot{\mathbf{u}}^{(s)} + \mathbf{C}^{(s)}\dot{\mathbf{u}}^{(s)} + \mathbf{K}^{(s)}\mathbf{u}^{(s)} = \mathbf{f}^{(s)} + \mathbf{g}^{(s)}, \tag{2.2}$$

where $\mathbf{C}^{(s)} = \partial \mathbf{f}_{nl}^{(s)}/\partial \dot{\mathbf{u}}^{(s)}$ and $\mathbf{K}^{(s)} = \partial \mathbf{f}_{nl}^{(s)}/\partial \mathbf{u}^{(s)}$ are the linearized (tangential) damping and stiffness matrices typically computed around equilibrium positions of the system (Géradin and Rixen 2015).

The unknowns $\mathbf{u}$ in Eqs. (2.1), (2.2) express the behavior of a substructure in terms of its physical displacements (discretized at nodes) in the time domain. Therefore, we will refer to them as the equations in the *physical and time domain*. The first qualification refers to the spatial meaning of the unknowns whereas the second refers to the spectral description of the problem. Another spatial and spectral description of the dynamic problem will now be introduced since the way in which the problem is described has an impact on what method can be applied for reduction or experimental assembly, as will be shown throughout these lecture notes. A summary of the different domains in which the substructure dynamics can be considered is given in Fig. 2.1. The substructure data in the different domains can be obtained either from numerical models, from experimentally measured data or from a mix of both. The different aspects are explained next.

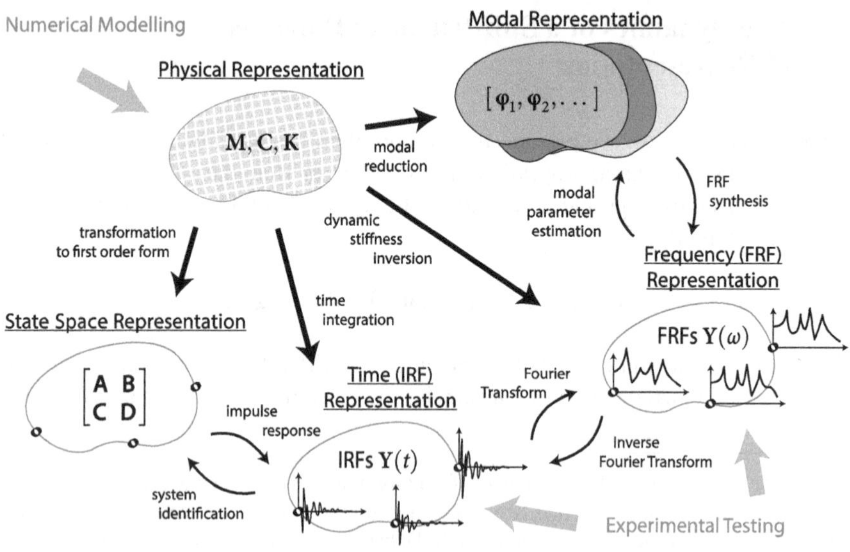

Fig. 2.1 Schematic overview of substructuring domains (van der Seijs 2016)

2.1.1 *Spatial Descriptions*

Physical Domain (Continuous and Discrete)

The description outlined in (2.2) expresses the dynamics of the substructure using the displacement at a specific nodal location and is referred to as a *physical representation*. Note that representation in the physical domain can be either discretized (as assumed in this text) or continuous. In the later case, the unknown is the physical response field and the associated equations are partial differential equations describing the (nonlinear) continuous dynamics in a body. This will not be considered here.

Modal Domain (Reduced and Unreduced)

It is often handy and useful to consider the unknowns of a dynamic problem as a combination of vectors of a (sub)space. The most well-known subspace representation is probably the one obtained by mode superposition. The free vibration modes of a substructure defined by the eigenvalue problem (Géradin and Rixen 2015)

$$\left(\mathbf{K}^{(s)} - \omega_i^{(s)2}\mathbf{M}^{(s)}\right)\boldsymbol{\phi}_i^{(s)} = \mathbf{0}, \tag{2.3}$$

where $\omega_i^{(s)}$ are the eigenfrequencies of the free–free substructure and $\boldsymbol{\phi}_i^{(s)}$ the associated eigenmodes that have the fundamental property of being mass and stiffness orthogonal, namely,

$$\boldsymbol{\phi}_i^{(s)T}\mathbf{K}^{(s)}\boldsymbol{\phi}_j^{(s)} = \omega_i^{(s)2}\delta_{ij}, \tag{2.4}$$

$$\boldsymbol{\phi}_i^{(s)T}\mathbf{M}^{(s)}\boldsymbol{\phi}_j^{(s)} = \delta_{ij}, \tag{2.5}$$

where we have assumed that the modes amplitudes have been chosen to be mass-normalized and where δ_{ij} is the Kronecker symbol such that

$$\delta_{ij} = 1 \quad \text{if} \quad i = j.$$
$$\delta_{ij} = 0 \quad \text{if} \quad i \neq j.$$

A modal representation of the substructure is then obtained by the change of variable

$$\mathbf{u}^{(s)} = \sum_{i=1}^{n^{(s)}} \boldsymbol{\phi}_i^{(s)}\eta_i^{(s)} = \boldsymbol{\Phi}^{(s)}, \boldsymbol{\eta}^{(s)} \tag{2.6}$$

where $n^{(s)}$ is the number of degrees of freedom in substructure (s). Here, $\eta_i^{(s)}$ are the amplitudes of the modal component of the response and are often called modal coordinate of substructure (s). Often, we will use a matrix notation as in the second equality of (2.6), where $\boldsymbol{\Phi}^{(s)}$ is a matrix containing in its columns the vibration modes and $\eta^{(s)}$ is a uni-column matrix containing all modal coordinates. Usually, only a subset of modes is considered in order to have an approximated but reduced representation of the substructure. This will be discussed in later chapters.

In general, the response $\mathbf{u}^{(s)}$ can be represented as a combination of $n^{(s)}$ independent vectors and we write

$$\mathbf{u}^{(s)} = \sum_{i=1}^{n^{(s)}} \mathbf{v}_i^{(s)} q_i^{(s)} = \mathbf{V}^{(s)} \mathbf{q}^{(s)}, \tag{2.7}$$

where $\mathbf{V}^{(s)}$ is a square matrix containing the basis vectors for the desired change of variables. Substituting in the dynamic equation (2.2) and premultiplying by ${\mathbf{V}^{(s)}}^T$ to project the equations onto the same space, we obtain

$${\mathbf{V}^{(s)}}^T \mathbf{M}^{(s)} \mathbf{V}^{(s)} \ddot{\mathbf{q}}^{(s)} + {\mathbf{V}^{(s)}}^T \mathbf{C}^{(s)} \mathbf{V}^{(s)} \dot{\mathbf{q}}^{(s)} + {\mathbf{V}^{(s)}}^T \mathbf{K}^{(s)} \mathbf{V}^{(s)} \mathbf{q}^{(s)} = {\mathbf{V}^{(s)}}^T \mathbf{f}^{(s)} + {\mathbf{V}^{(s)}}^T \mathbf{g}^{(s)}, \tag{2.8}$$

which is usually written as

$$\tilde{\mathbf{M}}^{(s)} \ddot{\mathbf{q}}^{(s)} + \tilde{\mathbf{C}}^{(s)} \dot{\mathbf{q}}^{(s)} + \tilde{\mathbf{K}}^{(s)} \mathbf{q}^{(s)} = \tilde{\mathbf{f}}^{(s)} + \tilde{\mathbf{g}}^{(s)}, \tag{2.9}$$

where the tilde superscript indicates that the matrices and vectors pertain now to a representation in a transformed space. The representation vectors stored in $\mathbf{V}^{(s)}$ can be any set of independent vectors, in particular, they can be chosen as the vibration modes $\boldsymbol{\Phi}^{(s)}$, in which case the transformed mass and stiffness matrices $\tilde{\mathbf{M}}^{(s)}$ and $\tilde{\mathbf{K}}^{(s)}$ will be diagonal.

The representation (2.9) will often be referred to as the *modal representation*, even when the base vectors are not vibration modes $\boldsymbol{\Phi}^{(s)}$ but general representation modes $\mathbf{V}^{(s)}$. The associated degrees of freedom $\dot{\mathbf{q}}^{(s)}$ are then called generalized degrees of freedom or modal coordinates and do in general not represent the solution at a particular physical location.

In case an incomplete basis is used for the representation, namely, when fewer modes than the number $n^{(s)}$ of degrees of freedom in the substructure are used, the modal representation represents the dynamics in a reduced subspace and in general only in an approximate way (this will be discussed in detail in Chap. 3). We will then call the representation a *reduced modal representation*.

2.1.2 Spectral Representation

Time Domain (Continuous and Discrete)

In the form of Eq. (2.2), the unknown dynamic response is considered a function of time and we say that, from a spectral point of view, the equations are expressed in the time domain.

The dynamic equation (2.2) considers the spatial unknowns to be a continuous function of time and the dynamics are expressed as ordinary differential equations in time (the space having already be discretized in that equation). As was done for the spatial domain, time can also be discretized using methods related to finite differences (typically variants of the Newmark time integration scheme in structural dynamics (Newmark 1959, Chung and Hulbert 1993, Brüls and Arnold 2008, Géradin and Rixen 2015), but other approaches can also be applied such as Finite Elements in time or variational-based approaches (Lew et al. 2004)). When discretized in time, we will still consider the resulting equations as being in the time domain since the unknowns are the responses at discrete time instances. The equations are then algebraic equations that are typically solved in a recursive form (time stepping), given the fact that the time problem is typically an initial value problem.[3]

Frequency Domain (Reduced and Unreduced)

Similar to the decomposition of the spatial response in component modes, the time dependency of the response can also be decomposed into a combination of time contributions. The most classical one is the Fourier decomposition[4] that writes the time function of the response in terms of harmonic functions. Using complex number notations, the Fourier decomposition can be written as

$$\mathbf{u}^{(s)}(t) = \int_{-\infty}^{\infty} \bar{\mathbf{u}}^{(s)}(\omega)\ \mathrm{e}^{-i\omega t} d\omega, \tag{2.10}$$

where i is to be understood as the imaginary unit number. This decomposition is very suitable for linear systems since replacing in (2.2) and using the orthogonality properties of harmonic functions, the harmonic component $\mathbf{u}(\omega)$ can be computed individually from the harmonic dynamic equation:

$$\left(-\omega^2\mathbf{M}^{(s)} + i\omega\mathbf{C}^{(s)} + \mathbf{K}^{(s)}\right)\bar{\mathbf{u}}^{(s)} = \bar{\mathbf{f}}^{(s)} + \bar{\mathbf{g}}^{(s)} \qquad \omega \in]-\infty, +\infty[, \tag{2.11}$$

where $\bar{\mathbf{f}}^{(s)}$ and $\bar{\mathbf{g}}^{(s)}$ are the Fourier components of the forces, for instance,

$$\bar{\mathbf{f}}^{(s)} = \frac{1}{2\pi}\int_{-\infty}^{\infty} \mathbf{f}^{(s)}(t)\ \mathrm{e}^{i\omega t} dt. \tag{2.12}$$

The dynamic equation in the frequency domain (2.14) is also often written as

$$\mathbf{Z}^{(s)}\bar{\mathbf{u}}^{(s)} = \bar{\mathbf{f}}^{(s)} + \bar{\mathbf{g}}^{(s)} \tag{2.13}$$

---

[3]For an interesting matrix description of time discretization see Rixen and van der Valk (2013), van der Valk and Rixen (2014).

[4]Note that other base functions in time can be used (such as wavelets), but this will not be discussed here.

where

$$\mathbf{Z}^{(s)}(\omega) = -\omega^2\mathbf{M}^{(s)} + i\omega\mathbf{C}^{(s)} + \mathbf{K}^{(s)} \tag{2.14}$$

$\mathbf{Z}^{(s)}$ is a *dynamic stiffness matrix* and is a function of the frequency ω. In this form, $\bar{\mathbf{u}}^{(s)}$ is the complex amplitude of the harmonic displacement response (or equivalently the Fourier component of the transient response). A similar equation can be written for the amplitude of the harmonic velocity or accelerations in which case the operator $\mathbf{Z}(\omega)$ is commonly called *the mechanical impedance or the apparent mass*, respectively. The dynamic relation can also be inverted and written

$$\bar{\mathbf{u}}^{(s)} = \mathbf{Y}^{(s)}\left(\bar{\mathbf{f}}^{(s)} + \bar{\mathbf{g}}^{(s)}\right), \tag{2.15}$$

where

$$\mathbf{Y}^{(s)}(\omega) = \mathbf{Z}^{(s)}(\omega)^{-1} = \left(-\omega^2\mathbf{M}^{(s)} + i\omega\mathbf{C}^{(s)} + \mathbf{K}^{(s)}\right)^{-1} \tag{2.16}$$

$\mathbf{Y}^{(s)}$ is a Frequency Response Function (FRF) matrix and is often called the *admittance or dynamic flexibility*, or more specifically *receptance, mobility, or accelerance/inertance* if $\bar{\mathbf{u}}^{(s)}$ are displacements, velocities or accelerations respectively.

Obviously, in practice, the harmonic components are calculated only for a finite discrete number of frequencies ω, and (2.10) is approximated by the Discrete Fourier Decomposition

$$\mathbf{u}^{(s)}(t) = \sum_{k=-N_\omega}^{N_\omega} \bar{\mathbf{u}}_k^{(s)}\ \mathrm{e}^{i\omega_k} \tag{2.17}$$

choosing a frequency range covering the spectral range of the excitation. It is noteworthy that the decomposition in (2.17) is comparable to the decomposition of the space function of the response in (2.6) and can also be seen as a reduction of the transient response in the time domain. It is an approximation unless the excitation can be exactly represented by a finite combination of harmonics.

Laplace Domain

Another often used representation of the time evolution of the dynamic response is in terms of the Laplace components. The idea is to look for the dynamic response when modulated with a decreasing exponential function, namely,

$$\bar{\mathbf{u}}^{(s)}(s) = \mathcal{L}\left(\mathbf{u}^{(s)}(t)\right) = \int_0^{\infty} \mathrm{e}^{-st}\mathbf{u}^{(s)}(t)\, dt \tag{2.18}$$

This transformation changes the differential equation in time into an algebraic equation in the Laplace variable s thanks to the fact that Laplace transforms of time derivatives of $\bar{\mathbf{u}}^{(s)}(t)$ can be written in terms of $\bar{\mathbf{u}}^{(s)}(s)$ using integration by parts. For instance,

$$\mathcal{L}\left(\ddot{\mathbf{u}}^{(s)}(t)\right) = s^2\bar{\mathbf{u}}^{(s)}(s) - s\mathbf{u}^{(s)}(t=0) - \dot{\mathbf{u}}^{(s)}(t=0)$$

Clearly, there is a similarity between Laplace and Fourier transforms since (2.18) becomes a Fourier transform if s is taken as imaginary. The main difference is that the inverse transform is trivial for the Fourier domain (leading to the frequency domain decomposition (2.10) or (2.17)), whereas finding the inverse Laplace transform is far more difficult and not general. In structural dynamics, Laplace transforms are used for highly transient problems that cannot efficiently be represented by harmonic superposition, such as impact responses and shock propagations (see for instance Sect. 4.3.2 in Géradin and Rixen (2015)).

2.1.3 *State Representation*

Displacement Space

In addition to representing the space and spectral behavior of the system in different domains as explained above, the very state of the system can be described in mainly two different manners: either one sees the displacements as the only independent unknowns (velocities and accelerations being dependent on the displacement through derivatives) or the velocities are seen as additional independent variables for which the derivative relation to the displacement is explicitly expressed in the formulation. For the first approach, often used in structural dynamics, the dynamics of the system are described by a single set of second-order equations as in (2.1) (for nonlinear structures) or (2.2) (for linear ones).

State-Space Representation

In the second case, namely, the velocities are seen as additional independent variables, the state of the system is described both by the displacements and the velocities:

$$\mathbf{x}^{(s)}(t) = \begin{bmatrix} \mathbf{u}^{(s)}(t) \\ \dot{\mathbf{u}}^{(s)}(t) \end{bmatrix} \tag{2.19}$$

and the associated linear dynamic equation can for instance be written as

$$\begin{bmatrix} \mathbf{I}^{(s)} & \mathbf{0} \\ \mathbf{0} & \mathbf{M}^{(s)} \end{bmatrix} \dot{\mathbf{x}}^{(s)} = \begin{bmatrix} \mathbf{0} & \mathbf{I}^{(s)} \\ -\mathbf{K}^{(s)} & -\mathbf{C}^{(s)} \end{bmatrix} \mathbf{x}^{(s)}(t) + \begin{bmatrix} \mathbf{0} \\ \mathbf{f}^{(s)} + \mathbf{g}^{(s)}. \end{bmatrix} \tag{2.20}$$

In this *state-space representation* the number of equations has doubled, but the order of the differential is now reduced to one. This representation is often used especially in control. This form is also commonly used in structural dynamics when strong damping is present since the concept of modes of vibration properly generalizes only when writing the system in the State Space (see, for instance, Sect. 3.3 in Géradin and Rixen (2015)).

2.1.4 Summary of Representation Domains

From the short summary of the formulation of the structural dynamics problem, it is clear that many variants to describe the problem, combining a spatial, spectral, and state representation, can be constructed. In Fig. 2.1, the different aspects are shown graphically and relations between them are summarized. Depending on the representation chosen, numerical and experimental techniques in substructuring can significantly differ as will be seen in these lecture notes.

2.2 Interface Conditions for Coupled Substructures

Let us consider again the linearized dynamic equilibrium equation (2.11) of a substructure in the physical space and in the frequency domain[5]

$$\mathbf{Z}^{(s)}\bar{\mathbf{u}}^{(s)} = \bar{\mathbf{f}}^{(s)} + \bar{\mathbf{g}}^{(s)} \qquad s = 1 \dots N^{sub}, \tag{2.21}$$

where N^{sub} is the total number of substructures in the system.

It is common to write the equilibrium of all substructures in a block matrix form as

$$\mathbf{Z}\bar{\mathbf{u}} = \bar{\mathbf{f}} + \bar{\mathbf{g}} \tag{2.22}$$

with the definitions

$$\mathbf{Z} = \begin{bmatrix} \mathbf{Z}^{(1)} & & \mathbf{0} \\ & \ddots & \\ \mathbf{0} & & \mathbf{Z}^{(N^{sub})} \end{bmatrix} \tag{2.23}$$

$$\bar{\mathbf{u}} = \begin{bmatrix} \bar{\mathbf{u}}^{(1)} \\ \vdots \\ \bar{\mathbf{u}}^{(N^{sub})} \end{bmatrix} \quad \bar{\mathbf{f}} = \begin{bmatrix} \bar{\mathbf{f}}^{(1)} \\ \vdots \\ \bar{\mathbf{f}}^{(N^{sub})} \end{bmatrix} \quad \bar{\mathbf{g}} = \begin{bmatrix} \bar{\mathbf{g}}^{(1)} \\ \vdots \\ \bar{\mathbf{g}}^{(N^{sub})} \end{bmatrix}.$$

The dimension of these block matrices and block vectors is $(\sum_s n^{(s)}) \times (\sum_s n^{(s)})$ and $(\sum_s n^{(s)}) \times 1$ respectively.

Since the substructures are part of a same assembly, two interface conditions need to be satisfied: *interface equilibrium and compatibility*.[6]

[5]Expressing the coupling of substructures in other domains (modal, time, state space ...) will be discussed in later chapters and use exactly the same approach.

[6]These lecture notes deal with structural problems. Nevertheless, the general theory is also applicable to the coupling of other physical domains such as acoustics or thermal problems.

2.2.1 Interface Equilibrium

The interface equilibrium requires that the interface forces, $\bar{\mathbf{g}}^{(s)}$, which are internal forces between the substructures, sum to zero when assembled. This is merely a manifestation of Newton's "action–reaction" principle. Considering, for instance, an interface $\Gamma^{(sr)}$ between two substructures s and r, one could express this condition as[7]

$$\mathbf{g}_b^{(s)} + \mathbf{g}_b^{(r)} = 0 \qquad \text{on} \quad \Gamma^{(sr)}, \tag{2.24}$$

$$\mathbf{g}_i^{(s)} = 0 \qquad\qquad \mathbf{g}_i^{(r)} = 0, \tag{2.25}$$

where the subscript b indicates a restriction of the DOF to the boundary and where we assumed that the DOF are numbered in the same manner on both sides of the interface. The subscript i denotes DOF that are not on a boundary and are thus internal DOF. On internal DOF, no connecting forces should exist.

In practice, the numbering of the DOF on the interface will not match across the interfaces and in addition more than two substructures can intersect on an interface (so-called *cross-points* in 2D and 3D, and edges in 3D). Hence in general, the interface equilibrium condition needs to be expressed using Boolean localization matrices $\mathbf{L}^{(s)^T}$ of dimension $n \times n^s$ that combine the forces on either side of the interface to satisfy force equilibrium. Interestingly, these localization matrices also map the DOF of substructure s to a global and unique set of n global DOF, as will be elaborated later. In general, the interface equilibrium thus is written as

$$\boxed{\sum_{s=1}^{N^{sub}} \mathbf{L}^{(s)^T} \bar{\mathbf{g}}^{(s)} = \mathbf{0}}. \tag{2.26}$$

This equilibrium condition can also be written using the block matrix $\mathbf{L}^T$ of dimension $n_b \times \sum_s n^s$ acting on the set of all substructure interface forces

$$\boxed{\mathbf{L}^T \bar{\mathbf{g}} = \mathbf{0}} \qquad \text{where} \quad \mathbf{L}^T = \left[\mathbf{L}^{(1)^T} \; \dots \; \mathbf{L}^{(N^{sub})^T}\right] \tag{2.27}$$

[7]In general, we will assume that the decomposition in substructures generates an interface discretization that is conforming and matching, namely, that the shape functions used on either side of the interface are identical and that the nodes coincide. In case of nonconforming or nonmatching interfaces, the theory used in these lecture notes are still generally applicable, but the assembly operators are then no longer Boolean. Indeed, going back to the variational principle underlying the discretized problem, the part related to the compatibility condition over an interface Γ can be written as

$$\int_\Gamma \boldsymbol{\mu}^T \mathbf{v} d\Gamma$$

where $\boldsymbol{\mu}$ and $\mathbf{v}$ are the field. More details about nonmatching interfaces can be found, for instance, in Rixen (1997).

► To illustrate the notation, consider the examples in Fig. 2.2. The green numbers indicate the global nodes for which the interface force conditions are written (i.e., the row number in the localization matrices).

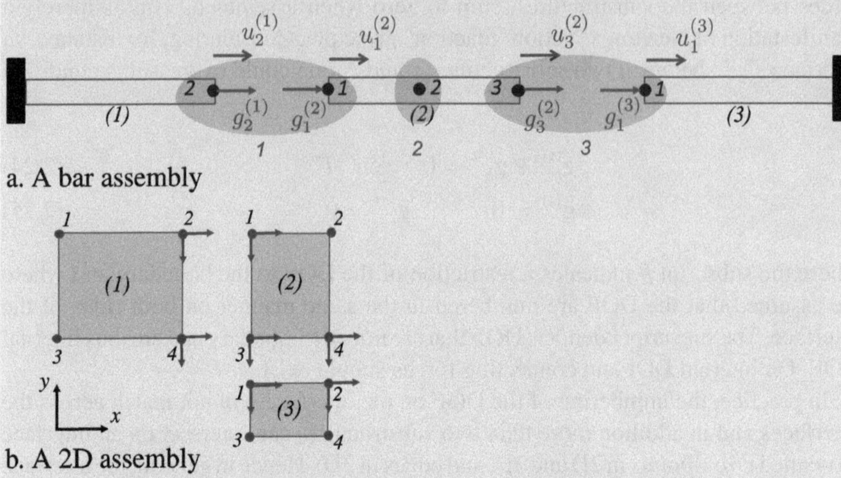

Fig. 2.2 Examples of assemblies: substructure DOFs and interface forces

For the beam example in Fig. 2.2a, the localization matrices are

$$\mathbf{L}^{(1)^T} = \begin{bmatrix} 1 \\ 0 \\ 0 \end{bmatrix} \quad \mathbf{L}^{(2)^T} = \begin{bmatrix} 1 & 0 & 0 \\ 0 & 1 & 0 \\ 0 & 0 & 1 \end{bmatrix} \quad \mathbf{L}^{(2)^T} = \begin{bmatrix} 0 \\ 0 \\ 1 \end{bmatrix}$$

and the interface equilibrium condition can be written as

$$\mathbf{L}^T \bar{\mathbf{g}} = \begin{bmatrix} \begin{bmatrix} 1 \\ 0 \\ 0 \end{bmatrix} \begin{bmatrix} 1 & 0 & 0 \\ 0 & 1 & 0 \\ 0 & 0 & 1 \end{bmatrix} \begin{bmatrix} 0 \\ 0 \\ 1 \end{bmatrix} \end{bmatrix} \begin{bmatrix} \begin{bmatrix} \bar{g}_2^{(1)} \end{bmatrix} \\ \begin{bmatrix} \bar{g}_1^{(2)} \\ \bar{g}_2^{(2)} \\ g_3^{(2)} \end{bmatrix} \\ \begin{bmatrix} \bar{g}_1^{(3)} \end{bmatrix} \end{bmatrix} = \mathbf{0}. \tag{2.28}$$

For the second example in Fig. 2.2b, we have two degrees of freedom per node and the localization matrix can be written as

$$\mathbf{L}^T = \begin{bmatrix} \mathbf{L}^{(1)} \ \mathbf{L}^{(2)} \ \mathbf{L}^{(3)} \end{bmatrix}$$

$$
\mathbf{L}^T = \left[\begin{bmatrix}
1&0&0&0&0&0&0&0\\
0&1&0&0&0&0&0&0\\
0&0&1&0&0&0&0&0\\
0&0&0&1&0&0&0&0\\
0&0&0&0&1&0&0&0\\
0&0&0&0&0&1&0&0\\
0&0&0&0&0&0&1&0\\
0&0&0&0&0&0&0&1\\
0&0&0&0&0&0&0&0\\
0&0&0&0&0&0&0&0\\
0&0&0&0&0&0&0&0\\
0&0&0&0&0&0&0&0\\
0&0&0&0&0&0&0&0\\
0&0&0&0&0&0&0&0\\
0&0&0&0&0&0&0&0\\
0&0&0&0&0&0&0&0
\end{bmatrix}
\begin{bmatrix}
0&0&0&0&0&0&0&0\\
0&0&0&0&0&0&0&0\\
1&0&0&0&0&0&0&0\\
0&1&0&0&0&0&0&0\\
0&0&0&0&0&0&0&0\\
0&0&0&0&0&0&0&0\\
0&0&0&0&1&0&0&0\\
0&0&0&0&0&1&0&0\\
0&0&1&0&0&0&0&0\\
0&0&0&1&0&0&0&0\\
0&0&0&0&0&0&1&0\\
0&0&0&0&0&0&0&1\\
0&0&0&0&0&0&0&0\\
0&0&0&0&0&0&0&0\\
0&0&0&0&0&0&0&0\\
0&0&0&0&0&0&0&0
\end{bmatrix}
\begin{bmatrix}
0&0&0&0&0&0&0&0\\
0&0&0&0&0&0&0&0\\
0&0&0&0&0&0&0&0\\
0&0&0&0&0&0&0&0\\
0&0&0&0&0&0&0&0\\
0&0&0&0&0&0&0&0\\
1&0&0&0&0&0&0&0\\
0&1&0&0&0&0&0&0\\
0&0&0&0&0&0&0&0\\
0&0&0&0&0&0&0&0\\
0&0&1&0&0&0&0&0\\
0&0&0&1&0&0&0&0\\
0&0&0&0&1&0&0&0\\
0&0&0&0&0&1&0&0\\
0&0&0&0&0&0&1&0\\
0&0&0&0&0&0&0&1
\end{bmatrix} \right]
$$

◀

These matrices are in fact identical to the localization matrices used in Finite Element codes to assemble elementary matrices in the global system, however here the localization is not written for one element but for one substructure (that one could consider as a super-element or macro-element). Obviously, it is not efficient to store these Boolean matrices as written above, but rather one should store them as sparse matrices, or even better one should construct the mapping tables based on the connectivity of the substructures over the interfaces.

2.2.2 *Interface Compatibility*

The second condition that needs to be satisfied on the interface is that DOF pertaining to some structural node have the same response on both sides of the interface, or in other words that the DOF are compatible on the interface. Considering the DOF of two substructures s and r coupled on the interface $\Gamma^{(sr)}$, the compatibility condition becomes

$$\bar{\mathbf{u}}_b^{(s)} - \bar{\mathbf{u}}_b^{(r)} = \mathbf{0} \qquad \text{on} \quad \Gamma^{(sr)}$$

where, as before, the subscript b indicates that the compatibility is written for the boundary DOF and where we assumed that the DOF are numbered identically on both sides of the interface.

In general, the numbering on the interface does not coincide and therefore the compatibility conditions are expressed using signed Boolean matrices $\mathbf{B}^{(s)}$. When operating on $\bar{\mathbf{u}}^{(s)}$, these operators extract the interfaces DOF and give them an opposite sign on each side of the interface. The interface compatibility can then be written in the following general form:

$$\boxed{\sum_{s=1}^{N^{sub}} \mathbf{B}^{(s)}\bar{\mathbf{u}}^{(s)} = \mathbf{0}} \tag{2.29}$$

One can use a block matrix notation to write this condition also in the form

$$\boxed{\mathbf{B}\bar{\mathbf{u}} = \mathbf{0}} \qquad \text{where} \quad \mathbf{B} = \left[\mathbf{B}^{(1)} \cdots \mathbf{B}^{(N^{sub})}.\right] \tag{2.30}$$

These equations can be understood as compatibility constraints imposed onto the independent sets of DOF in the substructures. The matrices $\mathbf{B}^{(s)}$ have dimension $n_\lambda \times n^{(s)}$, where n_λ is the number of interface compatibility constraints that need to be imposed.

► *Example: Boolean Compatibility Matrix*
To illustrate this notation, consider again the examples of Fig. 2.2.

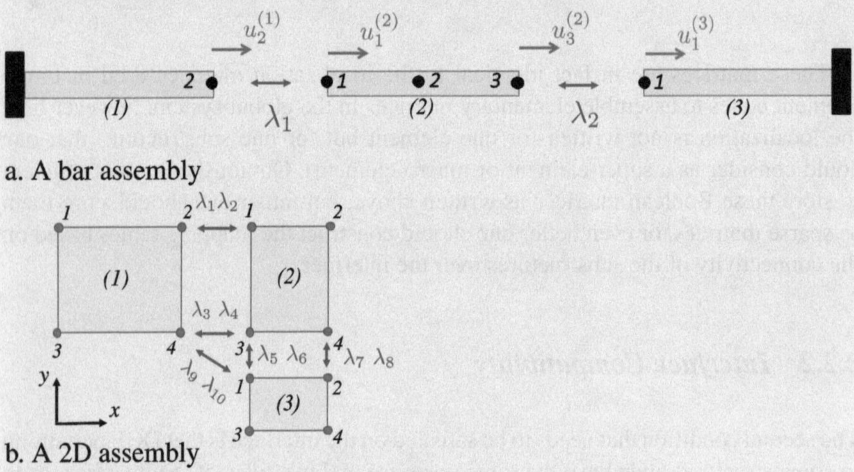

Fig. 2.3 Examples of assemblies: interpretation of the Lagrange multipliers

For the beam example, the compatibility condition can be written as

$$\mathbf{B}\bar{\mathbf{u}} = \left[\mathbf{B}^{(1)}\ \mathbf{B}^{(2)}\ \mathbf{B}^{(3)}\right]\bar{\mathbf{u}} = \left[\begin{bmatrix}1\\0\end{bmatrix}\begin{bmatrix}-1&0&0\\0&0&1\end{bmatrix}\begin{bmatrix}0\\-1\end{bmatrix}\right]\begin{bmatrix}\left[\bar{u}_2^{(1)}\right]\\ \begin{bmatrix}\bar{u}_1^{(2)}\\ \bar{u}_2^{(2)}\\ u_3^{(2)}\end{bmatrix}\\ \left[\bar{u}_1^{(3)}\right]\end{bmatrix} = \mathbf{0}. \tag{2.31}$$

The side on which the entry in **B** is positive and negative can be chosen freely. The interpretation of **B** and its associated Lagrange multipliers is depicted in Fig. 2.3a.

For the second example in Fig. 2.2, the constraint matrix can be written as

$$\mathbf{B} = \left[\mathbf{B}^{(1)}\ \mathbf{B}^{(2)}\ \mathbf{B}^{(3)}.\right]$$

$$= \left[\begin{bmatrix} 0&0&1&0&0&0&0&0\\ 0&0&0&1&0&0&0&0\\ 0&0&0&0&0&0&1&0\\ 0&0&0&0&0&0&0&1\\ 0&0&0&0&0&0&0&0\\ 0&0&0&0&0&0&0&0\\ 0&0&0&0&0&0&0&0\\ 0&0&0&0&0&0&0&0\\ 0&0&0&0&0&0&1&0\\ 0&0&0&0&0&0&0&1 \end{bmatrix} \begin{bmatrix} -1&0&0&0&0&0&0&0\\ 0&-1&0&0&0&0&0&0\\ 0&0&0&0&-1&0&0&0\\ 0&0&0&0&0&-1&0&0\\ 0&0&0&0&1&0&0&0\\ 0&0&0&0&0&1&0&0\\ 0&0&0&0&0&0&1&0\\ 0&0&0&0&0&0&0&1\\ 0&0&0&0&0&0&0&0\\ 0&0&0&0&0&0&0&0 \end{bmatrix} \begin{bmatrix} 0&0&0&0&0&0&0&0\\ 0&0&0&0&0&0&0&0\\ 0&0&0&0&0&0&0&0\\ 0&0&0&0&0&0&0&0\\ -1&0&0&0&0&0&0&0\\ 0&-1&0&0&0&0&0&0\\ 0&0&-1&0&0&0&0&0\\ 0&0&0&-1&0&0&0&0\\ -1&0&0&0&0&0&0&0\\ 0&-1&0&0&0&0&0&0 \end{bmatrix}\right]$$

The interpretation of **B** and its associated Lagrange multipliers is depicted in Fig. 2.3b. Note that the last two constraints are redundant. The compatibility of node 4 in $\Omega^{(1)}$ with node 3 in $\Omega^{(2)}$, and of node 1 in $\Omega^{(3)}$ with node 3 in $\Omega^{(2)}$ is imposed in lines 3, 4, 5, and 6 of **B**, so there is no need to impose in addition the compatibility between node 4 in $\Omega^{(1)}$ and node 1 in $\Omega^{(3)}$. Adding this redundant constraint does usually not harm the computation and can even be helpful for instance in the parallel computing algorithms. ◀

As for the localization matrix $\mathbf{L}^{(s)}$, the constraint matrices $\mathbf{B}^{(s)}$ are, in practice, not stored as full but as sparse, or only connectivity information is used to apply them to a vector.

To summarize this section, an assembly of substructures in the physical frequency domain (and similarly in other domains) is obtained by imposing compatibility and interface equilibrium conditions, leading to the set of equations

$$\boxed{\begin{cases} \mathbf{Z}^{(s)}\bar{\mathbf{u}}^{(s)} = \bar{\mathbf{f}}^{(s)} + \bar{\mathbf{g}}^{(s)} \qquad s = 1 \dots N^{sub} \\ \displaystyle\sum_{s=1}^{N^{sub}} \mathbf{B}^{(s)}\bar{\mathbf{u}}^{(s)} = \mathbf{0} \\ \displaystyle\sum_{s=1}^{N^{sub}} \mathbf{L}^{(s)^T}\bar{\mathbf{g}}^{(s)} = \mathbf{0} \end{cases}} \tag{2.32}$$

or in block matrix form

$$\boxed{\begin{cases} \mathbf{Z}\bar{\mathbf{u}} = \bar{\mathbf{f}} + \bar{\mathbf{g}} \\ \mathbf{B}\bar{\mathbf{u}} = \mathbf{0} \\ \mathbf{L}^T\bar{\mathbf{g}} = \mathbf{0} \end{cases}} \tag{2.33}$$

2.3 Primal and Dual Assembly

The form (2.32) (or equivalently (2.33)) of the coupled problem uses two interface fields, namely, the *primal unknowns* $\mathbf{u}$ per substructure (i.e., on each side of the interfaces) and the substructure interface forces $\mathbf{g}$ called *dual unknowns*.[8] Solving the dynamic problem of the assembly in the form (2.32) can be expensive since many interface unknowns need to be resolved. Hence, these equations can be rearranged in order to eliminate the interface forces and write the problem in terms of unique interface displacements (primal assembly) or by introducing interface forces satisfying the interface equilibrium (dual assembly).

2.3.1 Primal Assembly

We can define as primal unknowns for the interface a set of DOF $\mathbf{u}_g$ that are global and uniquely defined for the entire structure. The DOF of each substructure are then obtained by mapping the global set $\mathbf{u}_g$ to the local DOF of each substructure $\mathbf{u}^{(s)}$. Such a mapping was already introduced in the previous section, (2.26), to map the local interface forces to a global set. The same mapping, but now from the global DOF to the local ones can be used to write

$$\mathbf{u}^{(s)} = \mathbf{L}^{(s)}\mathbf{u}_g \qquad \text{or} \qquad \mathbf{u} = \mathbf{L}\mathbf{u}_g. \tag{2.34}$$

If the substructure DOF are obtained from a unique set as described above, they automatically satisfy the compatibility conditions that matching interface DOF must be equal. Hence

$$\mathbf{B}\mathbf{L}\mathbf{u}_g = \mathbf{0} \qquad \forall\, \mathbf{u}_g. \tag{2.35}$$

This relation mathematically means that $\mathbf{L}$ represents the nullspace of the constraint matrix $\mathbf{B}$:

$$\boxed{\mathbf{L} = \text{null}(\mathbf{B})} \tag{2.36}$$

[8]In these lecture notes, we will introduce the coupling conditions using basically a two field approach. A more general three-field approach can also be considered but will not be discussed here. See for instance Voormeeren (2012).

► In the example of the beam in Fig. 2.2a for which $\mathbf{L}$ was found in (2.28), this leads to

$$\begin{bmatrix} [\bar{u}_2^{(1)}] \\ \begin{bmatrix} \bar{u}_1^{(2)} \\ \bar{u}_2^{(2)} \\ \bar{u}_3^{(2)} \end{bmatrix} \\ [\bar{u}_1^{(3)}] \end{bmatrix} = \begin{bmatrix} [1\;0\;0] \\ \begin{bmatrix} 1&0&0 \\ 0&1&0 \\ 0&0&1 \end{bmatrix} \\ [0\;0\;1] \end{bmatrix} \begin{bmatrix} \bar{u}_1 \\ \bar{u}_2 \\ \bar{u}_3 \end{bmatrix} \tag{2.37}$$

Clearly, it can be seen that the substructure DOF are "drawn" from a global and unique set of DOF and thus automatically satisfy the compatibility constraint on the interface: with the compatibility constraint matrix $\mathbf{B}$ obtained in (2.31), one verifies that

$$\mathbf{Bu} = \mathbf{BLu}_g = \left[\begin{bmatrix} 1 \\ 0 \end{bmatrix} \begin{bmatrix} -1&0&0 \\ 0&0&1 \end{bmatrix} \begin{bmatrix} 0 \\ -1 \end{bmatrix} \right] \begin{bmatrix} [1\;0\;0] \\ \begin{bmatrix} 1&0&0 \\ 0&1&0 \\ 0&0&1 \end{bmatrix} \\ [0\;0\;1] \end{bmatrix} \mathbf{u}_g = \mathbf{0}$$

◄

When choosing a unique set of DOF as in (2.34), the coupled problem (2.33) can be simplified and written as

$$\begin{cases} \mathbf{ZLu}_g = \bar{\mathbf{f}} + \bar{\mathbf{g}} \\ \mathbf{L}^T \bar{\mathbf{g}} = \mathbf{0}, \end{cases} \tag{2.38}$$

where the compatibility condition is not present anymore since it is automatically verified. This expression is sometimes denoted as the Neumann–Dirichlet form of the coupled problem.

The interface forces $\mathbf{g}$ can be eliminated from this relation by premultiplying the dynamic equilibrium (first line of (2.38)) by $\mathbf{L}^T$, which builds the sum of the equilibrium equations on matching nodes. Considering the equilibrium of the internal forces (second line of (2.38)), the coupled problem can finally be expressed as

$$\boxed{\mathbf{Z}_g \mathbf{u}_g = \mathbf{f}_g \quad \text{with} \quad \mathbf{Z}_g = \mathbf{L}^T \mathbf{ZL} \quad \text{and} \quad \mathbf{f}_g = \mathbf{L}^T \bar{\mathbf{f}}.} \tag{2.39}$$

This form is generally referred to as the *primal assembly* of the coupled problem and matrix $\mathbf{L}^T\mathbf{ZL}$ is the primal-assembled impedance of the global system. This assembly is similar to the assembly of finite elements.

► To illustrate this process, consider again the first example in Fig. 2.2a, with the matrix **L** given in (2.28). The assembled impedance matrix for this example then writes

$$\mathbf{Z}_g = \mathbf{L}^T\mathbf{Z}\mathbf{L}$$

$$= \begin{bmatrix} \begin{bmatrix}1\\0\\0\end{bmatrix} & \begin{bmatrix}1&0&0\\0&1&0\\0&0&1\end{bmatrix} & \begin{bmatrix}0\\0\\1\end{bmatrix} \end{bmatrix} \begin{bmatrix} [Z_{22}^{(1)}] & \mathbf{0} & \mathbf{0} \\ \mathbf{0} & \begin{bmatrix} Z_{11}^{(2)} & Z_{12}^{(2)} & Z_{13}^{(2)} \\ Z_{21}^{(2)} & Z_{22}^{(2)} & Z_{23}^{(2)} \\ Z_{31}^{(2)} & Z_{32}^{(2)} & Z_{33}^{(2)} \end{bmatrix} & \mathbf{0} \\ \mathbf{0} & \mathbf{0} & [Z_{11}^{(3)}] \end{bmatrix} \begin{bmatrix} [1\;0\;0] \\ \begin{bmatrix}1&0&0\\0&1&0\\0&0&1\end{bmatrix} \\ [0\;0\;1] \end{bmatrix}$$

$$= \begin{bmatrix} Z_{22}^{(1)} + Z_{11}^{(2)} & Z_{12}^{(2)} & Z_{13}^{(2)} \\ Z_{21}^{(2)} & Z_{22}^{(2)} & Z_{23}^{(2)} \\ Z_{31}^{(2)} & Z_{32}^{(2)} & Z_{11}^{(2)} + Z_{33}^{(3)} \end{bmatrix}$$

and the dynamics of the beam are expressed by

$$\mathbf{Z}_g\mathbf{u}_g = = \mathbf{f}_g \qquad \Longrightarrow$$

$$\begin{bmatrix} Z_{22}^{(1)} + Z_{11}^{(2)} & Z_{12}^{(2)} & Z_{13}^{(2)} \\ Z_{21}^{(2)} & Z_{22}^{(2)} & Z_{23}^{(2)} \\ Z_{31}^{(2)} & Z_{32}^{(2)} & Z_{11}^{(2)} + Z_{33}^{(3)} \end{bmatrix} \begin{bmatrix} \bar{u}_1 \\ \bar{u}_2 \\ \bar{u}_3 \end{bmatrix} = \begin{bmatrix} \bar{f}_2^{(1)} + \bar{f}_1^{(2)} \\ \bar{f}_2^{(2)} \\ \bar{f}_3^{(2)} + \bar{f}_1^{(3)} \end{bmatrix}$$

◄

2.3.2 Dual Assembly

Let us start again from the coupled form (2.32) (or equivalently (2.33)), but instead of satisfying *a priori* the interface compatibility as in the primal assembly, we satisfy *a priori* the interface equilibrium.[9] This can be achieved by choosing coupling forces as

[9]For the description of the problem, it does obviously not manner if one first satisfies the equilibrium or the compatibility. This only determines which equation in the three-field formulation (2.32) is

$$\mathbf{g} = -\mathbf{B}^T \boldsymbol{\lambda} \tag{2.40}$$

Note that, since $\mathbf{BL} = \mathbf{0}$ and thus $\mathbf{L}^T\mathbf{B}^T = \mathbf{0}$, such coupling forces automatically satisfy the interface equilibrium:

$$\mathbf{L}^T\mathbf{B}^T\boldsymbol{\lambda} = \mathbf{0} \qquad \forall \boldsymbol{\lambda} \tag{2.41}$$

The minus in this definition is a choice to obtain a symmetric form later. The expression (2.40) can also be written substructure wise as

$$\begin{bmatrix} \mathbf{g}^{(1)} \\ \vdots \\ \mathbf{g}^{(N^{sub})} \end{bmatrix} = -\begin{bmatrix} \mathbf{B}^{(1)} \\ \vdots \\ \mathbf{B}^{(N^{sub})} \end{bmatrix} \boldsymbol{\lambda} \tag{2.42}$$

showing clearly that the coupling forces are generated from a unique set of interface forces, whose intensities are given by $\boldsymbol{\lambda}$ and are attributed, with the appropriate positive or negative sign, to the substructure interface. The interface force intensities $\boldsymbol{\lambda}$ are unknowns pertaining to the interface. The size n_λ of $\boldsymbol{\lambda}$ is equal to the number of interface compatibility conditions on the interface.

Substituting the coupling forces by their form (2.40), the coupled problem can be written as

$$\boxed{\begin{cases} \mathbf{Z}^{(s)}\bar{\mathbf{u}}^{(s)} = \bar{\mathbf{f}}^{(s)} + \mathbf{B}^{(s)T}\boldsymbol{\lambda} & s = 1 \ldots N^{sub} \\ \displaystyle\sum_{s=1}^{N^{sub}} \mathbf{B}^{(s)}\bar{\mathbf{u}}^{(s)} = \mathbf{0}, & \end{cases}} \tag{2.43}$$

where the interface equilibrium no longer appears since it is automatically satisfied given our choice (2.40). In block matrix form, we can write

$$\begin{cases} \mathbf{Z}\bar{\mathbf{u}} + \mathbf{B}^T\boldsymbol{\lambda} = \bar{\mathbf{f}} \\ \mathbf{B}\bar{\mathbf{u}} = \mathbf{0} \end{cases} \tag{2.44}$$

which is often put in the symmetric form

$$\boxed{\begin{bmatrix} \mathbf{Z} & \mathbf{B}^T \\ \mathbf{B} & \mathbf{0} \end{bmatrix} \begin{bmatrix} \bar{\mathbf{u}} \\ \boldsymbol{\lambda} \end{bmatrix} = \begin{bmatrix} \bar{\mathbf{f}} \\ \mathbf{0} \end{bmatrix}}. \tag{2.45}$$

One observes that $\boldsymbol{\lambda}$ can be interpreted as the Lagrange multipliers associated with the compatibility constraint (see the general theory of constrained systems in dynamics, for instance, in Géradin and Rixen (2015)).

---

eliminated. The end result will be identical, but the mathematical form and the approximations that can be applied will be depending on the form that will be solved.

► To illustrate this form, consider again the first example in Fig. 2.2a, with the matrix **B** given in (2.31). The dually assembled form is then written as

$$\begin{bmatrix} [Z_{22}^{(1)}] & \mathbf{0} & \mathbf{0} & \begin{bmatrix}1 & 0\end{bmatrix} \\ \mathbf{0} & \begin{bmatrix} Z_{11}^{(2)} & Z_{12}^{(2)} & Z_{13}^{(2)} \\ Z_{21}^{(2)} & Z_{22}^{(2)} & Z_{23}^{(2)} \\ Z_{31}^{(2)} & Z_{32}^{(2)} & Z_{33}^{(2)} \end{bmatrix} & \mathbf{0} & \begin{bmatrix} -1 & 0 \\ 0 & 0 \\ 0 & 1 \end{bmatrix} \\ \mathbf{0} & \mathbf{0} & [Z_{11}^{(3)}] & \begin{bmatrix}0 & -1\end{bmatrix} \\ \begin{bmatrix}1 \\ 0\end{bmatrix} & \begin{bmatrix} -1 & 0 & 0 \\ 0 & 0 & 1 \end{bmatrix} & \begin{bmatrix}0 \\ -1\end{bmatrix} & \mathbf{0} \end{bmatrix} \begin{bmatrix} [\bar{u}_2^{(1)}] \\ \begin{bmatrix} \bar{u}_1^{(2)} \\ \bar{u}_2^{(2)} \\ \bar{u}_3^{(2)} \end{bmatrix} \\ [\bar{u}_1^{(3)}] \\ \lambda_1 \\ \lambda_2 \end{bmatrix} = \begin{bmatrix} [\bar{f}_2^{(1)}] \\ \begin{bmatrix} \bar{f}_1^{(2)} \\ \bar{f}_2^{(2)} \\ \bar{f}_3^{(2)} \end{bmatrix} \\ [\bar{f}_1^{(3)}] \\ 0 \\ 0 \end{bmatrix}$$

Notice that this representation of the system is larger than the primally assembled counterpart, but in this representation the interface forces, λ are explicitly available. In future sections we shall see that this can be useful. ◄

2.3.3 *Usefulness of Different Assembly Formulations*

In this section, we have expressed the coupled problem in three different forms that are recalled in their block matrix form below for clarity:

$$\text{General form} \quad \boxed{\begin{cases} \mathbf{Z}\bar{\mathbf{u}} = \bar{\mathbf{f}} + \bar{\mathbf{g}} \\ \mathbf{B}\bar{\mathbf{u}} = \mathbf{0} \\ \mathbf{L}^T\bar{\mathbf{g}} = \mathbf{0} \end{cases}} \tag{2.33}$$

$$\text{Primal Assembly} \quad \boxed{\mathbf{Z}_g\mathbf{u}_g = \mathbf{f}_g \quad \text{with} \quad \mathbf{Z}_g = \mathbf{L}^T\mathbf{Z}\mathbf{L} \quad \text{and} \quad \mathbf{f}_g = \mathbf{L}^T\bar{\mathbf{f}}} \tag{2.39}$$

$$\text{Dual Assembly} \quad \boxed{\begin{bmatrix} \mathbf{Z} & \mathbf{B}^T \\ \mathbf{B} & \mathbf{0} \end{bmatrix} \begin{bmatrix} \bar{\mathbf{u}} \\ \boldsymbol{\lambda} \end{bmatrix} = \begin{bmatrix} \bar{\mathbf{f}} \\ \mathbf{0} \end{bmatrix}} \tag{2.45}$$

To illustrate the primal and dual assembly, Fig. 2.4 shows the assembly of components with forces or with displacements, and the corresponding matrices.

Each of these representations is mathematically equivalent and, hence, one might wonder why considering one form or the other might be advantageous. From a mechanical point of view, the interpretation of these forms is different, and so each provides different insights. The general form considers each substructure as independent (the DOF and coupling forces are defined separately per substructure) and therefore requires writing all of the interface conditions explicitly. In the primal form, coupling forces are no longer part of the problem: the only interface unknowns are

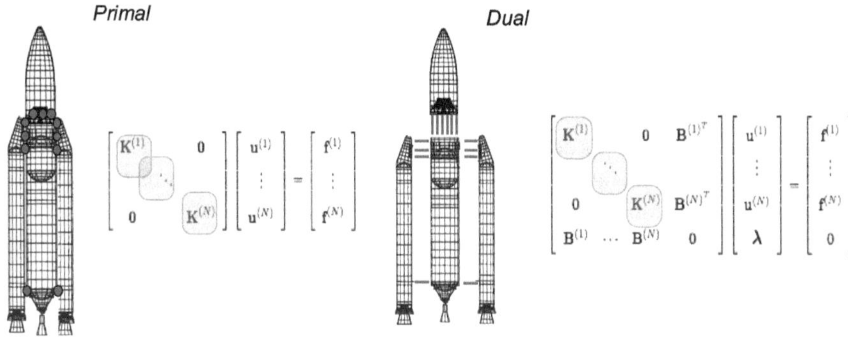

Fig. 2.4 Primal and dual assemble illustration

the interface DOF, which are unique for the entire structure. In the dual form, it is the interface force amplitudes $\boldsymbol{\lambda}$ that are unique for the entire structure and represent the interface unknowns (the interface DOF remaining defined per substructure).

The different mechanical interpretations of the assembly can be used to develop different numerical and experimental techniques in substructuring. This will hopefully become clear in the other chapters of these lecture notes, but here we provide a brief preview to the various uses of these forms.

Model Reduction (Chap. 3) In model reduction by substructuring, the dynamics in a substructure are approximated by representing the behavior of its physical DOF in a subspace of representative modes.

- When the primal-assembled form (2.39) is considered, a substructure is seen as receiving information from its neighboring substructures over the interface DOF they share. Hence, a natural set of modes for the reduction might correspond to the static response of the substructure to unit displacements at the interface, and these could be augmented by vibration modes in which the interface is fixed (i.e., the Hurty/Craig–Bampton family of methods naturally arise).
- When the dual assembly form (2.45) is considered, the substructure receives excitation from the neighbors as imposed interface forces. A natural set of representation modes, in that case, would consist of the static response of the substructure to unit interface forces and the additional dynamics would naturally be represented by vibration modes computed with the interface free.

The reduction methods will thus be based on different approximated representations depending on which mechanical interpretation of the coupling is considered.

Experimental Substructuring (Chap. 4) In experimental substructuring, the dynamics of (some) substructures are obtained by measurements. Since in experimental dynamics one typically measures a component with free interface, applying excitations and measuring the dynamic response, one typically measures dynamic flexibilities (FRFs), namely the response of the component to excitations. Therefore, considering the dually assembled form seems more natural. Obviously, the measured dynamic flexibility could be inverted and assembled in a primal way (impedance coupling) which would be mathematically equivalent. In practice however, since the measured dynamic flexibility of the substructure always contains imperfections or are not complete, an assembly should usually be performed only in a weak sense: the degrees of freedom on the interface should not be forced to match exactly, but rather only some simple deformation modes on the interface should be made compatible, allowing for some local incompatibility. This alleviates the effect of measurement errors in the assembly as will be explained later (Sect. 4.3).

References

Brüls, O., Arnold, M. (2008). The generalized-α scheme as a linear multistep integrator: Towards a general mechatronic simulator. *Transactions ASME, 3*, 041,007–1/041,007–10.

Chung, J., & Hulbert, G. M. (1993). A time integration algorithm for structural dynamics with improved numerical dissipation: The generalized-α method. *ASME Journal of Applied Mechanics, 60*, 371–375.

de Klerk, D., Rixen, D. J., & Voormeeren, S. N. (2008). General framework for dynamic substructuring: History, review and classifcation of techniques. *AIAA Journal, 46*(8), 1169–1181. https://doi.org/10.2514/1.33274.

Géradin, M., & Cardona, A. (2001). *Flexible multibody dynamics: A finite element approach.* Hoboken: Wiley.

Géradin, M., & Rixen, D. (2015). *Mechanical vibrations. Theory and application to structural dynamics* (3rd ed.). Chichester: Wiley.

Lew, A., Marsden, J., Ortiz, M., & West, M. (2004). Variational time integrators. *International Journal for Numerical Methods in Engineering, 60*(1), 153–212.

Newmark, N. (1959). Method of computation for structural dynamics. *Journal of Engineering Mechanics, 85*, 67–94.

Rixen, D. (1997). *Substructuring and dual methods in structural analysis*. Ph.D. thesis, Université de Liège, Belgium, Collection des Publications de la Faculté des Sciences appliquées, n° 175

Rixen, D. J., & van der Valk, P. L. (2013). An impulse based substructuring approach for impact analysis and load case simulations. *Journal of Sound and Vibration, 332*, 7174–7190.

van der Valk, P. L., & Rixen, D. J. (2014). An impulse based substructuring method for coupling impulse response functions and finite element models. *Computer Methods in Applied Mechanics and Engineering, 275*, 113–137.

van der Seijs, M.V. (2016). *Experimental dynamic substructuring: Analysis and design strategies for vehicle development*. Ph.D. thesis, Delft University of Technology. https://doi.org/10.4233/uuid:28b31294-8d53-49eb-b108-284b63edf670.

Voormeeren, S. (2012). *Dynamic substructuring methodologies for integrated dynamic analysis of wind turbines*. Ph.D. thesis, Delft University of Technology

Chapter 3
Model Reduction Concepts and Substructuring Approaches for Linear Systems

Abstract In this chapter, we give an overview of some of the most common reduction techniques based on substructuring. Although all techniques follow a similar approach, the main difference between the methods lies into the basis vectors used in the approximation subspace to represent the dynamics of each substructure and the manner in which the substructures are a couple. — *Chapter Authors: Daniel Rixen, Matt Allen, and Thomas Abrahamsson*

3.1 Model Reduction—General Concepts (Reduced Basis)

3.1.1 Reduction by Projection

Often, a finite element model is first built for static analysis in order to assess static deformations and stress levels which might be high especially in small details (stress concentration areas). Therefore, it is very common to find static models that have a very refined mesh, hence a high number of degrees of freedom (typically up to several million). Such very large static models can be solved by means of efficient solvers, but when it comes to computing free vibration modes, harmonic responses and/or transient responses, it is required to solve many static-like problems (i.e., in the inverse iterations of the eigensolvers or in the time-stepping loop) and the computing time required is often unacceptable.

Fortunately, for most dynamical problems, such highly refined meshes are not needed to capture the dynamic behavior. For instance, if we want to compute the fundamental modes, we know that the first free vibration modes have a rather smooth deformation and thus a coarser mesh would be sufficient to compute these modes. Similarly, when computing the dynamic response to external loads which are no shocks (i.e., which have no high spectral content compared to the eigenfrequency spectrum of the structure), a coarse mesh yields in most cases accurate results. In summary, *using a coarser representation of the displacement field in the dynamic model is often acceptable in terms of accuracy and is required in order to perform dynamic analyses in a reasonable time.*

M. S. Allen et al., *Substructuring in Engineering Dynamics*,
CISM International Centre for Mechanical Sciences 594,
https://doi.org/10.1007/978-3-030-25532-9_3

The structural analyst should thus in principle build two models, one for the static simulations, one for the dynamic analysis. Knowing that building a model is a significant part of the entire study, it would be very useful to device a procedure that reduces the size of the dynamic problem without modifying the mesh grid.

Such procedures indeed exist and are known as *reduction methods*. Similarly to the way one approximates continuous fields by a set of shape functions in a Rayleigh–Ritz approach (e.g., in Finite Elements), the driving idea in reduction techniques is to replace the DOF by a set of global variables representing the amplitudes of possible displacement modes (see also Sect. 2.1.1):

$$\mathbf{u}(t) \simeq \mathbf{T}\mathbf{q}(t), \tag{3.1}$$

where $\mathbf{u}$ are all the DOF of the system and $\mathbf{T}$ is a reduction matrix of dimension $n \times r, r < n$ and where $\mathbf{q}$ of dimension $r \times 1$ is a set of reduced generalized degrees of freedom. The columns of $\mathbf{T}$ define possible displacement shapes for the degrees of freedom $\mathbf{u}$, and $\mathbf{q}$ are the amplitude of these displacement shapes.

In general, replacing $\mathbf{u}$ by (3.1), only an approximate solution can be found and the accuracy of the approximation will depend on how well the assumed modes in $\mathbf{T}$ can represent the exact solution. Introducing (3.1) in the linear dynamic problem of motion

$$\mathbf{M}\ddot{\mathbf{u}} + \mathbf{C}\dot{\mathbf{u}} + \mathbf{K}\mathbf{u} = \mathbf{f}(t), \tag{3.2}$$

one obtains

$$\mathbf{M}\mathbf{T}\ddot{\mathbf{q}} + \mathbf{C}\mathbf{T}\dot{\mathbf{q}} + \mathbf{K}\mathbf{T}\dot{\mathbf{q}} = \mathbf{f}(t) + \mathbf{r}, \tag{3.3}$$

where $\mathbf{r}$ is a residual force that is a remainder for the dynamic equilibrium equation: since the reduction subspace $\mathbf{T}$ can in general not represent the exact solution, there will always be an equilibrium error, whatever the choice of $\mathbf{q}$. Indeed, one now has only r unknowns $\mathbf{q}$ to satisfy $n > r$ equations. Following then the idea of virtual work, one requires that the reduced DOF $\mathbf{q}$ be chosen such that the residual force $\mathbf{r}$ does not contribute to the dynamics in the representation space $\mathbf{T}$ or, in other words, that the residual force does not produce any work for the possible shapes of motion contained in $\mathbf{T}$:

$$\mathbf{T}^T\mathbf{r} = \mathbf{0} \tag{3.4}$$

With this condition, we can find the equations to determine $\mathbf{q}$ by projecting the dynamic equilibrium equations (3.3) onto the subspace $\mathbf{T}$:

$$\mathbf{T}^T\mathbf{M}\mathbf{T}\ddot{\mathbf{q}} + \mathbf{T}^T\mathbf{C}\mathbf{T}\dot{\mathbf{q}} + \mathbf{T}^T\mathbf{K}\mathbf{T}\mathbf{q} = \mathbf{T}^T\mathbf{f} \tag{3.5}$$

which is usually written as

$$\tilde{\mathbf{M}}\ddot{\mathbf{q}} + \tilde{\mathbf{C}}\dot{\mathbf{q}} + \tilde{\mathbf{K}}\mathbf{q} = \tilde{\mathbf{f}} \tag{3.6}$$

where the tilde superscript indicates that the matrices and vectors pertain now to a representation in a reduced space.

After having solved the reduced problem (3.6) for $\mathbf{q}$, one can build the solution $\mathbf{u}$ for the physical DOF by substituting in (3.1). The residual force $\mathbf{r}$ for the full problem can be computed by substitution in the original problem (3.3): it provides a way to monitor, a posteriori, the error on the equilibrium.

In what follows, we will drop the damping term in order to clarify the presentation. Reduction of the damping matrix, in particular for substructure reduction, is discussed in Gruber and Rixen (2018a, b) and the references therein.

3.1.2 The Guyan–Irons Method

The cornerstone of every reduction method consists of finding a representation space $\mathbf{T}$ of the solution that allows finding a good approximation and that can be obtained with a computational cost significantly smaller than the one involved in solving the full problem. One very common way to find a reasonably good representation space was proposed by Guyan (1965) and Irons (1965) as described next.

Let us consider the matrix equation that governs the system dynamics (3.2). To reduce the size of matrices $\mathbf{K}$ and $\mathbf{M}$, let us eliminate a subset of degrees of freedom. The remaining and condensed degrees of freedom are written, respectively, as $\mathbf{u}_2$ and $\mathbf{u}_1$. Assuming that no forces are applied on $\mathbf{u}_1$, the equation can be partitioned as follows:

$$\begin{bmatrix} \mathbf{M}_{22} & \mathbf{M}_{21} \\ \mathbf{M}_{12} & \mathbf{M}_{11} \end{bmatrix} \begin{bmatrix} \ddot{\mathbf{u}}_2 \\ \ddot{\mathbf{u}}_1 \end{bmatrix} + \begin{bmatrix} \mathbf{K}_{22} & \mathbf{K}_{21} \\ \mathbf{K}_{12} & \mathbf{K}_{11} \end{bmatrix} \begin{bmatrix} \mathbf{u}_2 \\ \mathbf{u}_1 \end{bmatrix} = \begin{bmatrix} \mathbf{f}_2 \\ \mathbf{0} \end{bmatrix} \tag{3.7}$$

or

$$\mathbf{M}_{22}\ddot{\mathbf{u}}_2 + \mathbf{M}_{21}\ddot{\mathbf{u}}_1 + \mathbf{K}_{22}\mathbf{u}_2 + \mathbf{K}_{21}\mathbf{u}_1 = \mathbf{f}_2, \tag{3.8}$$

$$\mathbf{M}_{12}\ddot{\mathbf{u}}_2 + \mathbf{M}_{11}\ddot{\mathbf{u}}_1 + \mathbf{K}_{12}\mathbf{u}_2 + \mathbf{K}_{11}\mathbf{u}_1 = \mathbf{0}. \tag{3.9}$$

One may imagine separating the condensed coordinates $\mathbf{u}_1$ into two contributions

$$\mathbf{u}_1 = \mathbf{u}_{1,stat} + \mathbf{u}_{1,dyn} \tag{3.10}$$

with the "static" part deduced from

$$\mathbf{u}_{1,stat} = -\mathbf{K}_{11}^{-1}\mathbf{K}_{12}\mathbf{u}_2. \tag{3.11}$$

This is obtained by neglecting the inertia forces in (3.9).

The static condensation algorithm consists in neglecting $\mathbf{u}_{1,dyn}$ and in building the reduction

$$\begin{bmatrix} \mathbf{u}_2 \\ \mathbf{u}_1 \end{bmatrix} = \mathbf{T}^{GI}\mathbf{u}_2 = \begin{bmatrix} \mathbf{I} \\ \mathbf{S} \end{bmatrix} \mathbf{u}_2 \tag{3.12}$$

where $\mathbf{T}^{GI}$ stands for the Guyan–Irons reduction matrix and where

$$\mathbf{S} = -\mathbf{K}_{11}^{-1}\mathbf{K}_{12} \tag{3.13}$$

is the static condensation matrix the columns of which contain the so-called *static modes* representing the static response of $\mathbf{u}_1$ for unit $\mathbf{u}_2$ displacements. They represent the static deformation induced in the inner part of the substructure when a unit displacement is given at an interface DOF.

The reduced matrices are then

$$\mathbf{K}^{GI} = \mathbf{T}^T\mathbf{K}\mathbf{T} = \mathbf{K}_{22} - \mathbf{K}_{21}\mathbf{K}_{11}^{-1}\mathbf{K}_{12}, \tag{3.14}$$

$$\mathbf{M}^{GI} = \mathbf{T}^T\mathbf{M}\mathbf{T} = \mathbf{M}_{22} - \mathbf{M}_{21}\mathbf{K}_{11}^{-1}\mathbf{K}_{12} - \mathbf{K}_{21}\mathbf{K}_{11}^{-1}\mathbf{M}_{12} + \mathbf{K}_{21}\mathbf{K}_{11}^{-1}\mathbf{M}_{11}\mathbf{K}_{11}^{-1}\mathbf{K}_{12}. \tag{3.15}$$

We observe that the reduced stiffness matrix is the stiffness matrix statically condensed on $\mathbf{u}_2$. The reduced mass matrix is the mass matrix associated to $\mathbf{u}_2$ and augmented by the inertia of $\mathbf{u}_1$ assumed to respond statically. The dynamic problem is then reduced to

$$\mathbf{M}^{GI}\ddot{\mathbf{u}}_2 + \mathbf{K}^{GI}\mathbf{u}_2 = \mathbf{f}_2^{GI}. \tag{3.16}$$

If the static condensation algorithm is applied to static problems, the exact solution is found. But when applied to dynamic problems, an approximation is introduced by neglecting the dynamic response of the interior of the substructure and thereby assuming that all internal nodes respond quasi-statically to the interface displacements. The validity of the condensation algorithm thus depends on the extent to which correction $\mathbf{u}_{1,dyn}$ is negligible. It is possible to show (see, for instance, Géradin and Rixen 2015) that the static condensation technique is valid if

$$\omega^2 \ll \mu_1^2, \tag{3.17}$$

where ω is the highest eigenfrequency that one wants to compute for the complete structure and μ_1 is the lowest eigenfrequency of the structure when $\mathbf{u}_2$ are clamped. The complete analysis allows to show that static condensation always leads to overestimating the eigenvalues compared to the full model. This is natural if we recall that a model obtained by applying consistently a Rayleigh–Ritz approach is always stiffer when further restrictions are introduced in the discretization field (see, for instance, Géradin and Rixen 2015).

The algorithm described above is very frequently used in the context of finite element structural analysis. Although originally proposed independently by Guyan and Irons, it is commonly known as the Guyan reduction method. In commercial codes, this method is often implemented and used to reduce the complexity of the problem when no substructuring approach is used. The choice of an appropriate set of master DOF $\mathbf{u}_2$ is sometimes done automatically by the software. One simple heuristics consists of ranking the DOF $\mathbf{u}$ according to their individual pseudo-frequency computed by the ratio of the diagonals of the stiffness and mass matrices, namely for a DOF k $\nu_k^2 = K_{kk}/M_{kk}$, and choosing as master DOF $\mathbf{u}_2$ the ones with the lowest ν_i^2.

3.1.3 Model Reduction Through Substructuring

A very interesting way of applying reduction technique is found by applying the following procedure.

1. Define subparts of the structure called *substructures*. These substructures correspond, for instance, to parts of the model that are analyzed and designed by different teams. For an aircraft, this might be the wings, fuselage, stabilizers, and tail. For a launcher system, it could be different stages of the rocket and the payload. For a vehicle, one could define as substructures the engine block, the accessories, the suspension and the car body. We will denote every substructure by $\Omega^{(s)}$. The stiffness and mass matrices corresponding to the non-assembled substructures will be denoted by $\mathbf{K}^{(s)}$ and $\mathbf{M}^{(s)}$, respectively. The degrees of freedom per substructure are denoted as $\mathbf{u}^{(s)}$.
2. For every subpart, define a reduction matrix $\mathbf{T}^{(s)}$ that retains the degrees of freedom on the interface boundary, called $\mathbf{u}_b^{(s)}$, such that

$$\mathbf{u}^{(s)} = \mathbf{T}^{(s)} \begin{bmatrix} \mathbf{u}_b^{(s)} \\ \boldsymbol{\zeta}^{(s)} \end{bmatrix} \tag{3.18}$$

   where $\boldsymbol{\zeta}^{(s)}$ are generalized degrees of freedom representing the amplitudes of representation modes additional to the ones governed by $\mathbf{u}_b^{(s)}$. The substructures are now seen as macro-elements (also called *super-element*) which stiffness and mass matrices are given by

$$\tilde{\mathbf{K}}^{(s)} = \mathbf{T}^{(s)^T} \mathbf{K}^{(s)} \mathbf{T}^{(s)} \qquad \tilde{\mathbf{M}}^{(s)} = \mathbf{T}^{(s)^T} \mathbf{M}^{(s)} \mathbf{T}^{(s)} \tag{3.19}$$

   These reduced matrices of the substructures can easily be shared between different design teams. Such methods are sometimes called *Component Mode Synthesis* or CMS.
3. The interface boundary degrees of freedom $\mathbf{u}_b^{(s)}$ are assembled on the interface of the substructures exactly as if one would assemble the macro-elements.

If one first assumes that the entire reduction of a substructure is performed by keeping only the interface DOF $\mathbf{u}_b^{(s)}$ (i.e., no additional reduced DOF $\boldsymbol{\zeta}^{(s)}$ per substructure), one can resort to the method of Guyan–Irons (see previous Section) where the interface DOF are the master DOF (previously called $\mathbf{u}_2^{(s)}$ when the Guayn–Irons reduction was applied to a non-decomposed problem). With a simple Guyan–Irons reduction on the interface DOF of the substructures, the dynamics of the DOF inside the substructures is neglected, which could be a crude assumption if the eigenfrequencies of the substructures fixed on the interface are not small (see criteria (3.17)). Hence, additional information about the substructure vibrations should be added to the reduction basis as explained in the next Section.

3.2 Numerical Techniques for Model Reduction of Substructures

3.2.1 The Hurty/Craig–Bampton Method

The substructure reduction method discussed in this section is one of the most commonly used substructuring technique in engineering practice. It was proposed by Roy Craig (1968), writing in a more intuitive form ideas previously published by Hurty (1965). Hence, although most commonly known as the Craig–Bampton method, we will call it the *Hurty/Craig–Bampton* method.

The FE discretized system of equations of motion for a substructure reads

$$\mathbf{M}^{(s)}\ddot{\mathbf{u}}^{(s)} + \mathbf{K}^{(s)}\mathbf{u}^{(s)} = \mathbf{f}^{(s)}, \tag{3.20}$$

where $\mathbf{M}^{(s)}$ and $\mathbf{K}^{(s)}$ are the mass and stiffness matrices, respectively, $\mathbf{u}^{(s)}$ is the displacement vector, $\mathbf{f}^{(s)}$ is the forcing vector and the superscript s denotes the sth substructure. The substructure equations of motion are partitioned into interior and boundary (or interface) DOF (denoted by subscripts i and b', respectively), as

$$\begin{bmatrix} \mathbf{M}_{ii}^{(s)} & \mathbf{M}_{ib}^{(s)} \\ \mathbf{M}_{bi}^{(s)} & \mathbf{M}_{bb}^{(s)} \end{bmatrix} \begin{Bmatrix} \ddot{\mathbf{u}}_i^{(s)} \\ \ddot{\mathbf{u}}_b^{(s)} \end{Bmatrix} + \begin{bmatrix} \mathbf{K}_{ii}^{(s)} & \mathbf{K}_{ib}^{(s)} \\ \mathbf{K}_{bi}^{(s)} & \mathbf{K}_{bb}^{(s)} \end{bmatrix} \begin{Bmatrix} \mathbf{u}_i^{(s)} \\ \mathbf{u}_b^{(s)} \end{Bmatrix} = \begin{Bmatrix} \mathbf{0} \\ \mathbf{f}_b^{(s)}, \end{Bmatrix} \tag{3.21}$$

where the subscripts b and i are indexes referring to the boundary and internal component of the matrices and vectors.

The representation modes, commonly called Component Modes (CMs) in substructuring, are formed by computing the static response of the interior of the substructure when one interface DOF is given a unit displacement and all the other DOF are held fixed. The CM set for the entire interface is expressed as

$$\boldsymbol{\Psi}^{(s)} = \begin{bmatrix} -\mathbf{K}_{ii}^{(s)^{-1}}\mathbf{K}_{ib}^{(s)} \\ \mathbf{I} \end{bmatrix}. \tag{3.22}$$

The resulting basis $\boldsymbol{\Psi}^{(s)}$ is used to statically eliminate all interface DOF from the model, retaining only the boundary DOF. The resulting reduced system is usually of small size since only boundary DOF are remaining unknowns. In order to capture the dynamics of the system, the static modes are augmented with a set of dynamic modes which are obtained by fixing the interface DOF and solving the following eigenvalue problem:

$$\left(\mathbf{K}_{ii}^{(s)} - \omega_r^2 \mathbf{M}_{ii}^{(s)}\right) \boldsymbol{\phi}_{i,r}^{(s)} = 0. \tag{3.23}$$

The eigenvectors obtained from this equation are referred to as fixed-interface modes. A truncated set of m of these mass normalized eigenvectors are collected into a fixed-interface mode matrix

$$\boldsymbol{\Phi}^{(s)} = \begin{bmatrix} \boldsymbol{\phi}_{i,1}^{(s)}, \dots, \boldsymbol{\phi}_{i,m}^{(s)} \\ \mathbf{0} \end{bmatrix} = \begin{bmatrix} \boldsymbol{\Phi}_i^{(s)} \\ \mathbf{0} \end{bmatrix}. \tag{3.24}$$

These modes provide a normal basis for the interior DOF of the substructure. The fixed-interface modes and the CMs are combined to form the HCB reduction matrix as

$$\mathbf{T}^{(s)\mathrm{HCB}} = \left[\boldsymbol{\Phi}^{(s)}\ \boldsymbol{\Psi}^{(s)}\right], \tag{3.25}$$

which provides a transformation from the substructure physical DOF to the HCB generalized DOF,

$$\begin{bmatrix} \mathbf{u}_i^{(s)} \\ \mathbf{u}_b^{(s)} \end{bmatrix} \approx \mathbf{T}^{\mathrm{HCB}^{(s)}} \begin{bmatrix} \mathbf{q}_i^{(s)} \\ \mathbf{u}_b^{(s)} \end{bmatrix}, \tag{3.26}$$

where $\mathbf{q}_i^{(s)}$ represents the modal coordinate vector associated with the fixed-interface modes. The meaning of the static modes and of the fixed-interface modes as used in the HCB method is illustrated in Fig. 3.1.

The uncoupled substructure-reduced mass and stiffness matrices are now formed by applying the HCB transformation to these matrices as

$$\mathbf{M}^{(s)\mathrm{HCB}} = \left(\mathbf{T}^{(s)\mathrm{HCB}}\right)^T \mathbf{M}^{(s)} \mathbf{T}^{(s)\mathrm{HCB}}, \qquad \mathbf{K}^{(s)\mathrm{HCB}} = \left(\mathbf{T}^{(s)\mathrm{HCB}}\right)^\mathrm{T} \mathbf{K}^{(s)} \mathbf{T}^{(s)\mathrm{HCB}}. \tag{3.27}$$

which, using (3.25), are given by

$$\mathbf{K}^{(s)HCB} = \begin{bmatrix} \boldsymbol{\Omega}_m^{(s)^2} & \mathbf{0} \\ \mathbf{0} & \tilde{\mathbf{K}}_{bb}^{(s)} \end{bmatrix} \quad \text{and} \quad \mathbf{M}^{(s)HBC} = \begin{bmatrix} \mathbf{I} & \tilde{\mathbf{M}}_{ib}^{(s)} \\ \tilde{\mathbf{M}}_{bi}^{(s)} & \tilde{\mathbf{M}}_{bb}^{(s)} \end{bmatrix} \tag{3.28}$$

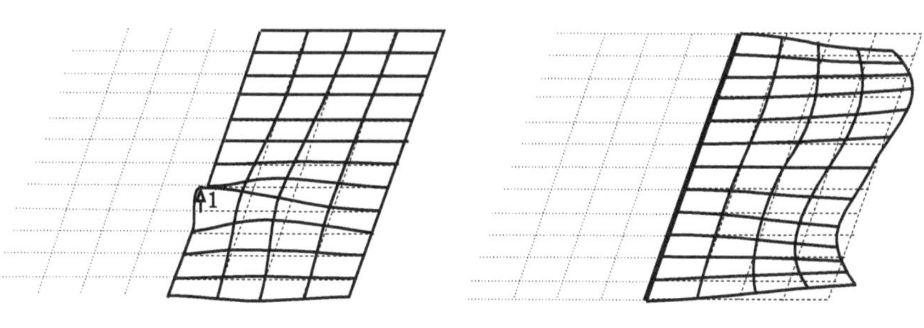

Fig. 3.1 Static modes and internal vibration modes of a substructure. In this example, the interface is the far left edge of the plate; the constraint modes show motion of a single node on the interface while the fixed-interface modes show the motion of the rest of the structure when the interface is fixed

with $\boldsymbol{\Omega}_m^{(s)^2}$ the diagonal matrix of the eigenfrequencies of the fixed- interface modes and with the full submatrices

$$\tilde{\mathbf{K}}_{bb}^{(s)} = \mathbf{K}_{bb}^{(s)} - \mathbf{K}_{bi}^{(s)} \mathbf{K}_{bb}^{(s)^{-1}} \mathbf{K}_{bi}^{(s)} \tag{3.29}$$

$$\begin{aligned} \tilde{\mathbf{M}}_{bb}^{(s)} &= \mathbf{M}_{bb}^{(s)} - \mathbf{M}_{bi}^{(s)} \mathbf{K}_{ii}^{(s)^{-1}} \mathbf{K}_{ib}^{(s)} - \mathbf{K}_{bi}^{(s)} \mathbf{K}_{ii}^{(s)^{-1}} \mathbf{M}_{ib}^{(s)} \\ &\quad + \mathbf{K}_{bi}^{(s)} \mathbf{K}_{ii}^{(s)^{-1}} \mathbf{M}_{ii}^{(s)} \mathbf{K}_{ii}^{(s)^{-1}} \mathbf{K}_{ib}^{(s)} \end{aligned} \tag{3.30}$$

$$\tilde{\mathbf{M}}_{ib} = \boldsymbol{\Phi}^{(s)^T} \left(\mathbf{M}_{ib}^{(s)} - \mathbf{M}_{ii}^{(s)} \mathbf{K}_{ii}^{(s)^{-1}} \mathbf{K}_{ib}^{(s)} \right) = \tilde{\mathbf{M}}_{bi}^T \tag{3.31}$$

The HCB reduced-order models are typically coupled using a primal assembly by defining a transformation between the uncoupled and coupled DOF that selects the substructure boundary DOF from a unique global set of boundary DOF (for an explanation of the Boolean assembly matrix, see Sect. 2.2). The assembled HCB mass and stiffness matrices are formed by adding the coupled contribution from each substructure as

$$\mathbf{M}^{\text{HCB}} = \sum_{j=1}^{n} \left(\mathbf{L}^{(\text{s})\text{HCB}} \right)^T \mathbf{M}^{(s)\text{HCB}} \mathbf{L}^{(\text{s})\text{HCB}} \tag{3.32}$$

$$\mathbf{K}^{\text{HCB}} = \sum_{j=1}^{n} \left(\mathbf{L}^{(\text{s})\text{HCB}} \right)^T \mathbf{K}^{(s)\text{HCB}} \mathbf{L}^{(\text{s})\text{HCB}}, \tag{3.33}$$

where the sth substructure coupling matrix, $\mathbf{L}^{(\text{s})\text{HCB}}$, is formed by extracting the corresponding rows of $\mathbf{L}^{\text{HCB}}$.

► *Example: W-bracket*

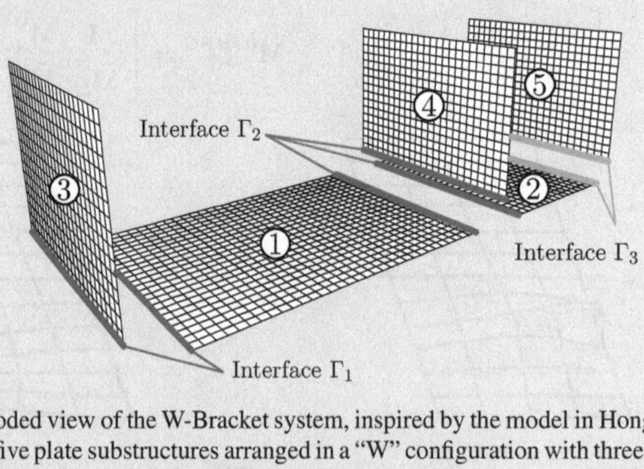

Fig. 3.2 Exploded view of the W-Bracket system, inspired by the model in Hong et al. (2013), consisting of five plate substructures arranged in a "W" configuration with three interface sets; the interface sets Γ_i, $i = 1, 2, 3$ are highlighted and labeled

For example, consider the W-bracket system from Krattiger et al. (2019) and shown in Fig. 3.2. This consists of five plates assembled into a shape that resembles the letter "W". For this example, the coupling matrix can be written as

$$
\begin{Bmatrix}
\mathbf{q}_{i,1} \\ \mathbf{u}_{\Gamma_1,1} \\ \mathbf{u}_{\Gamma_2,1} \\ \hdashline
\mathbf{q}_{i,2} \\ \mathbf{u}_{\Gamma_2,2} \\ \mathbf{u}_{\Gamma_3,2} \\ \hdashline
\mathbf{q}_{i,3} \\ \mathbf{u}_{\Gamma_1,3} \\ \hdashline
\mathbf{q}_{i,4} \\ \mathbf{u}_{\Gamma_2,4} \\ \hdashline
\mathbf{q}_{i,5} \\ \mathbf{u}_{\Gamma_3,5}
\end{Bmatrix}
=
\underbrace{\begin{bmatrix}
\mathbf{I}&\mathbf{0}&\mathbf{0}&\mathbf{0}&\mathbf{0}&\mathbf{0}&\mathbf{0}&\mathbf{0} \\
\mathbf{0}&\mathbf{0}&\mathbf{0}&\mathbf{0}&\mathbf{0}&\mathbf{I}&\mathbf{0}&\mathbf{0} \\
\mathbf{0}&\mathbf{0}&\mathbf{0}&\mathbf{0}&\mathbf{0}&\mathbf{0}&\mathbf{I}&\mathbf{0} \\ \hdashline
\mathbf{0}&\mathbf{I}&\mathbf{0}&\mathbf{0}&\mathbf{0}&\mathbf{0}&\mathbf{0}&\mathbf{0} \\
\mathbf{0}&\mathbf{0}&\mathbf{0}&\mathbf{0}&\mathbf{0}&\mathbf{0}&\mathbf{I}&\mathbf{0} \\
\mathbf{0}&\mathbf{0}&\mathbf{0}&\mathbf{0}&\mathbf{0}&\mathbf{0}&\mathbf{0}&\mathbf{I} \\ \hdashline
\mathbf{0}&\mathbf{0}&\mathbf{I}&\mathbf{0}&\mathbf{0}&\mathbf{0}&\mathbf{0}&\mathbf{0} \\
\mathbf{0}&\mathbf{0}&\mathbf{0}&\mathbf{0}&\mathbf{0}&\mathbf{I}&\mathbf{0}&\mathbf{0} \\ \hdashline
\mathbf{0}&\mathbf{0}&\mathbf{0}&\mathbf{I}&\mathbf{0}&\mathbf{0}&\mathbf{0}&\mathbf{0} \\
\mathbf{0}&\mathbf{0}&\mathbf{0}&\mathbf{0}&\mathbf{0}&\mathbf{0}&\mathbf{I}&\mathbf{0} \\ \hdashline
\mathbf{0}&\mathbf{0}&\mathbf{0}&\mathbf{0}&\mathbf{I}&\mathbf{0}&\mathbf{0}&\mathbf{0} \\
\mathbf{0}&\mathbf{0}&\mathbf{0}&\mathbf{0}&\mathbf{0}&\mathbf{0}&\mathbf{0}&\mathbf{I}
\end{bmatrix}}_{\mathbf{L}^{\mathrm{HCB}}}
\begin{Bmatrix}
\mathbf{q}_{i,1} \\ \mathbf{q}_{i,2} \\ \mathbf{q}_{i,3} \\ \mathbf{q}_{i,4} \\ \mathbf{q}_{i,5} \\ \mathbf{u}_{\Gamma_1} \\ \mathbf{u}_{\Gamma_2} \\ \mathbf{u}_{\Gamma_3}
\end{Bmatrix},
\tag{3.34}
$$

where the boundary DOF of each substructure are further partitioned according to the interface sets Γ_1, Γ_2, and Γ_3 as shown in Fig. 3.2.

◄

The choice of the number of fixed-interface modes kept in the reduction basis for each substructure can be made based on different criteria:

- A criterion often used in practice is based on the eigenfrequencies of the kept modes. Typically, one chooses all fixed-interface modes having an eigenfrequency lower than 1.8 or 2 times the highest frequency of interest in the assembly.
- One can also choose the fixed-interface modes based on how easily they can be excited through the support, for instance, by defining a measure of the interface reaction force associate to a mode. Following a similar reasoning, one evaluates how complete the base of the fixed- interface modes need to be by specifying how much of the mass of a substructure mass be represented in its reduced matrices. This can be evaluated using the concept of effective modal mass (see, for instance, Géradin and Rixen 2015).
- The number of modes can also be chosen based on *a posteriori* error estimators (Jakobsson et al. 2011), later reformulated in a less mathematical form and used for adaptive selection strategies in Voormeeren et al. (2013).

3.2.2 Substructure Reduction Using Free Interface Modes

The discussion in this Section is based on the overview in Gruber and Rixen (2016).

Considering the equation of motion (2.2) of substructure s,

$$
\mathbf{M}^{(s)}\ddot{\mathbf{u}}^{(s)} + \mathbf{C}^{(s)}\dot{\mathbf{u}}^{(s)} + \mathbf{K}^{(s)}\mathbf{u}^{(s)} = \mathbf{f}^{(s)} + \mathbf{g}^{(s)} \tag{3.35}
$$

every substructure can be seen as being excited by the interface connection forces and the external forces (contrary to the paradigm underlying the HCM method, where the substructures are considered as being excited by interface displacements).

This indicates that the displacements of each substructure $\mathbf{u}^{(s)}$ can be expressed in terms of local static solutions $\mathbf{u}^{(s)}_{stat}$ and in terms of eigenmodes associated to the entire substructure matrices $\mathbf{K}^{(s)}$ and $\mathbf{M}^{(s)}$ (hence the free interface modes):

$$\mathbf{u}^{(s)} = \mathbf{u}^{(s)}_{stat} + \sum_{j=1}^{n^{(s)}-r^{(s)}} \boldsymbol{\theta}^{(s)}_j \eta^{(s)}_j, \tag{3.36}$$

where $n^{(s)}$ and $r^{(s)}$ are the number of DOF and of rigid-body modes for substructure s, and where the free interface modes are solutions of the eigenvalue problem

$$\left(\mathbf{K}^{(s)} - \omega^{(s)^2}_j \mathbf{M}^{(s)}\right) \boldsymbol{\theta}^{(s)}_j = \mathbf{0}. \tag{3.37}$$

The static solution is written as follows (assuming no external forces are applied for simplicity):

$$\mathbf{u}^{(s)}_{stat} = -\mathbf{K}^{(s)^+} \mathbf{g}^{(s)} + \sum_{j=1}^{r^{(s)}} \mathbf{r}^{(s)}_j \alpha^{(s)}_j \tag{3.38}$$

The static solution arises from solving Eq. (3.35) under the assumption that there are no inertia forces and no external forces acting on the substructure. The notation $\mathbf{K}^{(s)^+}$ denotes the inverse of $\mathbf{K}^{(s)}$ when there are enough boundary conditions to prevent the substructure with a free interface from floating Rixen (2004). If a substructure is floating, $\mathbf{K}^{(s)^+}$ is the generalized inverse of $\mathbf{K}^{(s)}$ and $\mathbf{r}^{(s)}_j$ are the rigid-body modes of s (see, for instance, Géradin and Rixen 2015 on how to compute a generalized inverse and the rigid-body modes).

If only a limited number of free interface modes are used for the substructure dynamics, Eqs. (3.36), (3.38) results in the approximation

$$\mathbf{u}^{(s)} \approx -\mathbf{K}^{(s)^+} \mathbf{B}^{(s)^T} \boldsymbol{\lambda} + \mathbf{R}^{(s)} \boldsymbol{\alpha}^{(s)} + \boldsymbol{\Theta}^{(s)} \boldsymbol{\eta}^{(s)} \tag{3.39}$$

The vector $\boldsymbol{\alpha}^{(s)}$ contains the amplitudes of the rigid-body modes and the vector $\boldsymbol{\eta}^{(s)}$ contains the amplitudes of the retained $n^{(s)}_\theta$ local free interface modes eigenmodes. The matrices $\mathbf{R}^{(s)}$ and $\boldsymbol{\Theta}^{(s)}$ contain all rigid-body modes and the retained eigenmodes.

Since a part of the subspace spanned by $\boldsymbol{\Theta}^{(s)}$ is already included in $\mathbf{K}^{(s)^+}$, the residual flexibility matrix $\mathbf{G}^{(s)}_r$ can be used instead of $\mathbf{K}^{(s)^+}$, which is defined by

$$\mathbf{G}^{(s)}_r = \sum_{j=n^{(s)}_\theta+1}^{n^{(s)}-r^{(s)}} \frac{\boldsymbol{\theta}^{(s)}_j \boldsymbol{\theta}^{(s)^T}_j}{\omega^{(s)^2}_j} \tag{3.40}$$

$$= \mathbf{K}^{(s)+} - \sum_{j=1}^{n_\theta^{(s)}} \frac{\boldsymbol{\theta}_j^{(s)} \boldsymbol{\theta}_j^{(s)^T}}{\omega_j^{(s)^2}}.$$

Note that, by construction, $\mathbf{G}_r^{(s)} = \mathbf{G}_r^{(s)^T}$, which is computed using the second equality in Eq. (3.40). For further properties of $\mathbf{G}_r^{(s)}$, see Rixen (2004). As a result, the approximation of one substructure writes

$$\mathbf{u}^{(s)} \approx \underbrace{\left[\mathbf{R}^{(s)}\ \boldsymbol{\Theta}^{(s)}\ \mathbf{G}_r^{(s)}\mathbf{A}^{(s)^T}\right]}_{\mathbf{T}_1^{(s)}} \begin{bmatrix} \boldsymbol{\alpha}^{(s)} \\ \boldsymbol{\eta}^{(s)} \\ \mathbf{g}_b^{(s)} \end{bmatrix}. \tag{3.41}$$

$\mathbf{G}_r^{(s)}\mathbf{A}^{(s)^T}$ is the matrix containing the residual flexibility attachment modes of substructure s, since the Boolean localization matrix $\mathbf{A}^{(s)^T}$ simply picks the columns of $\mathbf{G}_r^{(s)}$ associated to the boundary DOF. In other words,

$$\mathbf{A}^{(s)^T}\mathbf{g}_b^{(s)} = \begin{bmatrix} \mathbf{0} \\ \mathbf{g}_b^{(s)} \end{bmatrix} = \mathbf{g}^{(s)} = \mathbf{B}^{(s)^T}\boldsymbol{\lambda}.$$

The approximation (3.41) can now be used to reduce the substructure DOF. Using the orthogonality properties of the modes in Eq. (3.41), the equation of motion of one substructure becomes

$$\mathbf{M}_{free}^{(s)} \begin{bmatrix} \ddot{\boldsymbol{\alpha}}^{(s)} \\ \ddot{\boldsymbol{\eta}}^{(s)} \\ \ddot{\mathbf{g}}_b^{(s)} \end{bmatrix} + \mathbf{K}_{free}^{(s)} \begin{bmatrix} \boldsymbol{\alpha}^{(s)} \\ \boldsymbol{\eta}^{(s)} \\ \mathbf{g}_b^{(s)} \end{bmatrix} = \mathbf{T}_1^{(s)^T}\left(\mathbf{f}^{(s)} + \mathbf{g}^{(s)}\right) \tag{3.42}$$

with the matrices

$$\mathbf{K}_{free}^{(s)} = \mathbf{T}_1^{(s)^T}\mathbf{K}^{(s)}\mathbf{T}_1^{(s)} = \begin{bmatrix} \mathbf{0} & \mathbf{0} & \mathbf{0} \\ \mathbf{0} & \boldsymbol{\Omega}^{(s)^2} & \mathbf{0} \\ \mathbf{0} & \mathbf{0} & \mathbf{G}_{r,bb}^{(s)} \end{bmatrix} \tag{3.43}$$

where

$$\mathbf{G}_{r,bb}^{(s)} = \mathbf{A}^{(s)}\mathbf{G}_r^{(s)}\mathbf{A}^{(s)^T} \tag{3.44}$$

and

$$\mathbf{M}_{free}^{(s)} = \mathbf{T}_1^{(s)^T}\mathbf{M}^{(s)}\mathbf{T}_1^{(s)} = \begin{bmatrix} \mathbf{I} & \mathbf{0} & \mathbf{0} \\ \mathbf{0} & \mathbf{I} & \mathbf{0} \\ \mathbf{0} & \mathbf{0} & \mathbf{M}_{r,bb}^{(s)} \end{bmatrix} \tag{3.45}$$

where

$$\mathbf{M}_{r,bb}^{(s)} = \mathbf{A}^{(s)}\mathbf{G}_r^{(s)}\mathbf{M}^{(s)}\mathbf{G}_r^{(s)}\mathbf{A}^{(s)^T}. \tag{3.46}$$

$\mathbf{G}_{r,bb}^{(s)}$ is the residual flexibility and $\mathbf{M}_{r,bb}^{(s)}$ is the interface inertia associated to the residual flexibility related to the boundary DOF, respectively, and $\boldsymbol{\Omega}^{(s)}$ being a diagonal matrix containing the retained $n_\theta^{(s)}$ eigenvalues $\omega_j^{(s)}$.

Rubin Method (RM)

The Rubin method was proposed by Rubin (1975) and, with the definitions above, can be explained as follows. In order to assemble in a primal manner the substructure equation of motion (3.42) in the global system, a second transformation is applied by the RM. The force DOF $\mathbf{g}_b^{(s)}$ are transformed back to the boundary displacements $\mathbf{u}_b^{(s)}$ using Eq. (3.41) (see for instance Voormeeren et al. 2011):

$$\mathbf{u}_b^{(s)} = \mathbf{A}^{(s)}\mathbf{u}^{(s)} = \mathbf{R}_b^{(s)}\boldsymbol{\alpha}^{(s)} + \boldsymbol{\Theta}_b^{(s)}\boldsymbol{\eta}^{(s)} + \mathbf{G}_{r,bb}^{(s)}\mathbf{g}_b^{(s)} \tag{3.47}$$

$\mathbf{R}_b^{(s)}$ and $\boldsymbol{\Theta}_b^{(s)}$ are the subparts of $\mathbf{R}^{(s)}$ and $\boldsymbol{\Theta}^{(s)}$ related to the boundary DOF, respectively. From this equation, the interface force DOF can be solved as

$$\mathbf{g}_b^{(s)} = \mathbf{K}_{r,bb}^{(s)}\left(\mathbf{u}_b^{(s)} - \mathbf{R}_b^{(s)}\boldsymbol{\alpha}^{(s)} - \boldsymbol{\Theta}_b^{(s)}\boldsymbol{\eta}^{(s)}\right) \tag{3.48}$$

with $\mathbf{K}_{r,bb}^{(s)} = \mathbf{G}_{r,bb}^{(s)^{-1}}$. The transformation matrix $\mathbf{T}_2^{(s)}$ from force DOF $\mathbf{g}_b^{(s)}$ back to the boundary displacements $\mathbf{u}_b^{(s)}$ leaving $\boldsymbol{\alpha}^{(s)}$ and $\boldsymbol{\eta}^{(s)}$ unchanged writes then

$$\mathbf{T}_2^{(s)} = \begin{bmatrix} \mathbf{I} & \mathbf{0} & \mathbf{0} \\ \mathbf{0} & \mathbf{I} & \mathbf{0} \\ -\mathbf{K}_{r,bb}^{(s)}\mathbf{R}_b^{(s)} & -\mathbf{K}_{r,bb}^{(s)}\boldsymbol{\Theta}_b^{(s)} & \mathbf{K}_{r,bb}^{(s)}. \end{bmatrix} \tag{3.49}$$

The RM approximation for one substructure writes, therefore,

$$\mathbf{u}^{(s)} = \mathbf{T}_R^{(s)} \begin{bmatrix} \boldsymbol{\alpha}^{(s)} \\ \boldsymbol{\eta}^{(s)} \\ \mathbf{u}_b^{(s)} \end{bmatrix} \quad \text{where} \quad \mathbf{T}_R^{(s)} = \mathbf{T}_1^{(s)}\mathbf{T}_2^{(s)}. \tag{3.50}$$

Application of this transformation to the matrices of Eqs. (3.45) and (3.43) gives the RM-reduced matrices of one substructure:

$$\mathbf{K}_{red,R}^{(s)} = \mathbf{T}_2^{(s)^T}\mathbf{K}_{free}^{(s)}\mathbf{T}_2^{(s)} = \mathbf{T}_R^{(s)^T}\mathbf{K}^{(s)}\mathbf{T}_R^{(s)}, \tag{3.51}$$

$$\mathbf{M}_{red,R}^{(s)} = \mathbf{T}_2^{(s)^T}\mathbf{M}_{free}^{(s)}\mathbf{T}_2^{(s)} = \mathbf{T}_R^{(s)^T}\mathbf{M}^{(s)}\mathbf{T}_R^{(s)}. \tag{3.52}$$

These matrices can be directly assembled using primal assembly to get the RM-reduced matrices $\mathbf{K}_{red,R}$ and $\mathbf{M}_{red,R}$ of the global system. The RM applies the reduction matrix $\mathbf{T}_R^{(s)}$ consistently to the mass and stiffness matrix resulting in a true Rayleigh–Ritz method as was observed in Craig and Chang (1977).

MacNeal Method (MNM)

The MNM MacNeal (1971) is nearly identical to the RM except for a small change. First, we will derive the preliminary MNM-reduced matrices $\tilde{\mathbf{K}}^{(s)}_{red,MN}$ and $\tilde{\mathbf{M}}^{(s)}_{red,MN}$ following the derivation of the RM to show the similarities between these two methods. The reduced stiffness matrix of both the RM and the MNM are identical (given in Eq. (3.51))

$$\tilde{\mathbf{K}}^{(s)}_{red,MN} = \mathbf{K}^{(s)}_{red,R} \tag{3.53}$$

but the MNM-reduced mass matrix $\tilde{\mathbf{M}}^{(s)}_{red,MN}$ is obtained differently. The residual mass term $\mathbf{M}^{(s)}_{r,bb}$ of the matrix $\mathbf{M}^{(s)}_{free}$ in Eq. (3.45) is neglected, resulting in a modified matrix labeled

$$\mathbf{M}^{(s)}_{free,MN} = \begin{bmatrix} \mathbf{I} & \mathbf{0} & \mathbf{0} \\ \mathbf{0} & \mathbf{I} & \mathbf{0} \\ \mathbf{0} & \mathbf{0} & \mathbf{0} \end{bmatrix} \tag{3.54}$$

instead of $\mathbf{M}^{(s)}_{free}$ for the MNM Voormeeren et al. (2011). The MNM-reduced mass matrix writes now

$$\tilde{\mathbf{M}}^{(s)}_{red,MN} = \mathbf{T}_2^{(s)^T} \mathbf{M}^{(s)}_{free,MN} \mathbf{T}_2^{(s)} = \mathbf{M}^{(s)}_{free,MN}. \tag{3.55}$$

This gives in fact inconsistent equations of motion since the mass and stiffness matrices are not reduced with the same basis. The assembly of the MNM-reduced matrices $\tilde{\mathbf{K}}^{(s)}_{red,MN}$ and $\tilde{\mathbf{M}}^{(s)}_{red,MN}$ in the global system proceeds in the same manner as for the RM. Observing that the boundary DOF $\mathbf{u}_b$ have no associated inertia in Eq. (3.55), those DOF can be condensed out of the equation of motion of the assembled problem and the final MNM-reduced matrices $\mathbf{K}_{red,MN}$ and $\mathbf{M}_{red,MN}$ are obtained (MacNeal 1971). Thus, the size of the assembled MNM system is reduced further by the number of DOF of $\mathbf{u}_b$.

Dual Craig–Bampton Method (DCBM)

The reduction transformation (3.41) can also be used directly, without transforming $\mathbf{A}^{(s)^T}\mathbf{g}_b^{(s)} = \mathbf{B}^{(s)^T}\boldsymbol{\lambda}$ in interface DOF (as done for the RM or for the MNM) if one considered the dually assembled problem (see Sect. 2.3.2). In the physical domain, the dual assembly writes

$$\begin{bmatrix} \ddots & \mathbf{0} & & \vdots \\ \mathbf{0} & \mathbf{M}^{(s)} & & \mathbf{0} \\ & & \ddots & \vdots \\ \dots & \mathbf{0} & \dots & \mathbf{0} \end{bmatrix} \begin{bmatrix} \vdots \\ \ddot{\mathbf{u}}^{(s)} \\ \vdots \\ \ddot{\boldsymbol{\lambda}} \end{bmatrix} + \begin{bmatrix} \ddots & \mathbf{0} & & \vdots \\ \mathbf{0} & \mathbf{K}^{(s)} & & \mathbf{B}^{(s)^T} \\ & & \ddots & \vdots \\ \dots & \mathbf{B}^{(s)} & \dots & \mathbf{0} \end{bmatrix} \begin{bmatrix} \vdots \\ \mathbf{u}^{(s)} \\ \vdots \\ \boldsymbol{\lambda} \end{bmatrix} = \begin{bmatrix} \vdots \\ \mathbf{f}^{(s)} \\ \vdots \\ \mathbf{0}. \end{bmatrix} \tag{3.56}$$

Assembling all substructures N in a dual fashion by keeping the interface forces $\boldsymbol{\lambda}$ as unknowns, the entire structure can consequently be approximated by

$$\begin{bmatrix} \mathbf{u} \\ \boldsymbol{\lambda} \end{bmatrix} \approx \mathbf{T}_{DCB} \begin{bmatrix} \boldsymbol{\alpha}^{(1)} \\ \boldsymbol{\eta}^{(1)} \\ \vdots \\ \boldsymbol{\alpha}^{(N)} \\ \boldsymbol{\eta}^{(N)} \\ \boldsymbol{\lambda} \end{bmatrix} \tag{3.57}$$

with the DCBM reduction matrix $\mathbf{T}_{DCB}$:

$$\mathbf{T}_{DCB} = \begin{bmatrix} \mathbf{R}^{(1)} & \boldsymbol{\Theta}^{(1)} & & \mathbf{0} & \mathbf{0} & -\mathbf{G}_r^{(1)}\mathbf{B}^{(1)^T} \\ & & \ddots \quad \ddots & & & \vdots \\ \mathbf{0} & \mathbf{0} & & \mathbf{R}^{(N)} & \boldsymbol{\Theta}^{(N)} & -\mathbf{G}_r^{(N)}\mathbf{B}^{(N)^T} \\ \mathbf{0} & \mathbf{0} & & \mathbf{0} & \mathbf{0} & \mathbf{I} \end{bmatrix}. \tag{3.58}$$

The approximation of the dynamic equations of the dual assembled system (3.56) is

$$\mathbf{M}_{red,DCB} \begin{bmatrix} \ddot{\boldsymbol{\alpha}}^{(1)} \\ \ddot{\boldsymbol{\eta}}^{(1)} \\ \vdots \\ \ddot{\boldsymbol{\alpha}}^{(N)} \\ \ddot{\boldsymbol{\eta}}^{(N)} \\ \ddot{\boldsymbol{\lambda}} \end{bmatrix} + \mathbf{K}_{red,DCB} \begin{bmatrix} \boldsymbol{\alpha}^{(1)} \\ \boldsymbol{\eta}^{(1)} \\ \vdots \\ \boldsymbol{\alpha}^{(N)} \\ \boldsymbol{\eta}^{(N)} \\ \boldsymbol{\lambda} \end{bmatrix} = \mathbf{T}_{DCB}^T \mathbf{f} \tag{3.59}$$

with the DCBM reduced mass and stiffness matrix

$$\mathbf{M}_{red,DCB} = \mathbf{T}_{DCB}^T \begin{bmatrix} \mathbf{M} & \mathbf{0} \\ \mathbf{0} & \mathbf{0} \end{bmatrix} \mathbf{T}_{DCB} = \begin{bmatrix} \mathbf{I} & \mathbf{0} \\ \mathbf{0} & \mathbf{M}_r \end{bmatrix} \tag{3.60}$$

$$\mathbf{K}_{red,DCB} = \mathbf{T}_{DCB}^T \begin{bmatrix} \mathbf{K} & \mathbf{B}^T \\ \mathbf{B} & \mathbf{0} \end{bmatrix} \mathbf{T}_{DCB} \tag{3.61}$$

with

$$\mathbf{M}_r = \sum_{s=1}^{N} \mathbf{B}^{(s)} \mathbf{G}_r^{(s)} \mathbf{M}^{(s)} \mathbf{G}_r^{(s)} \mathbf{B}^{(s)^T}. \tag{3.62}$$

$\mathbf{M}_{red,DCB}$ and $\mathbf{K}_{red,DCB}$ are diagonal for the parts related to the different substructures. The coupling between the substructures is only achieved by the rows and columns related to $\boldsymbol{\lambda}$. The DCBM applies the reduction matrix $\mathbf{T}_{DCB}$ consistently to the mass and stiffness matrix resulting in a true Rayleigh–Ritz method.

The DCBM enforces only a weak compatibility between the substructures and does not enforce a strong displacement compatibility between the interfaces compared to many other common reduction methods (Rixen 2004). Considering the system of Eqs. (3.56) and (3.35) multiplied by the reduction matrix $\mathbf{T}_{DCB}^T$, the last row of Eq. (3.59) results from

$$\begin{bmatrix} \mathbf{M}^{(1)}\ddot{\mathbf{u}}^{(1)} + \mathbf{K}^{(1)}\mathbf{u}^{(1)} + \mathbf{B}^{(1)^T}\boldsymbol{\lambda} = \mathbf{f}^{(1)} \\ \vdots \\ \mathbf{M}^{(N)}\ddot{\mathbf{u}}^{(N)} + \mathbf{K}^{(N)}\mathbf{u}^{(N)} + \mathbf{B}^{(N)^T}\boldsymbol{\lambda} = \mathbf{f}^{(N)} \\ \sum_{s=1}^{N} \mathbf{B}^{(s)}\mathbf{u}^{(s)} = \mathbf{0} \end{bmatrix} \tag{3.63}$$

multiplied from left by the last row of $\mathbf{T}_{DCB}^T$ which is

$$\left[-\mathbf{B}^{(1)}\mathbf{G}_r^{(1)} \ \cdots \ -\mathbf{B}^{(N)}\mathbf{G}_r^{(N)} \ \mathbf{I}\right]. \tag{3.64}$$

Replacing the strong interface compatibility condition of Eq. (3.56) by the weak form according to the multiplication of Eq. (3.63) by Eq. (3.64) can be interpreted as follows. Denote $\Delta\mathbf{f}^{(s)}$ the residual forces of substructure s resulting from the weak satisfaction of the local equilibrium of the substructure approximating the dynamics by a small number of free interface normal modes. Name $\Delta\mathbf{u}^{(s)} = \mathbf{G}_r^{(s)}\Delta\mathbf{f}^{(s)}$ the displacements these residual force $\Delta\mathbf{f}^{(s)}$ would create locally. Then the weak interface compatibility condition Eqs. (3.63) and (3.64) states that a compatibility error (i.e., an interface displacement jump) equal to the incompatibility of $\Delta\mathbf{u}^{(s)}$ is permitted Rixen (2004).

Compared to MacNeal's and Rubin's method (1971, 1975), the weak interface compatibility of the DCBM avoids locking problems occurring during the application of the aforementioned methods. Therefore, the approximation accuracy is improved Rixen (2004). But the fact that a weak interface compatibility is allowed in the DCBM implies that the infinite eigenvalues related to the Lagrange multipliers $\boldsymbol{\lambda}$ in the non-reduced problem (3.56) are now becoming finite and negative Rixen (2011). In practice those negative eigensolutions will appear only in the higher eigenvalue spectrum if the reduction space is rich enough (Rixen 2011). Nevertheless, the reduction basis has to be selected with care avoiding potential nonphysical effects of the possibly occurring negative eigenvalues.

If $\mathbf{M}_r$ in Eq. (3.60) is neglected strong interface compatibility is enforced again and the DCBM reduced system with $\mathbf{M}_r = \mathbf{0}$ is equivalent to the MNM (Rixen 2004). Then static condensation can be applied again to remove $\boldsymbol{\lambda}$ (as it was done for $\mathbf{u}_b$ at the end of the derivation of the MNM in section "MacNeal Method (MNM)") from the assembled system since no mass is associated. Thus, the size of the assembled system is reduced again by the number of DOF of $\boldsymbol{\lambda}$.

3.2.3 Numerical Examples of Different Substructure Reduction Techniques

The Benfield truss (Benfield and Hruda 1971) of Fig. 3.3 is used to compare the results obtainable by the HCB, the MNM, the RM, and the DCBM. The planar truss consists of two substructures having uniform bay section whereas all members have constant area and uniform stiffness and mass properties. The left component consists of five equal bays and has a total of 18 joints and the right component consists of four equal bays and has a total of 15 joints (Benfield and Hruda 1971). The lowest eigenfrequencies ω of the entire structure shall be approximated by the different methods.

The relative error $\varepsilon_{rel,j} = |\omega_{red,j} - \omega_{full,j}|/\omega_{full,j}$ of the jth eigenfrequency is used as a criterion to assess the accuracy of the different methods. Thereby, $\omega_{full,j}$ is the jth eigenfrequency of the full (non-reduced) system and $\omega_{red,j}$ represents the jth eigenfrequency of the reduced system obtained by each method.

Using 5 elastic (fixed or free interface normal modes) per substructure the relative errors ε_{rel} depicted in the semi-log graph in Fig. 3.4 are resulting. Since all methods give the correct rigid-body modes only the relative errors of the elastic modes are plotted. All methods give a relative error of less than 1% for the first six eigenfrequencies. Comparing the free interface methods for this example, the RM performs always better than the DCB and the DCB performs again always better as the MNM. The HCB and the DCB result in similar frequency errors.

The sparsity pattern of the reduced stiffness matrix $\mathbf{K}_{red}$ and reduced mass matrix $\mathbf{M}_{red}$ of the HCB (Fig. 3.5), the MNM (Fig. 3.7), the RM (Fig. 3.8) and the DCB (Fig. 3.6), respectively, illustrate the differences of the assembled reduced structures. Both the reduced stiffness matrix $\mathbf{K}_{red}$ and the reduced mass $\mathbf{M}_{red}$ matrix applying the HCB and the DCB, respectively, have only diagonal entries for the subparts of each substructure. On the one hand, the coupling between the substructures using the HCB is entirely achieved by the last rows and last columns in the mass matrix $\mathbf{M}_{red,CB}$ (Fig. 3.5b) and the remaining part is diagonal (Bampton and Craig 1968). On the other hand, the coupling applying the DCB is entirely achieved by the

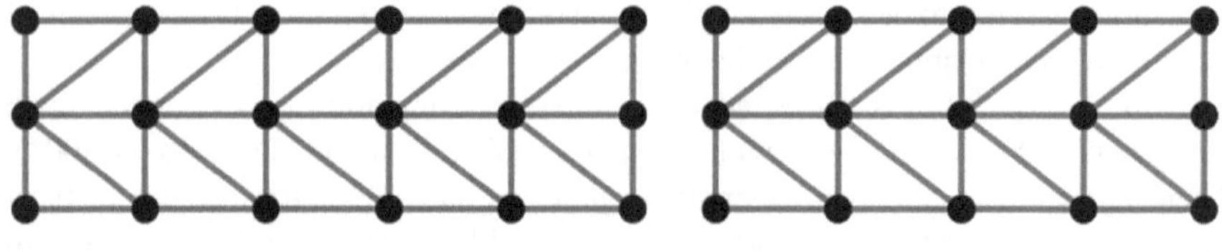

Fig. 3.3 Benfield truss (Benfield and Hruda 1971)

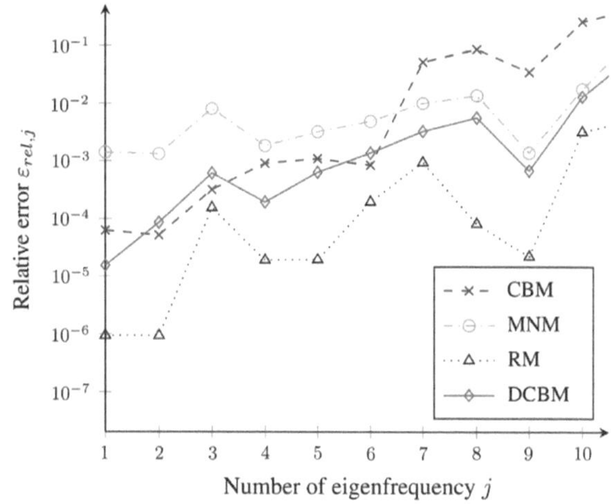

Fig. 3.4 Relative error $\varepsilon_{rel,j}$ of eigenfrequency j using 5 normal modes per substructure for the approximation of the lowest eigenfrequencies of the Benfield truss

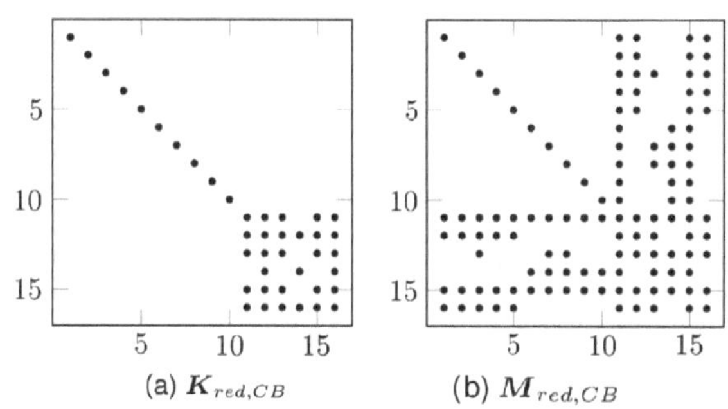

Fig. 3.5 Sparsity pattern of the reduced matrices applying the HCB using 5 normal modes per substructure

last rows and last columns in the stiffness matrix $\mathbf{K}_{red,DCB}$ (Fig. 3.6a) and again the remaining part is diagonal (Rixen 2004). The corresponding degrees of freedoms are either the interface displacements $\mathbf{u}_b$ or the interface forces $\boldsymbol{\lambda}$ but no direct coupling between the modal parameters of adjacent substructures occurs which ensures the sparse structure.

In contrast, the sparsity pattern of stiffness matrix $\mathbf{K}_{red}$ and the reduced mass matrix $\mathbf{M}_{red}$ obtained by the MNM and the RM, respectively, show full matrices. The MNM gives indeed an entirely diagonal reduced mass matrix $\mathbf{M}_{red,MN}$ (Fig. 3.7b) but causes always a full coupling between all DOF of all substructures via the reduced stiffness matrix $\mathbf{K}_{red,MN}$ (Fig. 3.7a). This makes the reusability of reduced models obtained by the MNM very inefficient and therefore nearly impossible from a practical point of view.

The RM also causes a coupling between the substructures via interface displacements $\mathbf{u}_b$ in the reduced stiffness matrix $\mathbf{K}_{red,R}$ (Fig. 3.8a) as well as in the reduced mass $\mathbf{M}_{red,R}$ (Fig. 3.8b). Moreover, all DOF belonging to one reduced substructure are coupled with all other DOF of the same substructure which is why the reduced matrices of the RM are full for the substructure blocks and not diagonal.

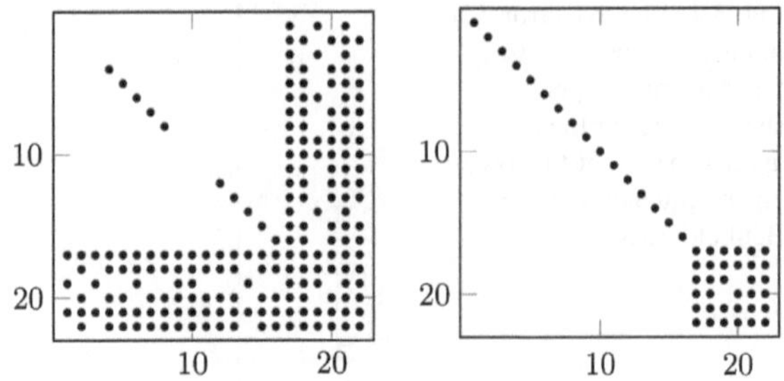

(a) $\boldsymbol{K}_{red,DCB}$ (b) $\boldsymbol{M}_{red,DCB}$

Fig. 3.6 Sparsity pattern of the reduced matrices applying the DCB using 5 normal modes per substructure

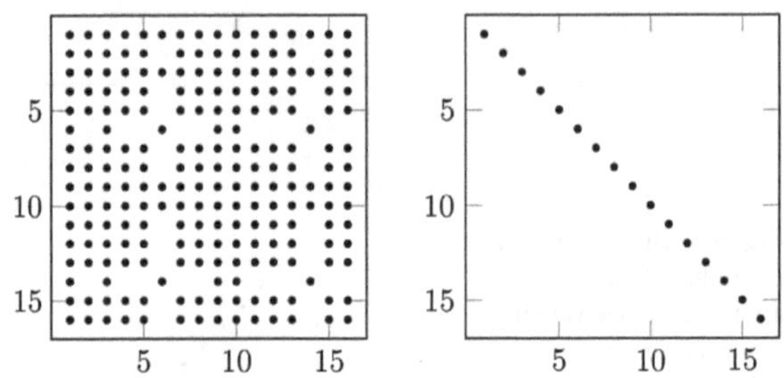

(a) $\boldsymbol{K}_{red,MN}$ (b) $\boldsymbol{M}_{red,MN}$

Fig. 3.7 Sparsity pattern of the reduced matrices applying the MNM using 5 normal modes per substructure

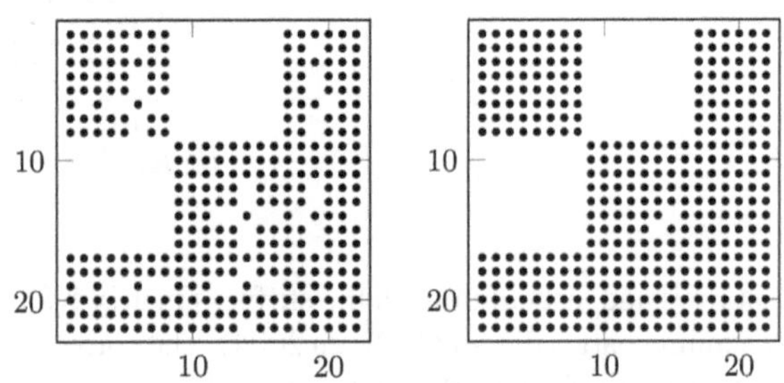

(a) $\boldsymbol{K}_{red,R}$ (b) $\boldsymbol{M}_{red,R}$

Fig. 3.8 Sparsity pattern of the reduced matrices applying the RM using 5 normal modes per substructure

This result concerning the sparsity of the reduced matrices is outlined in Table 3.1, which shows the number n of nonzero elements in the reduced matrices $\mathbf{K}_{red}$ and $\mathbf{M}_{red}$ and the sum n_{total} of both obtained by the different methods for this example. The reduced matrices of the HCB, the MNM, and the DCB contain a similar number of entries while the RM causes even for such a simple example a remarkable high number of entries. The number of entries of the MNM are comparable to the HCB and the DCB but will increase dramatically if the number of substructures is increased since $\mathbf{K}_{red}$ will always be completely full.

Table 3.1 Number n of nonzero elements in the reduced matrices obtained by the different methods for the Benfield truss using 5 normal modes per substructure

	HCB	MNM	RM	DCB
n in $\mathbf{K}_{red}$	40	216	314	196
n in $\mathbf{M}_{red}$	118	16	354	50
n_{total}	158	232	668	246

3.2.4 Other Reduction Techniques for Substructures

In the previous sections, we have outlined the classical substructure reduction methods (or Component Mode Synthesis) using either free or fixed interface modes.

Other methods, are not outlined here, like for instance the Craig–Chang approach which uses also free interface modes, but computes the interface forces from a global (assembled) problem (Craig and Chang 1977).

Many other variants of CMS methods were published over the past years and they can be classified as follows:

- **Loaded Interface Modes** Some authors have proposed to compute the quasi-static modes around a central frequency (Shyu et al. 1997) or to use vibration modes with an impedance attached to the interface. In certain cases, such modifications can improve the accuracy of the reduced model.
- **Modal Truncation Augmentation and Moment Matching** The vibration modes used, for instance, in the HCB or in the RM approaches are not specifically tuned for the excitations coming through the interface, whereas the static modes are. It is possible to enrich the static modes with higher order static contribution computed over a Krylov series. This leads to a method original called Modal Truncation Augmentation (or MTA) (Dickens and Stroeve 2000) and was generalized to higher orders corrections in the HCB in Rixen (2002). Later, techniques called Moment Matching (Su and Craig 1991), that basically used the same ideas as MTAs, were applied for model reduction mainly in the control community (see, for instance, (Antoulas 2005, Holzwarth and Eberhard 2015, Stoev and Möhring 2010). The idea of MTA was also applied for the DCBM (Rixen 2009, Kim et al. 2017).
- **Balanced Truncation** In the control community, reduction is seen from the point of view of controllability and observability. This leads to representation modes derived from so-called Grammians that were used for reducing substructure models in Holzwarth and Eberhard (2015), Stoev and Möhring (2010). Balancing of State-Space models will be discussed in Sect. 3.4.4.
- **Mixed Methods** Several methods were published where different types of vectors (attachment modes, static modes ...) were combined. One of the most commonly used in major Finite Element software (although not cited much in the community, maybe because the explanations in the publication are poor) is the one from Herting (Herting 1985). Note also that it is possible to mix primal and dual assembly

for different DOF on an interface (which can be advantageous depending on the stiffness ratios across the interface) (Voormeeren et al. 2011).

- Finally, let us mention that methods have been proposed where the reduction is performed in an iterative manner, computing the residual force on the full model after a reduction (see (3.3)) and enriching the reduction base with, for instance, the static response of the substructures to that residual force (Balmès 1997, Bathe and Dong 2014).

3.3 Interface Reduction with the Hurty/Craig–Bampton Method: Characteristic Constraint Modes

While the number of degrees of freedom in a model may be reduced dramatically using the Hurty/Craig–Bampton (HCB) method, that method retains all of the DOF at the interfaces between substructures, and so the resulting model may still be unacceptably large. Furthermore, since the minimum timestep for time integration is limited by the distance between the two closest nodes, the HCB model may still be almost as expensive to integrate as the original model. Several methods have been proposed over the years to address this problem, and while none has been widely implemented as has the HCB method, they do seem to be effective in many scenarios. This section presents a brief review of methods for interface reduction. For further details and a comparison of each method on the W-bracket, see the recent review in Krattiger et al. (2019).

Craig and Chang (1977) seem to have been the first to propose methods to reduce the interface DOF, presenting three methods in 1977. However, they do not appear to have been used subsequently until Castanier et al. (2001) rediscovered the modal method by applying a secondary eigenvalue analysis on the interface partition of the assembled CMS model and obtaining what they called the Characteristic Constraint (CC) modes. The assembled, system-level HCB matrices were used to compute these modes, so this method is referred to as the System-level Characteristic Constraint (S-CC) mode method. While the method is effective, it is often undesirable to have to assemble the system before reduction and so this has inspired the investigation of other methods that perform the reduction before assembly.

Hong et al. (2013) subsequently proposed an interface reduction technique that performs an eigenvalue analysis on the HCB interface DOF prior to assembly. They then concatenate the shapes obtained for each interface and enforce compatibility between the local-level characteristic constraint (L-CC) modes obtained from each interface. This method is referred to in Krattiger et al. (2019) as the "exact compatibility L-CC method," because it is possible to enforce exact compatibility using that method. However, in practice, some of the shapes are typically truncated and there is some level of approximation.

Alternatively, Kuether et al. (2017) proposed a method that weakens the compatibility at the interface. The L-CC modes of each substructure remain separate but are combined to minimize the compatibility error between connecting substructure interfaces. Aoyama et al. (2001) presented a method similar to S-CC except that each interface is assembled and then reduced separately. The recently presented Hybrid-level Characteristic Constraint (H-CC) mode method blends these concepts, performing an S-CC type analysis on each pair of interfaces. In a related work, Balmès (1996) explored a CMS basis that defined arbitrary interface deformations to describe a set of generalized DOF along an interface. A few other less commonly known methods are reviewed in (Krattiger et al. (2019)).

The following reviews three different interface reduction strategies: the S-CC method (Castanier et al. 2001), the Exact-Compatibility (EC) (Hong et al. 2013), the Weak-Compatibility (WC) method (Kuether et al. 2017) and the Virtual Node (VN) method. Additional methods are reviewed in Krattiger et al. (2019). The phrase "system-level" indicates that interface reduction is applied to the fully assembled HCB model, whereas "local-level" indicates that interface reduction is performed on the individual substructures before they are assembled.

3.3.1 System-Level Characteristic Constraint (S-CC) Modes

As done by Craig and Chang (1977), and detailed by Castanier et al. (2001), the HCB models for all subcomponents are first assembled, the interface partition of the assembled system is extracted and then an eigenvalue analysis is performed

$$\left(\mathbf{K}_{bb}^{\text{HCB}} - \omega_r^2 \mathbf{M}_{bb}^{\text{HCB}}\right) \boldsymbol{\phi}_r^{\text{S-CC}} = \mathbf{0}. \tag{3.65}$$

The eigenvectors obtained are called S-CC modes and the first n vectors are collected into a matrix,

$$\boldsymbol{\Phi}^{\text{S-CC}} = \left[\boldsymbol{\phi}_1^{\text{S-CC}}, \ldots, \boldsymbol{\phi}_n^{\text{S-CC}}\right]. \tag{3.66}$$

The matrix $\boldsymbol{\Phi}^{\text{S-CC}}$ is used to replace the interface DOF of the HCB model into amplitudes of the S-CC modes,

$$\mathbf{u}_b \approx \boldsymbol{\Phi}^{\text{S-CC}} \mathbf{q}_b. \tag{3.67}$$

The S-CC reduced mass (and similarly stiffness) matrices are obtained by pre- and post- multiplying the assembled HCB matrices in Eq. (3.33) with the S-CC transformation matrix to obtain

$$\mathbf{M}^{\text{S-CC}} = \begin{bmatrix} \mathbf{I} & \mathbf{0} \\ \mathbf{0} & \boldsymbol{\Phi}^{\text{S-CC}} \end{bmatrix}^{\text{T}} \begin{bmatrix} \mathbf{M}_{ii}^{\text{HCB}} & \mathbf{M}_{ib}^{\text{HCB}} \\ \mathbf{M}_{bi}^{\text{HCB}} & \mathbf{M}_{bb}^{\text{HCB}} \end{bmatrix} \underbrace{\begin{bmatrix} \mathbf{I} & \mathbf{0} \\ \mathbf{0} & \boldsymbol{\Phi}^{\text{S-CC}} \end{bmatrix}}_{\mathbf{T}^{\text{S-CC}}}. \tag{3.68}$$

This reduces the size of the interface partition of the HCB model from the number of physical DOF on the interface to the number of S-CC modes in the truncated eigenvector set, $n_{\boldsymbol{\Phi}_b}$. It is interesting to note that static reduction modes were proposed for the interface of a dual Craig–Bampton model in Rixen (2011).

3.3.2 *Local-Level Characteristic Constraint Modes*

As mentioned previously, the prior method requires that the complete HCB model of each component be assembled prior to reduction. This may not provide the desired computational gains in many applications, and it requires knowledge of what the component will connect to in order to perform the reduction. An alternative is the local-level interface reduction techniques, where the interface is reduced before the subsystems are assembled.

Exact Compatibility (Hong et al. 2013)

The EC L-CC interface reduction was first proposed by Hong et al. (2013) and is uses a similar eigenvalue analysis on the interface partition, only this is performed on each individual substructure prior to assembly. For the jth subcomponent, the following eigenvalue problem is solved:

$$\left(\mathbf{K}_{bb,j}^{\mathrm{HCB}} - \omega_r^2 \mathbf{M}_{bb,j}^{\mathrm{HCB}}\right) \left\{\boldsymbol{\phi}_j^{\mathrm{L\text{-}CC}}\right\}_r = \mathbf{0} \tag{3.69}$$

The L-CC modes are denoted $\boldsymbol{\Phi}_j^{\mathrm{L\text{-}CC}}$ for the jth substructure. The L-CC modes for each substructure are then used to reduce the interface DOF of that substructure.

Once again the W-bracket system in Fig. 3.2 can be used to illustrate the process. The L-CC modes in $\boldsymbol{\Phi}_1^{\mathrm{L\text{-}CC}}$ from Eq. (3.69) have interface DOF for boundary sets $\boldsymbol{\Gamma}_1$ and $\boldsymbol{\Gamma}_2$, while the modes in $\boldsymbol{\Phi}_2^{\mathrm{L\text{-}CC}}$ have interface DOF for boundary sets $\boldsymbol{\Gamma}_2$ and $\boldsymbol{\Gamma}_3$. Substructures 3, 4, and 5 only have one interface $\boldsymbol{\Gamma}_1$, $\boldsymbol{\Gamma}_2$, and $\boldsymbol{\Gamma}_3$, respectively. The L-CC modes for each subcomponent (1 through 5) are expressed then as

$$\begin{aligned} \boldsymbol{\Phi}_1^{\mathrm{L\text{-}CC}} &= \begin{bmatrix} \boldsymbol{\Phi}_{\Gamma_1,1}^{\mathrm{L\text{-}CC}} \\ \boldsymbol{\Phi}_{\Gamma_2,1}^{\mathrm{L\text{-}CC}} \end{bmatrix}, \quad \boldsymbol{\Phi}_2^{\mathrm{L\text{-}CC}} = \begin{bmatrix} \boldsymbol{\Phi}_{\Gamma_2,2}^{\mathrm{L\text{-}CC}} \\ \boldsymbol{\Phi}_{\Gamma_3,2}^{\mathrm{L\text{-}CC}} \end{bmatrix}, \\ \boldsymbol{\Phi}_3^{\mathrm{L\text{-}CC}} &= \boldsymbol{\Phi}_{\Gamma_1,3}^{\mathrm{L\text{-}CC}}, \quad \boldsymbol{\Phi}_4^{\mathrm{L\text{-}CC}} = \boldsymbol{\Phi}_{\Gamma_2,4}^{\mathrm{L\text{-}CC}}, \quad \boldsymbol{\Phi}_5^{\mathrm{L\text{-}CC}} = \boldsymbol{\Phi}_{\Gamma_3,5}^{\mathrm{L\text{-}CC}}. \end{aligned} \tag{3.70}$$

Following what was done in Krattiger et al. (2019), for each substructure the interface modes are computed for the entire interface set simultaneously, even if the set of interface nodes comprises multiple distinct interfaces. Then, for each distinct interface, the modes from every substructure are concatenated to create a complete basis for each interface. The result is called the augmented set, e.g., $\boldsymbol{\Phi}_{\boldsymbol{\Gamma}_1}^{\mathrm{aug}}$. For the W-bracket system in Fig. 3.2, we have the following augmented sets:

$$\begin{aligned}\boldsymbol{\Phi}_{\boldsymbol{\Gamma}_1}^{\text{aug}} &= \left[\boldsymbol{\Phi}_{\Gamma_{1,1}}^{\text{L-CC}}\ \boldsymbol{\Phi}_{\Gamma_{1,3}}^{\text{L-CC}}\right], \quad \boldsymbol{\Phi}_{\boldsymbol{\Gamma}_2}^{\text{aug}} = \left[\boldsymbol{\Phi}_{\Gamma_{2,1}}^{\text{L-CC}}\ \boldsymbol{\Phi}_{\Gamma_{2,2}}^{\text{L-CC}}\ \boldsymbol{\Phi}_{\Gamma_{2,4}}^{\text{L-CC}}\right], \\ \boldsymbol{\Phi}_{\boldsymbol{\Gamma}_3}^{\text{aug}} &= \left[\boldsymbol{\Phi}_{\Gamma_{3,2}}^{\text{L-CC}}\ \boldsymbol{\Phi}_{\Gamma_{3,5}}^{\text{L-CC}}\right].\end{aligned} \tag{3.71}$$

The augmented interface L-CC modes for each interface will generally not be linearly independent and this must be corrected before they can be used. Indeed, the matrices may be rank deficient to numerical precision such that they cannot be used as a basis. To address this, Hong et al. used a Singular Value Decomposition (SVD) on the augmented L-CC modes,

$$\mathbf{U}_{\Gamma_j}^{\text{aug}}\mathbf{S}_{\Gamma_j}^{\text{aug}}\left(\mathbf{V}_{\Gamma_j}^{\text{aug}}\right)^{\text{T}} = \boldsymbol{\Phi}_{\Gamma_j}^{\text{aug}}, \tag{3.72}$$

where $\mathbf{U}_{\Gamma_j}^{\text{aug}}$ and $\mathbf{S}_{\Gamma_j}^{\text{aug}}$ are the left singular vectors and the diagonal singular value matrix for the jth interface set. If some singular values are very small then the basis is ill-conditioned and the corresponding singular vectors must be truncated. In Hong's work (Hong et al. 2013), the truncation tolerance used was $\sigma_R = 1 \times 10^{-4} = 0.01\%$, where σ_R is the ratio of the minimum singular value to the maximum for the retained modes. In Krattiger et al. (2019), the effect of σ_R on the model performance was evaluated, revealing that $\sigma_R = 1 \times 10^{-4} = 0.01\%$ was optimal for the W-bracket, but this can be expected to be problem dependent.

The resulting matrices $\mathbf{U}_{\boldsymbol{\Gamma}_j}$ can then be grouped for each subsystem j to form a block-diagonal matrix that contains all the interface sets for substructure j. Take substructures 1 and 3 as examples. The orthogonal L-CC basis is

$$\mathbf{U}_1 = \begin{bmatrix}\mathbf{U}_{\Gamma_1} & \mathbf{0} \\ \mathbf{0} & \mathbf{U}_{\Gamma_2}\end{bmatrix}, \quad \mathbf{U}_3 = \mathbf{U}_{\Gamma_1}. \tag{3.73}$$

Then, the reduced mass and stiffness matrices for each subcomponent j are

$$\mathbf{M}_j^{\text{EC L-CC}} = \begin{bmatrix}\mathbf{I} & \mathbf{0} \\ \mathbf{0} & \mathbf{U}_j\end{bmatrix}^{\text{T}} \begin{bmatrix}\mathbf{M}_{ii,j}^{\text{HCB}} & \mathbf{M}_{ib,j}^{\text{HCB}} \\ \mathbf{M}_{bi,j}^{\text{HCB}} & \mathbf{M}_{bb,j}^{\text{HCB}}\end{bmatrix} \begin{bmatrix}\mathbf{I} & \mathbf{0} \\ \mathbf{0} & \mathbf{U}_j\end{bmatrix}. \tag{3.74}$$

Further details are provided in Krattiger et al. (2019).

Weak Compatibility

The WC L-CC method also begins by computing local interface modes for each substructure using Eq. (3.69). This transformation expresses the physical interface DOF as a linear combination of the interface modes,

$$\mathbf{u}_{b,j} \approx \boldsymbol{\Phi}_j^{\text{L-CC}}\mathbf{q}_{b,j}. \tag{3.75}$$

Then the reduced-order mass matrix for each substructure is

$$\mathbf{M}_j^{\text{WC L-CC}} = \begin{bmatrix} \mathbf{I} & \mathbf{0} \\ \mathbf{0} & \boldsymbol{\Phi}_j^{\text{L-CC}} \end{bmatrix}^{\mathrm{T}} \begin{bmatrix} \mathbf{M}_{ii,j}^{\text{HCB}} & \mathbf{M}_{ib,j}^{\text{HCB}} \\ \mathbf{M}_{bi,j}^{\text{HCB}} & \mathbf{M}_{bb,j}^{\text{HCB}} \end{bmatrix} \begin{bmatrix} \mathbf{I} & \mathbf{0} \\ \mathbf{0} & \boldsymbol{\Phi}_j^{\text{L-CC}} \end{bmatrix}. \tag{3.76}$$

The models for each substructure are not statically complete at the interface, so they cannot simply be assembled. Furthermore, because the basis for each interface may not span the same space, they could lock if strict constraints are enforced. Instead, the following approach is used. The constraint equations for the interface are written as

$$\mathbf{B}\mathbf{q}_{cat} = \mathbf{0} \tag{3.77}$$

where $\mathbf{q}_{cat}$ contains the interface modal coordinates concatenated from each uncoupled substructure, e.g., $\mathbf{q}_{cat} = \left[\mathbf{u}_{\Gamma_1,1}, \mathbf{u}_{\Gamma_1,3}\right]$. A transformation is sought such that, $\mathbf{q}_{cat} = \mathbf{L}_b\mathbf{q}_{asm}$, and this can be obtained if $\mathbf{L}_b$ lies in the null space of $\mathbf{B}$. Hence, the matrix $\mathbf{B}$ is formed and its null space is computed and used to define $\mathbf{L}_b$.

For the W-bracket example, the following constraint equations need to be satisfied:

$$\mathbf{u}_{\Gamma_1,1} = \mathbf{u}_{\Gamma_1,3}, \qquad \mathbf{u}_{\Gamma_2,1} = \mathbf{u}_{\Gamma_2,2}, \qquad \mathbf{u}_{\Gamma_2,1} = \mathbf{u}_{\Gamma_2,4}, \qquad \mathbf{u}_{\Gamma_3,2} = \mathbf{u}_{\Gamma_3,5}. \tag{3.78}$$

Collecting all of the constraint equations into a single matrix and expressing them in modal coordinates gives

$$\underbrace{\begin{bmatrix} \boldsymbol{\Phi}_{\Gamma_1,1}^{\text{L-CC}} & \mathbf{0} & -\boldsymbol{\Phi}_{\Gamma_1,3}^{\text{L-CC}} & \mathbf{0} & \mathbf{0} \\ \boldsymbol{\Phi}_{\Gamma_2,1}^{\text{L-CC}} & -\boldsymbol{\Phi}_{\Gamma_2,2}^{\text{L-CC}} & \mathbf{0} & \mathbf{0} & \mathbf{0} \\ \boldsymbol{\Phi}_{\Gamma_2,1}^{\text{L-CC}} & \mathbf{0} & \mathbf{0} & -\boldsymbol{\Phi}_{\Gamma_2,4}^{\text{L-CC}} & \mathbf{0} \\ \mathbf{0} & \boldsymbol{\Phi}_{\Gamma_3,2}^{\text{L-CC}} & \mathbf{0} & \mathbf{0} & -\boldsymbol{\Phi}_{\Gamma_3,5}^{\text{L-CC}} \end{bmatrix}}_{\mathbf{B}} \begin{Bmatrix} \mathbf{q}_{b,1} \\ \mathbf{q}_{b,2} \\ \mathbf{q}_{b,3} \\ \mathbf{q}_{b,4} \\ \mathbf{q}_{b,5} \end{Bmatrix} = \mathbf{0}. \tag{3.79}$$

Then, the transformation $\mathbf{L}_b$ is sought that minimizes $\mathbf{B}\mathbf{L}_b$. Specifically, the SVD, $\mathbf{B} = \mathbf{U}_\mathbf{B}\mathbf{S}_\mathbf{B}\mathbf{V}_\mathbf{B}^{\mathrm{T}}$ is computed and the columns of $\mathbf{V}_\mathbf{B}$ are retained if their singular values are less than a null space tolerance, σ_N. The compatibility will be good (although typically not exact) as long as the singular values are small. Those motions at the interface that are compatible will govern the motion of the coupled system, while those displacements that are deemed incompatible are discarded.

Once an approximate null space has been obtained, it is placed into the boundary partitions of the overall coupling matrix. For the W-bracket model, this gives

$$\begin{Bmatrix} \mathbf{q}_{i,1} \\ \mathbf{q}_{b,1} \\ \mathbf{q}_{i,2} \\ \mathbf{q}_{b,2} \\ \mathbf{q}_{i,3} \\ \mathbf{q}_{b,3} \\ \mathbf{q}_{i,4} \\ \mathbf{q}_{b,4} \\ \mathbf{q}_{i,5} \\ \mathbf{q}_{b,5} \end{Bmatrix} = \underbrace{\begin{bmatrix} \mathbf{I} & \mathbf{0} & \mathbf{0} & \mathbf{0} & \mathbf{0} & \mathbf{0} \\ \mathbf{0} & \mathbf{0} & \mathbf{0} & \mathbf{0} & \mathbf{0} & \mathbf{N}_1 \\ \mathbf{0} & \mathbf{I} & \mathbf{0} & \mathbf{0} & \mathbf{0} & \mathbf{0} \\ \mathbf{0} & \mathbf{0} & \mathbf{0} & \mathbf{0} & \mathbf{0} & \mathbf{N}_2 \\ \mathbf{0} & \mathbf{0} & \mathbf{I} & \mathbf{0} & \mathbf{0} & \mathbf{0} \\ \mathbf{0} & \mathbf{0} & \mathbf{0} & \mathbf{0} & \mathbf{0} & \mathbf{N}_3 \\ \mathbf{0} & \mathbf{0} & \mathbf{0} & \mathbf{I} & \mathbf{0} & \mathbf{0} \\ \mathbf{0} & \mathbf{0} & \mathbf{0} & \mathbf{0} & \mathbf{0} & \mathbf{N}_4 \\ \mathbf{0} & \mathbf{0} & \mathbf{0} & \mathbf{0} & \mathbf{I} & \mathbf{0} \\ \mathbf{0} & \mathbf{0} & \mathbf{0} & \mathbf{0} & \mathbf{0} & \mathbf{N}_5 \end{bmatrix}}_{\mathbf{L}^{\text{WC L-CC}}} \begin{Bmatrix} \mathbf{q}_{i,1} \\ \mathbf{q}_{i,2} \\ \mathbf{q}_{i,3} \\ \mathbf{q}_{i,4} \\ \mathbf{q}_{i,5} \\ \mathbf{q}_{b} \end{Bmatrix}, \tag{3.80}$$

where $\mathbf{N}_j$ denotes the rows of $\mathbf{N}$ corresponding to the interface DOF of the jth substructure. Finally, the reduced-order system is assembled by summing over the substructures.

$$\mathbf{M}^{\text{WC L-CC}} = \sum_{j=1}^{n} \left(\mathbf{L}_j^{\text{WC L-CC}}\right)^{\mathrm{T}} \mathbf{M}_j^{\text{WC L-CC}} \mathbf{L}_j^{\text{WC L-CC}}, \tag{3.81}$$

where $\mathbf{L}_j^{\text{WC L-CC}}$ contains the rows of $\mathbf{L}^{\text{WC L-CC}}$ corresponding to the jth uncoupled substructure.

Virtual Node

Lindberg et al. (2013) presented an interface reduction method that formalizes some concepts that have been used in industry for many years. This method was proposed as a way to connect stiff and soft substructures at a single Virtual Node (VN) having six DOF. In Krattiger et al. (2019), this approach was generalized to allow n virtual nodes along an interface. Each virtual node is connected to neighboring nodes with multi-point constraints, and so the node set can collectively move in six shapes defined by three orthogonal translations and three rotations around the VN's position.

This type of approach is used very commonly in industry, although there are various flavors. For example, the RBE3 elements that are commonly used in Nastran perform a similar operation, although they do not eliminate the interface nodes merely compute the motion of a virtual node that tracks the average motion of the interface. *The experimental counterpart, operating on measured FRFs, can be found in the virtual point transformation, which is discussed in Sect.* 4.6.

Other Approaches

Coppolino proposed a complementary approach called Modified Guyan Reduction (MGR) in which forces are applied in the shape of a certain vector, such as the displacement patches used in the VN approach, thus allowing the interfaces to deform in a natural way but reducing the number of interface DOF (Coppolino 2018).

Other extensions to the interface reduction methods proposed here are certainly possible. One notable addition is the Hybrid method (H-CC), which is described in detail in Krattiger et al. (2019).

3.4 State-Space Model Reduction

A state-space model representation gives a unified approach to the modeling of finite- dimensional dynamical systems. It is often used in control theory and in system identification theory alike but less so in classical structural dynamics theory that rely more on second-order system descriptions. However, it also suits well in structural dynamics and as such gives a good bridging link to the system identification that estimates mathematical models from experimental data. The state-space system formulation for nonlinear or linear dynamical systems provides a general first-order differential form description that relates the system responses to known and unknown system stimuli. In testing, the system responses are measured with sensors using a data acquisition system that unavoidably mixes the true system response with noise. Since the unknown stimuli and sensor noise are often best seen as being stochastic, a so-called stochastic state-space description that embeds deterministic and stochastic parts can be expressed by the state transition differential equation, also called the *dynamic equation* together with the *output equation* as

$$\begin{aligned} \mathbf{x} &= \mathbf{A}\mathbf{x} + \mathbf{B}\mathbf{s} + \mathbf{E}\mathbf{w}, \\ \mathbf{r} &= \mathbf{C}\mathbf{x} + \mathbf{D}\mathbf{s} + \mathbf{v}. \end{aligned} \tag{3.82}$$

Here $\mathbf{x} \in \mathbb{R}^{n}$ is the state vector, $\mathbf{s} \in \mathbb{R}^{n_s}$ is the vector of known stimuli and $\mathbf{r} \in \mathbb{R}^{n_r}$ is the output vector. The system order is n, the number of independent stimuli elements of the excitation vector is n_s, and the number of response elements of the output vector is n_r. The column vectors $\mathbf{w} \in \mathbb{R}^{n_w}$ and $\mathbf{v} \in \mathbb{R}^{n_v}$ represent unknown process noise and output noise, respectively. The matrices **A, B, C, D,** and **E** are state-space coefficient matrices out of which $\mathbf{B} \in \mathbb{R}^{n \times n_s}$ is the *output matrix*, $\mathbf{C} \in \mathbb{R}^{n_r \times n}$ is the *input matrix*, and $\mathbf{D} \in \mathbb{R}^{n_r \times n_s}$ is the *direct throughput matrix*. In general, for a nonlinear and time-varying system, the system coefficients are state and time dependent. However, for linear time-invariant systems the matrices are constant coefficient.

In vibration testing, the system of interest is often put on soft supports to isolate it from ambient vibrations. The system is excited by some measured stimuli **s** and its responses **r** are picked up as sensor signals which are filtered, digitized and processed in a data acquisition system. In reality, the registered stimuli **s** is not the true system stimuli since sensors are not ideal and the support system does not fully isolate the test object from ambient vibrations which lead to unknown vibration stimuli **w** transmitted to the system. However, in many cases, the stochastic terms may rightfully be deemed negligable which leads to the deterministic state-space description

$$\begin{aligned} \mathbf{x} &= \mathbf{A}\mathbf{x} + \mathbf{B}\mathbf{s} \\ \mathbf{r} &= \mathbf{C}\mathbf{x} + \mathbf{D}\mathbf{s} \end{aligned} \tag{3.83}$$

for which there is an initial value time domain solution given by a convolution integral expression, see Eq. (3.89) below.

The frequency domain counterpart for the steady-state dynamics of the deterministic time domain model follows from the assumption that a harmonic stimulus with amplitude $\hat{\mathbf{s}}$ and frequency ω, i.e., $\mathbf{s}(t) = \mathfrak{Re}\{\hat{\mathbf{s}}e^{i\omega t}\}$, results in a harmonic response $\mathfrak{Re}\{\hat{\mathbf{r}}e^{i\omega t}\}$. This gives the relation between input and output as

$$\mathfrak{Re}\{\hat{\mathbf{r}}e^{i\omega t}\} = \mathfrak{Re}\{[\mathbf{C}(i\omega\mathbf{I} - \mathbf{A})^{-1}\mathbf{B} + \mathbf{D}]\hat{\mathbf{s}}e^{i\omega t}\} \tag{3.84}$$

which defines the system's *transfer function* $\mathbf{H}$, also known as its *frequency response function* as

$$\mathbf{H}(\omega) = \mathbf{C}(i\omega\mathbf{I} - \mathbf{A})^{-1}\mathbf{B} + \mathbf{D} \tag{3.85}$$

A state-space model, on the form given by Eqs. (3.83) or (3.85), attempting to mimic the behavior of a real system may be called a *realization* within a model structure $\mathfrak{M}$ of a system. For a multi-input-multi-output system, the coefficient matrix quadruple $\mathfrak{M} = \{\mathbf{A}, \mathbf{B}, \mathbf{C}, \mathbf{D}\}$ generally holds $n^2 + nn_r + nn_s + n_r n_s$ elements in total. However, for a given input/output relation, there is no unique state-space description. Unique realizations can be obtained after adding model form constraints. Four such realizations; the balanced realization, the modal realization, the Jordan form realization, and the coupling realization are of particular interest here. For the balanced realization, the observability and controllability Grammians (see below) are forced to coincide. The modal form is applicable for most systems and is a realization in which its states are fully decoupled. Some systems with esoteric behavior cannot be transformed into a system with fully decoupled states. For these, the Jordan normal form gives the realization with the minimal coupling between states. The coupling realization is particularly useful for coupling of state-space models for which force equilibrium and displacement compatibility need to be enforced at subsystem interfaces.

A change of variables $\mathbf{x} = \mathbf{T}\mathbf{z}$ with a nonsingular transformation matrix $\mathbf{T}$ leads to a proof of the nonuniqueness. Setting out from Eq. (3.83) and using the transformation $\mathbf{x} = \mathbf{T}\mathbf{z}$ give

$$\begin{aligned} \mathbf{T}\dot{\mathbf{z}} &= \mathbf{A}\mathbf{T}\mathbf{z} + \mathbf{B}\mathbf{s}, \\ \mathbf{r} &= \mathbf{C}\mathbf{T}\mathbf{z} + \mathbf{D}\mathbf{s}. \end{aligned} \tag{3.86}$$

Since **T** is nonsingular this leads to

$$\begin{aligned}\dot{\mathbf{z}} &= \mathbf{T}^{-1}\mathbf{A}\mathbf{T}\mathbf{z} + \mathbf{T}^{-1}\mathbf{B}\mathbf{s} \equiv \bar{\mathbf{A}}\mathbf{z} + \bar{\mathbf{B}}\mathbf{s} \\ \mathbf{r} &= \mathbf{C}\mathbf{T}\mathbf{z} + \mathbf{D}\mathbf{s} \equiv \bar{\mathbf{C}}\mathbf{z} + \mathbf{D}\mathbf{s},\end{aligned} \tag{3.87}$$

which is a similar realization with the state-space matrix quadruple $\bar{\mathfrak{M}} = \{\bar{\mathbf{A}}, \bar{\mathbf{B}}, \bar{\mathbf{C}}, \mathbf{D}\}$ and states $\mathbf{z}$ that gives the same input/output relation as the realization $\{\mathbf{A}, \mathbf{B}, \mathbf{C}, \mathbf{D}\}$ with states $\mathbf{x}$. A similarity transformation, such as $\mathbf{x} = \mathbf{T}\mathbf{z}$, which is known to preserve the eigenvalues of the system, thus also preserves its input/ output relation.

Markov Parameters and the Hankel Matrix. Associated to the dynamic equation of the state-space description (3.83) is the matrix integral solution of the initial value problem

$$\mathbf{x} = \mathbf{A}\mathbf{x} + \mathbf{B}\mathbf{s} \quad \text{with} \quad \mathbf{x}(0) = \mathbf{x}_0 \tag{3.88}$$

for which the exact solution is given by the convolution

$$\mathbf{x}(t) = \Phi(t)\mathbf{x}_0 + \int_0^t \Phi(t-\tau)\mathbf{B}\mathbf{s}(\tau)d\tau. \tag{3.89}$$

Here $\Phi(t)$ is the state transition matrix, which is

$$\Phi(t) = e^{\mathbf{A}t} = \mathbf{I} + t\mathbf{A}/1! + t^2\mathbf{A}^2/2! + \ldots \tag{3.90}$$

An important special case is for the impact stimuli $\mathbf{s}(t) = [1\ 1 \ldots 1]^T\delta(t)$ of a system at rest with $\mathbf{x}_0 = 0$ and $\delta(t)$ being the Dirac delta function. For such an ideal impact (hit), the convolution integral in (3.89) evaluates to

$$\mathbf{x}(t) = [\mathbf{A}^0\mathbf{B} + t\mathbf{A}\mathbf{B}/1! + t^2\mathbf{A}^2\mathbf{B}/2! + \ldots][1\ 1 \ldots 1]^T \tag{3.91}$$

and the system response $\mathbf{r} \equiv \mathbf{r}_{\text{hit}}$ is thus given by

$$\mathbf{r}_{\text{hit}}(t) = [\mathbf{D}\delta(t) + \mathbf{C}\mathbf{A}^0\mathbf{B} + t\mathbf{C}\mathbf{A}\mathbf{B}/1! + t^2\mathbf{C}\mathbf{A}^2\mathbf{B}/2! + \ldots][1\ 1 \ldots 1]^T. \tag{3.92}$$

The terms of the system's impact response series are associated to its Markov parameters defined as

$$\mathbf{h}_0 = \mathbf{D} \quad \mathbf{h}_k = \mathbf{C}\mathbf{A}^{k-1}\mathbf{B}, \quad k = 1, 2, \ldots \tag{3.93}$$

from which it can be noted that the impact response is

$$\mathbf{r}_{\text{hit}}(t) = [\mathbf{h}_0\delta(t) + \sum_{k=0} \mathbf{h}_{k+1}t^k/k!][1\ 1 \ldots 1]^T \tag{3.94}$$

The infinite-size Hankel matrix $\mathscr{H}$ of the system is constructed from its Markov parameters as follows:

$$\mathscr{H} = \begin{bmatrix} \mathbf{h}_1 \ \mathbf{h}_2 \ \mathbf{h}_3 \ \mathbf{h}_4 \ \ldots \\ \mathbf{h}_2 \ \mathbf{h}_3 \ \mathbf{h}_4 \ \mathbf{h}_5 \ \vdots \\ \mathbf{h}_3 \ \mathbf{h}_4 \ \mathbf{h}_5 \ \mathbf{h}_6 \ \vdots \\ \vdots \ \ \vdots \ \ \vdots \ \ \vdots \ \ddots \end{bmatrix} = \begin{bmatrix} \mathbf{CB} & \mathbf{CAB} & \mathbf{CA^2B} & \mathbf{CA^3B} & \ldots \\ \mathbf{CAB} & \mathbf{CA^2B} & \mathbf{CA^3B} & \mathbf{CA^4B} & \vdots \\ \mathbf{CA^2B} & \mathbf{CA^3B} & \mathbf{CA^4B} & \mathbf{CA^5B} & \vdots \\ \vdots & \vdots & \vdots & \vdots & \ddots \end{bmatrix} \tag{3.95}$$

Although the Hankel matrix is of infinite size, its rank is bounded to be rank$(\mathscr{H}) \leq n$. This follows from the Cayley–Hamilton theorem stating that $\mathbf{A}$ raised to any power can be expressed as a linear combination with a finite number of terms as $\mathbf{A}^k = \sum_{j=0}^{n-1} \alpha_{jk} \mathbf{A}^k$. This, in turn, leads to that the number of linearly independent columns (and rows, and thus the rank) of the Hankel matrix is limited. The number of positive singular values of the Hankel matrix is thus also bounded. Its largest singular value $\sigma_1(\mathscr{H})$ is associated with the largest impulse response of the system.

Discrete-Time State-Space Modeling. The convolution solution (3.89) is utilized for deriving efficient computational schemes for discrete-time solutions. Consider the solutions at two consequent discrete times kT and $kT + T$, where $T = 1/f_s$ and f_s is referred to as the sampling frequency. The solutions are

$$\mathbf{x}_k = e^{\mathbf{A}kT}\mathbf{x}_0 + \int_0^{kT} e^{\mathbf{A}(kT-\tau)}\mathbf{Bs}(\tau)d\tau \tag{3.96}$$

and

$$\mathbf{x}_{k+1} = e^{\mathbf{A}(kT+T)}\mathbf{x}_0 + \int_0^{kT} e^{\mathbf{A}(kT+T-\tau)}\mathbf{Bs}(\tau)d\tau \tag{3.97}$$

$$= e^{\mathbf{A}T}(e^{\mathbf{A}kT}\mathbf{x}_0 + \int_0^{kT} e^{\mathbf{A}(kT-\tau)}\mathbf{Bs}(\tau)d\tau) + \int_{kT}^{kT+T} e^{\mathbf{A}(kT+T-\tau)}\mathbf{Bs}(\tau)d\tau$$

$$= e^{\mathbf{A}T}\mathbf{x}_k + \int_{kT}^{kT+T} e^{\mathbf{A}(kT+T-\tau)}\mathbf{Bs}(\tau)d\tau.$$

It is interesting to note that after a sample interval $[kT \ kT + T]$ in which no excitation take place the state $\mathbf{x}_{k+1}$ can be computed without approximation using the discrete- time transition matrix and the previous state $\mathbf{x}_k$ for an arbitrary slow sampling rate T.

Using so-called *zero-order-hold* assumption, i.e., assuming that the stimulus $\mathbf{s}_k$ is constant over the sampling period T, the excitation term can be approximated as

$$\int_{kT}^{kT+T} e^{\mathbf{A}(kT+T-\tau)}\mathbf{Bs}(\tau)d\tau \approx (\int_{kT}^{kT+T} e^{\mathbf{A}(kT+T-\tau)}d\tau)\mathbf{Bs}_k. \tag{3.98}$$

With the change of variables $kT + T - \tau = \tau'$, the integral expression can be evaluated into

$$\int_{kT}^{kT+T} e^{\mathbf{A}(kT+T-\tau)} d\tau = -\int_{T}^{0} e^{\mathbf{A}\tau'} d\tau' = \int_{0}^{T} e^{\mathbf{A}\tau'} d\tau' \tag{3.99}$$

$$= \int_{0}^{T} (\mathbf{I} + \tau'\mathbf{A} + \tau'^2\mathbf{A}^2/2! + \ldots) d\tau' = T\mathbf{I} + T^2\mathbf{A}/2! + T^3\mathbf{A}^2/3! + \ldots$$
$$= \mathbf{A}^{-1}(T\mathbf{A} + T^2\mathbf{A}^2/2! + T^3\mathbf{A}^3/3! + \ldots) = \mathbf{A}^{-1}(e^{\mathbf{A}T} - \mathbf{I}).$$

Which leads to the explicit time-stepping algorithm for the states

$$\mathbf{x}_{k+1} = \hat{\mathbf{A}}\mathbf{x}_k + \hat{\mathbf{B}}\mathbf{s}_k \tag{3.100}$$

and the response

$$\mathbf{y}_k = \mathbf{C}\mathbf{x}_k + \mathbf{D}\mathbf{s}_k \tag{3.101}$$

with the discrete-time state-space matrices

$$\hat{\mathbf{A}} = e^{\mathbf{A}T} \text{ and } \hat{\mathbf{B}} = \mathbf{A}^{-1}(e^{\mathbf{A}T} - \mathbf{I})\mathbf{B}. \tag{3.102}$$

It can be shown that for stable systems $||\hat{\mathbf{A}}|| < 1$, which is a requirement for the time-stepping numerical integration scheme (3.100) to be stable.

3.4.1 Structural Dynamics Equations in State Space

It is interesting to relate the state-space model structure to the second-order linear mechanical system that is so often used in finite element (FE) structural dynamic modeling that in this setting reads

$$\mathbf{M}\ddot{\mathbf{u}} + (\mathbf{V} + \mathbf{G})\dot{\mathbf{u}} + (\mathbf{K} + \lambda\mathbf{K}_g + \mathbf{L})\mathbf{u} = \mathbf{F}(t) \tag{3.103}$$

with the displacement and load vectors $\mathbf{u}, \mathbf{F} \in \Re^N$. Here $\mathbf{M}, \mathbf{V}, \mathbf{G}, \mathbf{K}, \mathbf{K}_g, \mathbf{L}$ are the mass, viscous damping, gyroscopic, stiffness, geometric stiffness, and circulatory matrices, respectively, and N is the number of degrees of freedom of the model. The λ is the buckling load parameter with $\lambda = \lambda_{cr}$ giving static buckling. Let $\mathbf{F}(t)$ relate to the nonzero independent stimuli $\mathbf{s}(t)$ with the linear relation $\mathbf{F} = \mathbf{P}_s\mathbf{s}$. Here $\mathbf{P}_s \in \mathbb{R}^{N \times n_s}$ is the so-called picking matrix that, however general, most often is a Boolean matrix used in the procedure to pick out the nonzero loads of $\mathbf{F}$, bringing that to the nonzero stimuli $\mathbf{s}$ and by $\mathbf{P}_s$ redistribute the nonzeros back into $\mathbf{F}$. The mass matrix is assumed to be symmetric positive definite and therefore Eq. (3.103) can be rewritten as

$$\ddot{\mathbf{u}} + \mathbf{M}^{-1}(\mathbf{V}+\mathbf{G})\dot{\mathbf{u}} + \mathbf{M}^{-1}(\mathbf{K}+\lambda\mathbf{K}_g+\mathbf{L})\mathbf{u} = \mathbf{M}^{-1}\mathbf{P}_s\mathbf{s}(t) \tag{3.104}$$

Combined with the trivial equation $\mathbf{I}\dot{\mathbf{q}} - \mathbf{I}\dot{\mathbf{q}} = 0$, Eq. (3.104) can be written as the state-space dynamic equation

$$\dot{\mathbf{x}} = \begin{bmatrix} \mathbf{0} & \mathbf{I} \\ -\mathbf{M}^{-1}(\mathbf{V}+\mathbf{G}) & -\mathbf{M}^{-1}(\mathbf{K}+\lambda\mathbf{K}_g+\mathbf{L}) \end{bmatrix}\mathbf{x} + \begin{bmatrix} \mathbf{0} \\ \mathbf{M}^{-1}\mathbf{P}_s \end{bmatrix}\mathbf{s} \equiv \mathbf{A}\mathbf{x}+\mathbf{B}\mathbf{s} \tag{3.105}$$

where the state vector $\mathbf{x}$ has been introduced so that

$$\mathbf{x} = \begin{pmatrix} \mathbf{u} \\ \dot{\mathbf{u}} \end{pmatrix} \text{ and thus } \dot{\mathbf{x}} = \begin{pmatrix} \dot{\mathbf{u}} \\ \ddot{\mathbf{u}} \end{pmatrix} \tag{3.106}$$

The quantities of interest in mechanical analyses is often a selected set of displacements, velocities, accelerations or quantities that can be derived from these, *e.g.*, stresses and strains. Let the quantities of interest be the outputs $\mathbf{y}$ of the system. In finite element vibrational analysis, the outputs of the model are often obtained by post-processing the analysis results. The analysis results, in this case, are the nodal displacements $\mathbf{u}$ and velocities $\dot{\mathbf{u}}$ of the model. The analyst then specifies what quantities are of his interest and lets the post-processor calculate these using analysis results together with complementary model data. These quantities are often linearly related to the displacements and velocities given by the analysis, and are therefore natural ingredients of in a linear state-space model. Displacement and velocity output elements may easily be extracted. This can be made by letting a picking matrix $\mathbf{P}$ operate on the state vector $\mathbf{x}$ containing all nodal displacement and velocities of the FE model. The time derivative of the state vector $\dot{\mathbf{x}}$ holds the acceleration data. Let $\mathbf{P}_d$ be the picking matrix that picks out the displacements of interest $\mathbf{r}_d$ from the displacement partition of the state vector and $\mathbf{P}_v$ be the one that picks out the velocities of interest $\mathbf{r}_d$ from the velocity part. Then the output equation becomes

$$\mathbf{r} \equiv \begin{pmatrix} \mathbf{r}_d \\ \mathbf{r}_v \end{pmatrix} = \begin{bmatrix} \mathbf{P}_d & \mathbf{0} \\ \mathbf{0} & \mathbf{P}_v \end{bmatrix}\mathbf{x} \equiv \begin{bmatrix} \mathbf{C}_d \\ \mathbf{C}_v \end{bmatrix}\mathbf{x}. \tag{3.107}$$

Let further $\mathbf{P}_a$ be the picking matrix that picks out the accelerations of interest $\mathbf{r}_a$ from the acceleration partition of the state vector's time derivative $\dot{\mathbf{x}}$. Then the output equation for accelerations becomes

$$\mathbf{r} \equiv \mathbf{r}_a = \begin{bmatrix} \mathbf{0} & \mathbf{P}_a \end{bmatrix}\dot{\mathbf{x}} = \begin{bmatrix} \mathbf{0} & \mathbf{P}_a \end{bmatrix}[\mathbf{A}\mathbf{x}+\mathbf{B}\mathbf{s}] \equiv \mathbf{C}_a\mathbf{x}+\mathbf{D}\mathbf{s}. \tag{3.108}$$

In summary, for a combined output

$$\mathbf{r} \equiv \begin{pmatrix} \mathbf{r}_d \\ \mathbf{r}_v \\ \mathbf{r}_a \end{pmatrix} \tag{3.109}$$

the corresponding state-space model quadruple {**A, B, C, D**} can be identified to be

$$\mathbf{A} = -\begin{bmatrix} \mathbf{0} & -\mathbf{I} \\ \mathbf{M}^{-1}(\mathbf{V}+\mathbf{G}) & \mathbf{M}^{-1}(\mathbf{K}+\lambda\mathbf{K}_g+\mathbf{L}), \end{bmatrix} \qquad \mathbf{B} = \begin{bmatrix} \mathbf{0} \\ \mathbf{M}^{-1}\mathbf{P}_s, \end{bmatrix} \tag{3.110}$$

$$\mathbf{C} = \begin{bmatrix} \mathbf{C}_d \\ \mathbf{C}_v \\ \mathbf{C}_a, \end{bmatrix} = \begin{bmatrix} \mathbf{P}_d & \mathbf{0} \\ \mathbf{0} & \mathbf{P}_v \\ [\mathbf{0}\ \mathbf{P}_a]\mathbf{A} \end{bmatrix} \qquad \mathbf{D} = \begin{bmatrix} \mathbf{D}_d \\ \mathbf{D}_v \\ \mathbf{D}_a \end{bmatrix} = \begin{bmatrix} \mathbf{0} \\ \mathbf{0} \\ [\mathbf{0}\ \mathbf{P}_a]\mathbf{B}. \end{bmatrix}$$

Here it may be noted that only acceleration output has static contribution through the direct throughput term $\mathbf{D}_a\mathbf{s}$, i.e., direct contribution from the excitation besides the dynamics of the system via the dynamic equation $\dot{\mathbf{x}} = Ax + Bs$. This is consistent with Newton's second law that directly relates acceleration contribution to force. The velocities and displacements, on the other hand, need to be obtained as the integral solution to the initial value problem $\dot{\mathbf{x}} = Ax + Bs$, $\mathbf{x}(0) = \mathbf{x}_0$. The velocity output is also related to displacement output by $\mathbf{r}_v = d\mathbf{r}_d/dt$ and therefore it is also given that

$$\mathbf{r}_v = \mathbf{C}_d[\mathbf{A}\mathbf{x} + \mathbf{B}\mathbf{s}] = \mathbf{C}_d\mathbf{A}\mathbf{x} + \mathbf{C}_d\mathbf{B}\mathbf{s} = \mathbf{C}_d\mathbf{A}\mathbf{x} \tag{3.111}$$

since it can be easily verified from above that $\mathbf{C}_d\mathbf{B} = 0$ and thus no direct throughput from **s** occur to velocity in this formulation either.

For other output quantities which relate linearly to the above quantities, e.g., strains and stresses, the corresponding state-space description is straightforward.

3.4.2 *State Observability and State Controllability*

State observability and state controllability play important roles in experimental vibration engineering. Loosely speaking, the state observability condition tells whether the states of the system can be uniquely determined from recorded response **r**. Similarly, the state controllability tells whether the model's states may be excited independently from each other by the stimuli **s**. In vibration testing, often targeting eigenmodes of a structure, the observability and controllability of the modes given by the chosen sensor-actuator configuration are critical for success. In the sensor placement, the observability criterion should, therefore, be fulfilled such that no critical states are non-observable. Likewise, state controllability of critical states should be maintained by proper actuator placement such that these states are excited and therefore contribute to the measured response.

State Observability. The concept of state observability is linked to the outputs and states of the system. Given a state-space model and the set of responses **r** and stimuli **s**, the question of observability is whether the model states **x** are deducible from the set. This is more rigorously formulated in the following definition: *A linear system is said to be observable at time t_0 if the state $\mathbf{x}(t_0)$ can be uniquely determined* *from the response* $\mathbf{r}(t)$ *when* $t \geq t_0$*. If the system is observable for all times, then the system is said to be completely observable.*

The observability of a state-space model $\{\mathbf{A}, \mathbf{B}, \mathbf{C}, \mathbf{D}\}$ can be investigated provided the model and its inputs and outputs are known. When the output is given at times $t \geq t_0$, also the time derivatives of the output at t may be determined. Using that $\mathbf{r} = \mathbf{C}\mathbf{x} + \mathbf{D}\mathbf{s}$, and therefore the sequence

$$\begin{aligned} \dot{\mathbf{r}} &= \mathbf{C}\dot{\mathbf{x}} + \mathbf{D}\dot{\mathbf{s}} = \mathbf{C}\mathbf{A}\mathbf{x} + \mathbf{C}\mathbf{B}\mathbf{s} + \mathbf{D}\dot{\mathbf{s}} \\ \ddot{\mathbf{r}} &= \mathbf{C}\mathbf{A}\dot{\mathbf{x}} + \mathbf{C}\mathbf{B}\dot{\mathbf{s}} + \mathbf{D}\ddot{\mathbf{s}} = \mathbf{C}\mathbf{A}^2\mathbf{x} + \mathbf{C}\mathbf{A}\mathbf{B}\mathbf{s} + \mathbf{C}\mathbf{B}\dot{\mathbf{s}} + \mathbf{D}\ddot{\mathbf{s}} \\ &\vdots \end{aligned} \tag{3.112}$$

leads to maximum n independent equations (not more according to the Cayley–Hamilton theorem, see below) for $\mathbf{x}$. On matrix form these equations are

$$\mathbf{R} = \mathfrak{O}\mathbf{x} + \mathbf{G}\mathbf{S} \tag{3.113}$$

with $\mathbf{R}$ being the vectorial concatenation of $\mathbf{r}, \dot{\mathbf{r}}, \ddot{\mathbf{r}}, \ldots$ and $\mathbf{S}$ being the vectorial concatenation of $\mathbf{s}, \dot{\mathbf{s}}, \ddot{\mathbf{s}}, \ldots$. The *observability matrix* of the system $\mathfrak{O}$ is

$$\mathfrak{O} = \begin{bmatrix} \mathbf{C} \\ \mathbf{C}\mathbf{A} \\ \mathbf{C}\mathbf{A}^2 \\ \vdots \\ \mathbf{C}\mathbf{A}^{n-1} \end{bmatrix} \tag{3.114}$$

and matrix $\mathbf{G}$ is irrelevant for observability. If the $n_r n$-by-n observability matrix has a rank of less then n, some linear combination of the n columns add to zero, and therefore there are (transformed) states that do not contribute to $\mathbf{R}$. It is thus necessary for observability of all system states that the observability matrix is of full rank. Is the full rank condition also sufficient for observability? To examine this, we start with multiplying equation (3.113) with the transpose of $\mathfrak{O}$ to obtain

$$\mathfrak{O}^{\mathrm{T}}\mathbf{R} = \mathfrak{O}^{\mathrm{T}}\mathfrak{O}\mathbf{x} + \mathfrak{O}^{\mathrm{T}}\mathbf{G}\mathbf{S}. \tag{3.115}$$

If $\mathfrak{O}$ is of full rank then $\mathfrak{O}^T\mathfrak{O}$ is nonsingular and thus the state vector $\mathbf{x}$ can be determined as its unique solution, which is

$$\mathbf{x} = (\mathfrak{O}^{\mathrm{T}}\mathfrak{O})^{-1}\mathfrak{O}^{\mathrm{T}}(\mathbf{R} - \mathbf{G}\mathbf{S}). \tag{3.116}$$

Is this also the solution to Eq. (3.113)? If it is not, for the two different solutions $\mathbf{x}_1$ and $\mathbf{x}_2$ we would have $\mathfrak{O}(\mathbf{x}_1 - \mathbf{x}_2) = \mathbf{0}$ which means that some linear combination of the columns of $\mathfrak{O}$ is zero, which contradicts the assumption that $\mathfrak{O}$ is of full column rank. In conclusion, we may thus state that: A realization is uniquely observable if and only if the observability matrix $\mathfrak{O}$ has full rank n.

It is important to notice that observability of states or lack of observability of states is not a given system property but can be actively affected by the selection of sensor configuration. In the planning of a vibration test, the observability issue can be addressed and various sensor configurations can be compared with respect to observability. Since the selection of sensor configuration affects the **C** matrix, different configurations can be evaluated for best observability in a pretest planning phase using FE analyses if an FE model is available. It is further interesting to note that, for high-order models, the time-continuous observability matrix Eq. (3.114) involve large entries when n is large and $||\mathbf{A}|| > 1$. Its numerical evaluation may then be troublesome since that may involve high powers of **A**. The observability matrix of the stable discrete-time model

$$\mathfrak{O}^{\mathrm{T}} = [\mathbf{C}^{\mathrm{T}}\ (\mathbf{C}\hat{\mathbf{A}})^{\mathrm{T}}\ \ (\mathbf{C}\hat{\mathbf{A}}^{2})^{\mathrm{T}}\ \ldots\ (\mathbf{C}\hat{\mathbf{A}}^{\mathrm{n-1}})^{\mathrm{T}}] \tag{3.117}$$

with $||\hat{\mathbf{A}}|| \leq 1$ does not suffer from this problem and is thus better suited for numerical evaluation.

State Controllability. The concept of controllability relates to the input and the states of a system. The state-space first-order differential equation can be used to examine the concept. The controllability is defined by the following: *The system* $\dot{\mathbf{x}} = \mathbf{A}\mathbf{x} + \mathbf{B}\mathbf{s}$ *is said to be state controllable at time* $t = t_0$ *if there exists a piecewise continuous input*$\mathbf{s}(t)$ *that will drive the initial state*$\mathbf{x}(t_0)$ *to any final state*$\mathbf{x}(t_f)$ *within a finite time interval*$t_f - t_0$*. If this is true for all initial times and all initial states, the system is said to be completely state controllable.*

If a time-invariant system is state controllable, it is thus also completely state controllable. For such systems, a quantitative test of controllability can be derived. To this end, let the initial time be $t_0 = 0$, for which the solution to the state-space equation is

$$\mathbf{x}(t_f) = e^{\mathbf{A}t_f}x(0) + \int_0^{t_f} e^{\mathbf{A}(t_f-\tau)}\mathbf{B}\mathbf{s}(\tau)d\tau, \tag{3.118}$$

which, for nonsingular $e^{\mathbf{A}t_f}$, can be reduced to

$$e^{-\mathbf{A}t_f}x(t_f) - \mathbf{x}(0) \equiv \Delta\mathbf{x}(t_f) = \int_0^{t_f} e^{-\mathbf{A}\tau}\mathbf{B}\mathbf{s}(\tau)d\tau. \tag{3.119}$$

The Cayley–Hamilton theorem giving $\mathbf{A}^k = \sum_{J=0}^{n-1}\alpha_{jk}\mathbf{A}^k$ together with the definition $\exp(\mathbf{A}t) = \mathbf{I} + t\mathbf{A} + t^2\mathbf{A}^2/2! + \ldots$ gives

$$e^{-\mathbf{A}\tau} = \sum_{j=0}^{n-1}\alpha_j(\tau)\mathbf{A}^j \tag{3.120}$$

with

$$\alpha_j(\tau) = \sum_{k=0}^{\infty}(-1)^k \alpha_{jk}\tau^k/k! \tag{3.121}$$

Hence, Eqs. (3.119) through (3.121) give

$$\Delta\mathbf{x}(t_f) = \sum_{j=0}^{n-1}\mathbf{A^jB}\int_0^{t_f}\alpha_j(\tau)\mathbf{s}(\tau)d\tau \equiv \sum_{j=0}^{n-1}\mathbf{A^jBc}_j(t_f) \tag{3.122}$$

with

$$\mathbf{c}_j(t_f) = \int_0^{t_f}\alpha_j(\tau)\mathbf{s}(\tau)d\tau \tag{3.123}$$

Equation (3.122) can then be written in matrix form as

$$\Delta\mathbf{x}(t_f) = [\mathbf{B\ AB\ A^2B\ \ldots\ A^{n-1}B}][\mathbf{c}_0^T\ \mathbf{c}_1^T\ \ldots\ \mathbf{c}_{n-1}^T]^T \tag{3.124}$$

and represents a set of n equations and nn_u unknowns. The matrix equation has a solution for any $\Delta\mathbf{x}(t_f)$ provided that the $n \times nn_s$ matrix

$$\mathfrak{C} = [\mathbf{B\ AB\ A^2B\ \ldots\ A^{n-1}B}] \tag{3.125}$$

has n independent columns. The matrix $\mathfrak{C}$ is known as the *controllability matrix* and thus the system is completely state controllable if $\mathfrak{C}$ has rank n.

Similar to the observability matrix of continuous-time systems, the controllability matrix has bad numerical properties when the system order gets large and $||\mathbf{A}|| > 1$. The observability matrix of the associated discrete-time model

$$\mathfrak{C} = [\hat{\mathbf{B}}\ \hat{\mathbf{A}}\hat{\mathbf{B}}\ \hat{\mathbf{A}}^2\hat{\mathbf{B}}\ldots\hat{\mathbf{A}}^{n-1}\hat{\mathbf{B}}] \tag{3.126}$$

of the stable system is then better suited. This matrix related to discrete-time entities is sometimes called the *reachability matrix*.

Controllability and Observability Grammians. Suppose a realization is given by $\dot{\mathbf{x}}= \mathbf{Ax} + \mathbf{Bs}$, $\mathbf{x}(t_0) = \mathbf{x}_0$ and the quest is to bring the state to zero by a finite amount of energy input $\mathbf{u}$ in a fixed time t_f. Is that possible? To answer that question, the initial value solution for a given $\mathbf{s}$

$$\mathbf{x}(t_f) = e^{\mathbf{A}(t_f-t_0)}x(0) + \int_{t_0}^{t_f} e^{\mathbf{A}(t_f-\tau)}\mathbf{Bs}(\tau)d\tau \tag{3.127}$$

can be used. For a nonsingular $e^{\mathbf{A}(t_f-t_0)}$, we thus have

$$-\mathbf{x}_0 = \int_{t_0}^{t_f} e^{\mathbf{A}(t_f-t_0)} e^{\mathbf{A}(t_f-\tau)} \mathbf{B}\mathbf{s}(\tau) d\tau = \int_{t_0}^{t_f} e^{\mathbf{A}(t_0-\tau)} \mathbf{B}\mathbf{s}(\tau) d\tau. \tag{3.128}$$

This is an integral equation for the sought stimuli **s**. Its solution can be shown to be

$$\mathbf{s}(\tau) = -\mathbf{B}^T e^{\mathbf{A}^T(t_0-\tau)} \mathfrak{G}_c^{-1}(t_0, t_f)\mathbf{x}_0, \tag{3.129}$$

where

$$\mathfrak{G}_c(t_0, t_f) = \int_{t_0}^{t_f} e^{\mathbf{A}(t_0-\tau)} \mathbf{B}\mathbf{B}^T e^{\mathbf{A}^T(t_0-\tau)} d\tau \tag{3.130}$$

is the *controllability Grammian*. To see that Eq. (3.130) really gives the solution, let it enter into the integral of Eq. (3.128) which renders

$$\begin{aligned} &-\int_{t_0}^{t_f} e^{\mathbf{A}(t_0-\tau)} \mathbf{B}\mathbf{B}^T e^{\mathbf{A}^T(t_0-\tau)} \mathfrak{G}_c^{-1}(t_0, t_f)\mathbf{x}_0 d\tau = \\ &-[\int_{t_0}^{t_f} e^{\mathbf{A}(t_0-\tau)} \mathbf{B}\mathbf{B}^T e^{\mathbf{A}^T(t_0-\tau)} d\tau]\mathfrak{G}_c^{-1}(t_0, t_f)\mathbf{x}_0 = \\ &\qquad -\mathfrak{G}_c(t_0, t_f)\mathfrak{G}_c^{-1}(t_0, t_f)\mathbf{x}_0 = -\mathbf{x}_0 \end{aligned} \tag{3.131}$$

since $\mathfrak{G}_c(t_0, t_f)$ is constant. The process requires that the controllability Grammian is nonsingular and therefore invertible, which is, therefore, the Grammian-related condition for controllability.

Similarly, for the realization $\dot{\mathbf{x}} = \mathbf{A}\mathbf{x} + \mathbf{B}\mathbf{s}$, $\mathbf{r} = \mathbf{C}\mathbf{x}$, the corresponding observability Grammian

$$\mathfrak{G}_o(t_0, t_f) = \int_{t_0}^{t_f} e^{\mathbf{A}^T(t_0-\tau)} \mathbf{C}^T \mathbf{C} e^{\mathbf{A}(t_0-\tau)} d\tau \tag{3.132}$$

must be nonsingular (Kailath 1980) for the realization to be observable from the output during over times $t_0 \leq t \leq t_f$.

The Grammian singularity test is known to be a test for linear dependence of functions l_j and l_k for which the Grammian $G_{jk} = \int_{t_0}^{t_1} l_j l_k d\tau$ should be nonzero for linear independence. In the case here, the functions for independence test are the system's state sequence $e^{\mathbf{A}t}\mathbf{B}$ to impulse excitation. Let the start time for control stimuli vary, i.e., let $t_0 = t$, and observe the following property of the controllability matrix

$$\begin{aligned} &\frac{d}{dt}\mathfrak{G}_c(t, t_f) = \frac{d}{dt}\int_{t}^{t_f} e^{\mathbf{A}(t-\tau)} \mathbf{B}\mathbf{B}^T e^{\mathbf{A}^T(t-\tau)} d\tau = \\ &\int_{t}^{t_f} \frac{d}{dt}[e^{\mathbf{A}(t-\tau)} \mathbf{B}\mathbf{B}^T e^{\mathbf{A}^T(t-\tau)}] d\tau - e^0 \mathbf{B}\mathbf{B}^T e^0 = \\ &\qquad \mathbf{A}\mathfrak{G}_c(t, t_f) + \mathfrak{G}_c(t, t_f)\mathbf{A}^T - \mathbf{B}\mathbf{B}^T. \end{aligned} \tag{3.133}$$

For large time ranges $t_f - t$ and damped (asymptotically stable) systems, the state impulses eventually die out asymptotically and do not contribute more to the Grammian. Thus, for large control times the Gramian derivative with respect to initial time variation is zero. The infinitely long time Grammian $\mathfrak{G}_c^\infty$ is thus governed by the Lyapunov equation

$$\mathbf{A}\mathfrak{G}_c^\infty(t, t_f) + \mathfrak{G}_c^\infty(t, t_f)\mathbf{A}^T - \mathbf{B}\mathbf{B}^T = \mathbf{0}. \tag{3.134}$$

On the other hand, the corresponding long time observability Grammian $\mathfrak{G}_o^\infty$ can be shown (see Kailath 1980) to be the solution of another Lyapunov equation

$$\mathbf{A}^T\mathfrak{G}_o^\infty(t, t_f) + \mathfrak{G}_o^\infty(t, t_f)\mathbf{A} - \mathbf{C}^T\mathbf{C} = \mathbf{0}. \tag{3.135}$$

Numerical methods for solving the Lyapunov equations exist but are computationally expensive. For large order systems (say $n > 10.000$), the solution may take hours to obtain even with a fast computer (as of year 2018). Recall that the Grammians obtained by the Lyapunov equation are for infinitely long control and observation times. The general controllability Grammian, as defined by Eq. (3.130), is seen to be dependent on control time and is thus not unique. Also, observe that the Grammians are not invariant to similarity transformations. Therefore, a similar realization as in Eq. (3.86) normally gives other Gramians $\mathfrak{G}_c$ and $\mathfrak{G}_o$. This is used to the advantage for model reduction using the balancing realization, one important realization form discussed below.

3.4.3 *State-Space Realizations*

State-Space Realization on Diagonal Form. The *diagonal form*, also known as the *modal decomposition form*, is a particular state-space realization that is strongly linked to the free (i.e., $\mathbf{s}(t) = 0$) decaying system state from an initial state $\mathbf{x}(0) = \Phi \neq \mathbf{0}$ for which it holds that

$$\dot{\mathbf{x}}(t) = \mathbf{A}\mathbf{x}(t), \quad \mathbf{r}(t) = \mathbf{C}\mathbf{x}(t), \quad \mathbf{x}(0) = \Phi. \tag{3.136}$$

Under the assumption that the free decay is governed by the solution $\mathbf{x}(t) = \Phi e^{\lambda t}$ one has

$$\mathbf{A}\Phi = \lambda\Phi, \quad \mathbf{r}(t) = \mathbf{C}\Phi e^{\lambda t} \tag{3.137}$$

for which there are nontrivial solutions (λ, Φ) provided that λ is a root of the characteristic polynomial $det(\mathbf{A} - \lambda\mathbf{I}) = \mathbf{0}$. Such roots are also called *system poles* and the associated solution vectors Φ are the *eigenvectors* of $\mathbf{A}$. Assuming that $\mathbf{A}$ and $\mathbf{C}$ stem from a physically realizable system (for which the response $\mathbf{r}$ to a real-valued stimulus $\mathbf{r}$ is real), a real form of $\mathbf{A} \in \mathbb{R}^{n \times n}$ and $\mathbf{C} \in \mathbb{R}^{n_r \times n}$ is possible. Thus the characteristic polynomial has real-valued polynomial coefficients and the poles are thus either real or appear in complex-conjugate pairs. The eigenvalue problem (3.137) for all eigensolutions combined is

$$\mathbf{AT} = \mathbf{T}\Lambda, \tag{3.138}$$

where the *modal matrix* $\mathbf{T}$ has all system eigenvectors as columns and the *eigenvalue matrix* Λ has the associated system poles, normally sorted in increasing magnitude order, as elements along its diagonal. It has been shown, see e.g., Golub and van Loan (2012), that Λ is fully diagonal provided that all system poles are unique. That state-space realization thus has fully decoupled states. For systems with system poles that are not all unique, but for which some or all appear in clusters of coalescing poles, it has been shown that a minimal-form Λ has a 2×2 block-diagonal form for the associated poles and is otherwise diagonal. Such systems, with a so-called *deficient system matrix* $\mathbf{A}$, are treated in the next section that treats the Jordan normal form while this section is devoted to systems that can be brought to a fully diagonal form for which the system matrix $\mathbf{A}$ is non-deficient. That includes systems that has repeated eigenvalues but for which $\mathbf{A}$ is still non-deficient.

Using that $\mathbf{T}^{-1}\mathbf{AT} = \mathbf{T}^{-1}\mathbf{T}\Lambda = \Lambda$ together with the state transformation $\mathbf{x} = \mathbf{Tz}$, the realization (3.87) becomes

$$\begin{aligned} \dot{\mathbf{z}} &= \Lambda\mathbf{z} + \mathbf{T}^{-1}\mathbf{Bs} \equiv \Lambda\mathbf{z} + \bar{\mathbf{B}}\mathbf{s} \\ \mathbf{r} &= \mathbf{CTz} + \mathbf{Ds} \equiv \bar{\mathbf{C}}\mathbf{z} + \mathbf{Ds} \end{aligned} \tag{3.139}$$

and since Λ is diagonal the first-order differential equation system thus becomes fully decoupled. In free vibration in one mode only, i.e., $z_k(t) \neq 0$, $z_m = 0\ \forall m \neq k$, one notes that the response is

$$\mathbf{r}(t) = \bar{\mathbf{C}}_{.k} z_k(t) \tag{3.140}$$

and the k:th column of $\bar{\mathbf{C}}$ (denoted $\bar{\mathbf{C}}_{.k}$) is thus the k:th eigenvector of $\mathbf{A}$ as seen by the sensors through the projection of $\mathbf{C}$.

Since the eigenvalues may either be real with real-valued eigenvectors, or appear in complex-conjugate pairs with associated complex-conjugate eigenvectors, a block-diagonal real form of the generally complex-valued realization $\{\Lambda, \bar{\mathbf{B}}, \bar{\mathbf{C}}, \mathbf{D}\}$ is possible. For each complex-conjugate pair of eigenvalues $\lambda_k = \Re\{\lambda_k\} \pm \Im\{\lambda_k\}$, the corresponding 2×2 block of the system matrix becomes

$$\begin{bmatrix} \Re\{\lambda_k\} & -\Im\{\lambda_k\} \\ \Im\{\lambda_k\} & \Re\{\lambda_k\} \end{bmatrix} \tag{3.141}$$

and the associated two columns of $\mathbf{T}$ become $\Re\{\Phi\}$ and $\Im\{\Phi\}$.

For the modal form realization (3.139) the frequency domain transfer function, see Eq. (3.85), can be expressed as

$$\mathbf{H}(\omega) = \bar{\mathbf{C}}(i\omega\mathbf{I} - \Lambda)^{-1}\bar{\mathbf{B}} + \mathbf{D} \tag{3.142}$$

Since the state matrix A is diagonal, with the kth eigenvalue λ_i on its kth row and column, one notes in particular that a transfer function element H_{ij} can be expressed as

$$H_{ij}(\omega) = \bar{\mathbf{C}}_{i.}diag(\frac{1}{i\omega - \lambda_k})\bar{\mathbf{B}}_{.j} + D_{ij} \tag{3.143}$$

or using the so-called *pole-residue form* as

$$H_{ij}(\omega) = \sum_{k=1}^{n} \frac{\bar{C}_{ik}\bar{B}_{kj}}{i\omega - \lambda_k} + D_{ij}, \tag{3.144}$$

where $\bar{C}_{ik}\bar{B}_{kj}$ are the n residues of the transfer function H_{ij}. Since the eigenvectors of $\mathbf{A}$ are generally complex-valued, the residue may be a complex-valued scalar.

► *A two-degree-of-freedom example*

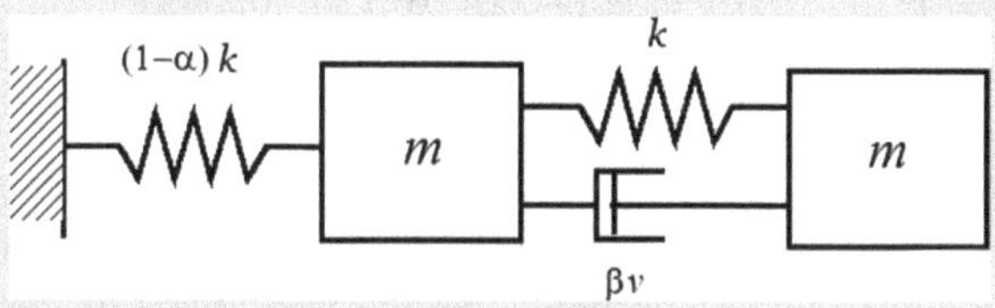

Fig. 3.9 A two-degree-of-freedom system

Let the parameters of the system in Fig. 3.9 be $\alpha = 0$, $\beta = 1$, $k = 100$ N/m, $v = 10$ Ns/m and $m = 1$ kg. Let further the output of the system be the displacement of the rightmost mass and the input be the force applied on the other mass. A numerical state-space representation, see Eq. (3.85), is then

$$\mathbf{A} = \begin{bmatrix} 0 & 0 & 1 & 0 \\ 0 & 0 & 0 & 1 \\ -200 & 100 & -1 & 1 \\ 100 & -100 & 1 & -1 \end{bmatrix} \quad \mathbf{B} = \begin{bmatrix} 0 \\ 0 \\ 1 \\ 0 \end{bmatrix} \quad \mathbf{C} = \begin{bmatrix} 0 & 1 & 0 & 0 \end{bmatrix}$$

for which the eigenvalues to three significant digits are $\lambda_{1,2} = -.0527 \pm 6.18i$ and $\lambda_{3,4} = -.947 \pm 16.14i$ [rad/s] and the transformed system on diagonal form is

$$\bar{\mathbf{A}} = \begin{bmatrix} -.0527 - 6.18i & 0 & 0 & 0 \\ 0 & -.0527 + 6.18i & 0 & 0 \\ 0 & 0 & -.947 - 16.1i & 0 \\ 0 & 0 & 0 & -.947 + 16.1i \end{bmatrix}$$

$$\bar{\mathbf{B}} = \begin{bmatrix} .269 - .00555i \\ .269 + .00555i \\ -.427 + .00779i \\ -.427 - .00779i \end{bmatrix}$$

$$\bar{\mathbf{C}} = \begin{bmatrix} -.00116 + .136i & -.00116 - .136i & .000435 + .0326i & .000435 - .0326i \end{bmatrix}$$

The corresponding real-valued 2 × 2 block-diagonal realization is

$$\bar{\mathbf{A}} = \begin{bmatrix} -.0527 & -6.18 & 0 & 0 \\ 6.18 & -.0527 & 0 & 0 \\ 0 & 0 & -.947 & -16.1 \\ 0 & 0 & 16.1 & -.947 \end{bmatrix}$$

$$\bar{\mathbf{B}} = \begin{bmatrix} .533 \\ .0111 \\ -.853 \\ -.0158 \end{bmatrix} \quad \bar{\mathbf{C}} = \begin{bmatrix} -.00116 & .136 & .000435 & .0326 \end{bmatrix}$$

A root locus plot of pole positions for various combinations of parameters α_1 and α_2 can be seen in Fig. 3.10. From that, it can be noted that some parameter combinations render coalescing poles that are the subject of the succeeding numerical example.

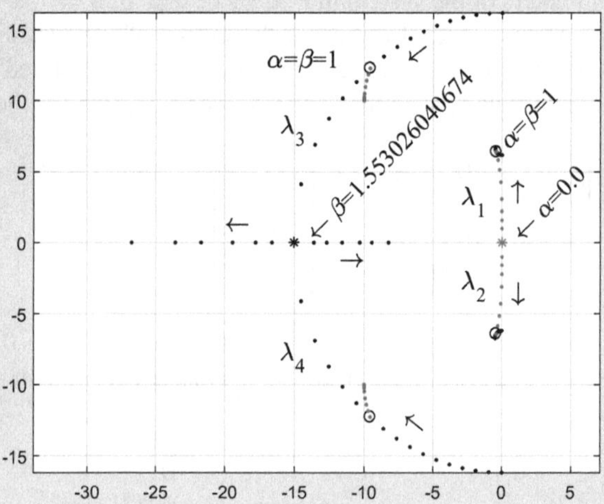

Fig. 3.10 Root locus of four system poles for discrete step variation of α with fixed $\beta = 1$ (red) and of varying β with fixed $\alpha = 1$ (black). Arrows indicate increasing parameters α and β. Asterisks (with values of α and β) indicate where poles coalesce which result in deficient systems

◄

State-Space Realization on Jordan Normal Form. For some systems with repeated eigenvalues, i.e., $\lambda_k = \lambda_{k+1} = \ldots = \lambda_{k+m_k}$, with multiplicity $m_k + 1$ it is impossible to form the same number of eigenvectors to diagonalize the system. Such systems have a deficient system matrix **A** with lesser than n unique eigenvectors and its *principal vectors* (sometimes called *generalized eigenvectors*) need to be found as the missing columns of **T** to form the minimal system, called the *Jordan normal form* Λ. That system is minimal in the sense that it gives a minimal number of couplings between its states and at the maximum couples states two-by-two. Systems with rigid-body modes or critical viscous damping are examples of such, as are illustrated by three examples below. The structure of the minimal coupling form is given by the following theorem.

Jordan Normal Form Theorem: *If* $\mathbf{A} \in \mathbb{R}^{n\times n}$, *then there exists a full rank* $\mathbf{T} \in \mathbb{C}^{n\times n}$ *such that* $\mathbf{T}^{-1}\mathbf{A}\mathbf{T} = blockdiag(\mathbf{J}_1, \ldots, \mathbf{J}_t)$ *is block diagonal with Jordan blocks* $\mathbf{J}_k$ *related to eigenvalues* λ_k *with multiplicity* m_k and $\sum_{k=1}^{t} m_k = n$. *The kth such block is*

$$\mathbf{J}_k = \begin{bmatrix} \lambda_k & 1 & & & 0 \\ & \lambda_k & 1 & & \\ & & \ddots & \ddots & \\ & & & \lambda_k & 1 \\ 0 & & & & \lambda_k \end{bmatrix}_{m_k \times m_k} \tag{3.145}$$

Proof See Halmos (1958). □

► *Three variants of the two-degree-of-freedom example*
Let the parameters of the system in Fig. 3.9 be such that (a) $\alpha = 1$ and $\beta = 0$, (b) $\alpha = 0$ and $\beta = 1.553$, and (c) $\alpha = 1$ and $\beta = \sqrt{2}$. In all three cases $k = 100$ N/m, $v = 10$ Ns/m and $m = 1$ kg. This will render; (a) an undamped system with one rigid-body mode, (b) a system with one critically damped mode, and (c) a system with both one rigid-body mode and one critically damped mode. Principal value calculation will then give the following transformation matrices and resulting system matrices on Jordan normal form:

(a)

$$\mathbf{T} = \begin{bmatrix} .5 & 0 & +.0177i & -.0177i \\ .5 & 0 & -.0177i & +.0177i \\ 0 & 0 & +.25 & +.25 \\ 0 & 0 & -.25 & -.25 \end{bmatrix} \qquad \bar{\mathbf{A}} = \begin{bmatrix} 0 & 1 & 0 & 0 \\ 0 & 0 & 0 & 0 \\ 0 & 0 & -14.1421i & 0 \\ 0 & 0 & 0 & 14.1421i \end{bmatrix}$$

(b)

$$\mathbf{T} = \begin{bmatrix} -.0378 & -.0015 & +.0222 - 0.1153i & +.0222 + .1153i \\ +.0545 & +.0043 & -.0111 - 0.1500i & -.0111 + .1500i \\ +.5685 & -.0158 & +.7537 + 0.2039i & +.7537 - .2039i \\ -.8200 & -.0106 & 1 & 1 \end{bmatrix}$$

$$\bar{\mathbf{A}} = \begin{bmatrix} -15.0372 & 1 & 0 & 0 \\ 0 & -15.0372 & 0 & 0 \\ 0 & 0 & -.4931 + 6.6319i & 0 \\ 0 & 0 & 0 & -.4931 - 6.6319i \end{bmatrix}$$

(c)

$$\mathbf{T} = \begin{bmatrix} +.0499 & +.0035 & -.7071 & 0 \\ -.0499 & -.0035 & -.7071 & 0 \\ -.7053 & +.0002 & 0 & -.7071 \\ & -.0002 & 0 & -.7071 \end{bmatrix}$$

$$\bar{\mathbf{A}} = \begin{bmatrix} -14.1421 & 1 & 0 & 0 \\ 0 & -14.1421 & 0 & 0 \\ 0 & 0 & 0 & 1 \\ 0 & 0 & 0 & 0 \end{bmatrix}$$

◀

State-Space Realization on Balanced Form. Moore (1981) has showed that it is possible to find a similarity transformation $\mathbf{x} = \mathbf{T}_b\mathbf{z}$ such that the controllability Grammian of Eq. (3.134) $\mathfrak{G}_c^\infty$ and the observability Grammian $\mathfrak{G}_o^\infty$ of Eq. (3.135) simultaneously become diagonal and equal (balanced), i.e., $\mathfrak{G}_c^\infty = \mathfrak{G}_c^\infty$. The corresponding state-space model

$$\begin{aligned} \dot{\mathbf{z}} &= \mathbf{T}_b^{-1}\mathbf{A}\mathbf{T}_b\mathbf{z} + \mathbf{T}_b^{-1}\mathbf{B}\mathbf{s} \\ \mathbf{r} &= \mathbf{C}\mathbf{T}_b\mathbf{z} + \mathbf{D}\mathbf{s} \end{aligned} \tag{3.146}$$

is called a balanced realization, with Grammians balanced over the control and observation range $[0, \infty]$.

The procedure to obtain the balancing transformation involves the solution of two Lyapunov equations and goes as the following.

Balancing procedure:

1. Solve two Lyapunov equations for matrices **P** and **Q**

$$\mathbf{A}^T\mathbf{P}\mathbf{A} - \mathbf{P} + \mathbf{C}^T\mathbf{C} = 0 \qquad \mathbf{A}\mathbf{Q}\mathbf{A}^T - \mathbf{Q} + \mathbf{B}\mathbf{B}^T = 0. \tag{3.147}$$

2. Make a Cholesky decomposition of the symmetric and positive definite **Q**

$$\mathbf{R}^T\mathbf{R} = \mathbf{Q}. \tag{3.148}$$

3. Compute a singular value decomposition to obtain the singular value matrix Σ for singular values in decreasing order in the relation

$$\mathbf{U}\Sigma\mathbf{U}^T = \mathbf{RPR}^T. \tag{3.149}$$

4. Then the balancing transformation $\mathbf{T}_b$ is given by

$$\mathbf{T}_b = \mathbf{R}^T\mathbf{U}\Sigma^{-1/2}. \tag{3.150}$$

It should be noted that, although the controllability and observability Grammians are both diagonalized and equal, the realization $\bar{\mathfrak{M}} = \{\mathbf{T}_b^{-1}\mathbf{AT}_b, \mathbf{T}_b^{-1}\mathbf{B}, \mathbf{CT}_b, \mathbf{D}\}$ related to Eq. (3.146) is generally not taken to diagonal form by the transformation $\mathbf{z} = \mathbf{T}_b\mathbf{x}$. For large-scale models with many states (say >10.000) it should also be noted that the solution to the Lyapunov equations is very computationally demanding. However, the balanced form is very well suited for model reduction as states that contribute little to the input/output relation can be singled out by the associated small elements of the diagonalized Grammians and by that reduced from the model. State reduction schemes will be described next.

3.4.4 State-Space Reduction Based on Transfer Strength

Many model order reduction techniques have been developed in order to balance the accuracy and simplicity of the systems, see Refs. (Antoulas 2005, Noor 1994, Ersal et al. 2008). Eigenvalue-based model reduction techniques, such as dynamic condensation, component mode synthesis, and modal truncation, continue to receive attention due to their low computational cost and applicability for very large systems (Varga 1995, Qu 2004). These approaches produce reduced-order models under the assumption that the eigenmodes with eigenfrequencies in the vicinity or within the dominant spectrum of the loading are the system's dominant eigenmodes to be kept in a reduced-order model.

One of the key features of the system is its input–output behavior and selection metrics for reduction have been developed to best conserve this feature of the original system. Skelton (1980, 1983) proposed a component cost analysis method for first-order continuous-time systems which assigns a cost function associated with the mean-squared system output under white noise stimuli. This method is best suited for small-scale systems since it requires the computationally expensive solution of a Lyapunov equation for the evaluation of the cost function. Another metric based on the unit-step response was proposed by Aguirre (1993) and is based on a simple formulation for either transfer function or state-space representation of the system.

Balancing Truncation. A reduction scheme that is popular in control engineering is based on the balanced input–output relation $\bar{\mathfrak{M}} = \{\mathbf{T}_b^{-1}\mathbf{AT}_b, \mathbf{T}_b^{-1}\mathbf{B}, \mathbf{CT}_b, \mathbf{D}\} \equiv \{\bar{\mathbf{A}}, \bar{\mathbf{B}}, \bar{\mathbf{C}}, \mathbf{D}\}$. It has the drawback of requiring the costly balancing operation but has the beneficial properties given by the balancing reduction theorem below. Let $\bar{\mathfrak{M}}$ be balanced with diagonal controllability and observability Grammians $\mathfrak{G}_c = \mathfrak{G}_o \equiv \mathfrak{G} \equiv \text{diag}(\mathfrak{g})$. Partition the matrices related to the $\bar{n}$ separated larger Grammians in

$\mathfrak{g}_1$ from the $\underline{n}$ smaller in $\mathfrak{g}_1$ of $\mathfrak{G}$ as

$$\mathfrak{G} = \begin{bmatrix} \mathfrak{g}_1 & 0 \\ 0 & \mathfrak{g}_2 \end{bmatrix} \quad \bar{\mathbf{A}} = \begin{bmatrix} \bar{\mathbf{A}}_{11} & \bar{\mathbf{A}}_{12} \\ \bar{\mathbf{A}}_{21} & \bar{\mathbf{A}}_{22} \end{bmatrix} \quad \bar{\mathbf{B}} = \begin{bmatrix} \bar{\mathbf{B}}_1 \\ \bar{\mathbf{B}}_2 \end{bmatrix} \quad \bar{\mathbf{C}} = \begin{bmatrix} \bar{\mathbf{C}}_1 & \bar{\mathbf{C}}_2 \end{bmatrix} \tag{3.151}$$

Let further the systems $\mathfrak{S} \in \mathbb{R}^{(n+n_r)\times(n+n_s)}$ and $\mathfrak{S}_r \in \mathbb{R}^{(\bar{n}+n_r)\times(\bar{n}+n_s)}$ be

$$\mathfrak{S} = \begin{bmatrix} \bar{\mathbf{A}} & \bar{\mathbf{B}} \\ \bar{\mathbf{C}} & \mathbf{0} \end{bmatrix} \quad \mathfrak{S}_r = \begin{bmatrix} \bar{\mathbf{A}}_{11} & \bar{\mathbf{B}}_1 \\ \bar{\mathbf{C}}_1 & \mathbf{0} \end{bmatrix} \tag{3.152}$$

The model $\bar{\mathfrak{M}}_r = \{\bar{\mathbf{A}}_{11}, \bar{\mathbf{B}}_1, \bar{\mathbf{C}}_1, \mathbf{D}\}$ is then a reduced-order system obtained from $\{\bar{\mathbf{A}}, \bar{\mathbf{B}}, \bar{\mathbf{C}}, \mathbf{D}\}$ by balanced truncation. It has some guaranteed properties related to stability, controllability, observability, and impulse response which are given by the following theorem.

Balancing Truncation Theorem: *Given the controllable, observable, and stable system $\bar{\mathfrak{M}}$, the reduced-order system $\bar{\mathfrak{M}}_r$ obtained by balanced truncation have the following properties:*

1. *$\bar{\mathfrak{M}}_r$ is balanced and has no unstable poles.*
2. *If $\lambda_p = \text{eig}(\mathbf{A}_{11}) \neq \lambda_q = \text{eig}(\mathbf{A}_{22})/\forall p, q$ then $\bar{\mathfrak{M}}_r$ is controllable and observable.*
3. *Let the ordered singular values of $\mathfrak{S}$ be σ_i, $i = 1, 2, \ldots, q$ and let further the number of singular values of $\mathfrak{S}_r$ be k with $k < q$. The Hankel $\mathscr{H}_\infty$-norm of the difference between the full-order system $\mathfrak{S}$ and the reduced-order system $\mathfrak{S}_r$ is then twice the sum of neglected Hankel singular values as: $||\mathfrak{S} - \mathfrak{S}_r||_{\mathscr{H}_\infty} \leq 2(\sigma_{k+1} + \ldots + \sigma_q)$*

Proof See Antoulas (2005). □

The last part of the above theorem implies that if the neglected singular values are small, the frequency response functions of $\mathfrak{M}$ and $\mathfrak{M}_r$ are close. The error bounds of continuous-time and discrete-time state-space model reduction were established by Glover (1984) and Enns (1984).

Modal Truncation. To obtain accurate reduced-order modal state-space models, it is necessary to determine the dominant eigenmode subset which keeps the most important features of the system. In the simplest form given by Davison (1966), the lowest damped modes has been considered as the dominant subset to be kept while the complement set was eliminated. The shortcoming of this reduction approach is that the retained modes can be selected from the uncontrollable or unobservable eigenvalue subspace of the original system. Such modes do not contribute to the input–output relation of the system at all. Thus, the created reduced-order model is not minimal. In Varga (1995), the original system is divided to several subsystems which are in the balanced form. In each subsystem, eigenmodes with largest Hankel singular values (the square root of the eigenvalues of the product of the observability and controllability gramians) are considered as dominant modes to retain.

A frequency-limited interval modal dominancy index for continuous-time systems with both under-damped and over-damped eigenvalues was introduced by Vakilzadeh et al. (2014, 2014). The index quantify the two-norm contribution of each eigenmode to the system output deviation. Thus, the nondominant modes with less output contribution can be identified and excluded to obtain the appropriate reduced-order model. The main advantage of this index is that it is on explicit form in terms of modal contribution and the frequency bound of interest. In addition, the retained low-order model does not contain any uncontrollable and unobservable modes.

Consider the diagonal (i.e., modal) realization of a linear, time invariant, and stable continuous-time system given by Eq. (3.139). Let the partitioned diagonalized state-space representation be

$$\begin{bmatrix} \dot{\mathbf{z}}_1 \\ \dot{\mathbf{z}}_2 \end{bmatrix} = \begin{bmatrix} \Lambda_1 & 0 \\ 0 & \Lambda_2 \end{bmatrix} \begin{bmatrix} \mathbf{z}_1 \\ \mathbf{z}_2 \end{bmatrix} + \begin{bmatrix} \bar{\mathbf{B}}_1 \\ \bar{\mathbf{B}}_2 \end{bmatrix} \mathbf{s}$$

$$\mathbf{r} = \begin{bmatrix} \bar{\mathbf{C}}_1 & \bar{\mathbf{C}}_2 \end{bmatrix} \begin{bmatrix} \mathbf{z}_1 \\ \mathbf{z}_2 \end{bmatrix} + \mathbf{D}\mathbf{s}, \tag{3.153}$$

where $\mathbf{z}_1$ contains the n_r modal coordinates to be retained in the low-order system and Λ_1 is a diagonal matrix which involves the n_r dominant eigenvalues of the full system $\mathfrak{S} = (\boldsymbol{\Lambda}, \bar{\mathbf{B}}, \bar{\mathbf{C}}, \mathbf{D})$. Thus, the truncated system can be written as $\mathfrak{S}_r = (\boldsymbol{\Lambda}_1, \bar{\mathbf{B}}_1, \bar{\mathbf{C}}_1, \mathbf{D})$.

The low-order model obtained by modal truncation has some guaranteed properties. First, the H_∞ norm of the difference between the full model and the low-order model has an a priori upper bound. In diagonalized form, the difference between the transfer functions of the full model $G(\mathfrak{S}, \omega)$ and the reduced-order model, $G_r(\mathfrak{S}_r, \omega)$, can be written as (see Eq. 3.144)

$$G(\mathfrak{S}, \omega) - G_r(\mathfrak{S}_r, \omega) = \sum_{k=n_r+1}^{n_s} \frac{\bar{c}_k \bar{b}_k}{i\omega - \lambda_k} \tag{3.154}$$

Thus, the H_∞ norm of the error system is upper bounded by the following expression

$$||G(\mathfrak{S}, \omega) - G_r(\mathfrak{S}_r, \omega)||_\infty = \sup_{\omega \in \Re}[\bar{\sigma}(G(\mathfrak{S}, \omega) - G_r(\mathfrak{S}_r, \omega))] \leq - \sum_{k=n_r+1}^{n_s} \frac{\bar{\sigma}(\bar{c}_k \bar{b}_k)}{\Re\mathfrak{e}(\lambda_k)} \tag{3.155}$$

where $\bar{\sigma}(.)$ is the largest singular value of matrix (.). Secondly, the eigenvalues of the low-order model is a subset of the eigenvalues of the original model and therefore they keep their physical interpretations (Ersal et al. 2008), e.g., the modal truncation preserves the stability property of the full system.

The upper bound of Eq. (3.155) hints to a reduction strategy to yield the low-order model. It remains to be determined which set of eigensolutions to be used in the reduced-order model in order to conserve the input–output behavior of the full system to as large extent as possible. To this end, let the k:th modal coordinate of the diagonalized state-space equation be described by

$$\begin{aligned} \dot{z}_k &= \lambda_k z_k + \bar{b}_k \mathbf{s} \\ \Delta r_k &= \bar{c}_k z_k, \end{aligned} \tag{3.156}$$

where the stimuli vector $\mathbf{s} \in \Re^{n_s}$ is unit-impulse at $t = 0$ and the response vector $\Delta r_k \in \Re^{n_r}$ are the modal contributions to the system output. The following metric (or dominance index) quantify the contribution of the k:th mode to the full system output

$$M_k = \int_0^{\infty} (\Delta r_k^H \Delta r_k) dt. \tag{3.157}$$

According to Parseval's theorem this dominance index can be transformed into the frequency domain as

$$M_k = \int_{-\infty}^{+\infty} (\Delta \mathbf{R}_k(\omega)^H \Delta \mathbf{R}_k(\omega)) d\omega. \tag{3.158}$$

Utilizing the Laplace transformation, the input–output relation can be written as

$$\Delta \mathbf{R}_k = \bar{c}_k (i\omega - \lambda_k)^{-1} \bar{b}_k \mathbf{S}(\omega) \tag{3.159}$$

With a unit-impulse signal ($\mathbf{S}(\boldsymbol{\omega}) = 1 \forall \omega$) and a substitution of Eq. (3.159) into Eq. (3.158) then leads to

$$M_k = \int_{-\infty}^{+\infty} \bar{b}_k^H (i\omega - \lambda_k)^{-1} \bar{c}_k^H \bar{c}_k (i\omega - \lambda_k)^{-1} \bar{b}_k d\omega = -\frac{\pi \bar{b}_k^H \bar{c}_k^H \bar{c}_k \bar{b}_k}{\Re\mathfrak{e}(\lambda_k)}. \tag{3.160}$$

However, most loading situations of relevance here are such that the (one-sided) load spectrum is dominated by the content in a specific frequency region from ω_1 thru ω_2. For the reduced-order model to more accurately represent the behavior in this frequency region, the frequency information of the input can be taken into the account by assuming that $\mathbf{S}(\boldsymbol{\omega}) = 1 \ \forall \omega = [\omega_1, \omega_2]$ and $\mathbf{S}(\boldsymbol{\omega}) = 0$ elsewhere. That leads to the following frequency-weighted metric

$$\bar{M}_k = \int_{\omega_1}^{\omega_2} (\Delta R_k(\omega)^H \Delta R_k(\omega)) d\omega. \tag{3.161}$$

With Eq. (3.159) this then leads to

$$\bar{M}_k = \frac{\bar{b}_k^H \bar{c}_k^H \bar{c}_k \bar{b}_k}{\Re\mathfrak{e}(\lambda_k)} (\text{atan}(\omega_1 - \Im\mathfrak{m}(\lambda_k)) - \text{atan}(\omega_2 - \Im\mathfrak{m}(\lambda_k)) \tag{3.162}$$

Based on this metric, a frequency interval dominancy based definition of modal dominancy can be introduced; **Definition**: For a given threshold value $\epsilon \geq 0$ and frequency interval $[\omega_1, \omega_2]$, let n_r be the number of metrics for which $\bar{M}_k > \epsilon$ for $k = 1, 2, \ldots, n_r$. The modes associated to this are the dominant modes.

As a consequence, the full model has n_r dominant and $n_x - n_r$ nondominant eigensolutions. An important feature of this metric is that the retained dominant eigenvalue subspace is both controllable and observable since each non-observable or noncontrollable mode k would render $\bar{M}_k = 0$.

A fuller description of the reduction procedure, also treating the case with sets of nonunique coalescing eigenvalues, together with a numerical example that illustrates its performance is given in Vakilzadeh et al. (2015).

References

Aguirre, L. A. (1993). Quantitative measure of modal dominance continuous systems. In Anon (ed.) *Proceedings of the 32nd IEEE conference on decision and control* (vol. 3). IEEE. https://doi.org/10.1109/CDC.1993.325629.

Antoulas, A. C. (2005). *Approximation of large-scale dynamical systems*. Philadelphia: Society for Industrial and Applied Mathematics.

Aoyama, Y., & Yagawa, G. (2001). Component mode synthesis for large-scale structural eigenanalysis. *Computers and Structures*, *79*(6), 605–615.

Balmès, E. (1996). Use of generalized interface degrees of freedom in component mode synthesis. In *Proceedings of the 14th International Modal Analysis Conference*, **1**.

Balmès, E. (1997). *Modèles expérimentaux complets et modèles analytiques réduits en dynamique des structures*. Ph.D. thesis, Université Pierre et Marie Curie, Paris VI, ONERA, Directions des Structure, DRET/AC.

Bampton, M. C. C., & Craig, J. R. R. (1968). Coupling of substructures for dynamic analyses. *AIAA Journal*, *6*(7), 1313–1319.

Bathe, K. J., & Dong, J. (2014). Component mode synthesis with subspace iterations for controlled accuracy of frequency and mode shape solutions. *Computers and Structures*, *139*, 28–32.

Benfield, W., & Hruda, R. (1971). Vibration analysis of structures by component mode substitution. *AIAA Journal*, *9*(7), 1255–1261.

Castanier, M. P., Tan, Y. C., & Pierre, C. (2001). Characteristic constraint modes for component mode synthesis. *AIAA Journal*, *39*(6), 1182–1187.

Coppolino, R.N. (2018). Consolidation of weakly coupled experimental system modes. In *Proceedings of IMAC-XXXVI: A Conference on Structural Dynamics*. Bethel, CT: Society for Experimental Mechanics.

Craig, R. R., & Chang, C. J. (1977). On the use of attachement modes in substructure coupling for dynamics analysis. *18st Structures* (pp. 89–99). Structural Dynamics and Material Conference San Diego: AIAA/ASME.

Craig, R. R, Jr., & Chang, C. J. (1977). *Substructure Coupling for Dynamic Analysis and Testing*. Report, National Aeronautics and Space Administration.

Davison, E. (1966). A method for simplifying linear dynamic systems. *EEE Transactions on Automatic Control*, *11*(1), 93–101. https://doi.org/10.1109/tac.1966.1098264.

Dickens, J.M., & Stroeve, A. (2000). Modal truncation vectors for reduced dynamic substructure models. In *Structures, Structural Dynamics and Material Conference. 41st AIAA/ASME/ASCE/AHS/ASC Structures, Structural Dynamics, and Materials Conference and Exhibit*, Atlanta

Enns, D. F. (1984). Model reduction with balanced realizations: An error bound and frequency weighted generalization. In *Proceedings of the IEEE Conference on Decision and Control* (pp. 127–132).

Ersal, T., Fathy, H.K., Rideout, D.G., Louca, L.S. & Stein, J.L. (2008). A review of proper modeling techniques. *Journal of Dynamic Systems, Measurement, and Control* **130**(6). https://doi.org/10.1115/1.2977484.

Géradin, M., & Rixen, D. (2015). *Mechanical Vibrations: Theory and Application to Structural Dynamics* (3rd ed.). Chichester: Wiley.

Glover, K. (1984). All optimal hankel-norm approximations of linear multivariable system and their L_∞-error bounds. *International Journal of Control*, *39*, 1115–1193.

Golub, G.H., van Loan, C.F. (2012). *Matrix computations* (Vol. 3). John Hopkins University Press.

Gruber, F., & Rixen, D. (2018a). A comparison of craig-bampton approaches for systems with arbitrary viscous damping in dynamic substructuring. In *IMAC-XXXVI: International Modal Analysis Conference, Orlando, FL*. Bethel, CT: Society for Experimental Mechanics.

Gruber, F. M., & Rixen, D. J. (2016). Evaluation of substructure reduction techniques with fixed and free interfaces. *Strojniški vestnik - Journal of Mechanical Engineering*, *62*(7–8), 452–462.

Gruber, F. M., & Rixen, D. J. (2018b). Dual craig-bampton component mode synthesis method for model order reduction of nonclassically damped linear systems. *Mechanical Systems and Signal Processing*, *111*, 678–698.

Guyan, R. (1965). Reduction of mass and stiffness matrices. *AIAA Journal*, *3*(2), 380.

Halmos, P. (1958). *Finite dimensional vector spaces*. New York: Van Nostrand.

Herting, D. (1985). A general purpose, multi-stage, component modal synthesis method. *Finite Elements in Analysis and Design*, *1*(2), 153–164.

Holzwarth, P., & Eberhard, P. (2015). Svd-based improvements for component mode synthesis in elastic multibody systems. *European Journal of Mechanics-A/Solids*, *49*, 408–418.

Hong, S. K., Epureanu, B. I., & Castanier, M. P. (2013). Next-generation parametric reduced-order models. *Mechanical Systems and Signal Processing*, *37*(12), 403–421.

Hurty, W. C. (1965). Dynamic analysis of structural systems using component modes. *AIAA Journal*, *3*(4), 678–685.

Irons, B. (1965). Structural eigenvalue problems-elimination of unwanted variables. *AIAA Journal*, *3*(5), 961–962.

Jakobsson, H., Bengzon, F., & Larson, M. G. (2011). Adaptive component mode synthesis in linear elasticity. *International Journal for Numerical Methods in Engineering*, *86*, 829–844.

Kailath, T. (1980). *Linear systems*. Wiley.

Kim, J. H., Kim, J., & Lee, P. S. (2017). Improving the accuracy of the dual craig-bampton method. *Computers and Structures*, *191*, 22–32.

Krattiger, D., Wu, L., Zacharczuk, M., Buck, M., Kuether, R. J., Allen, M. S., et al. (2019). Interface reduction for hurty/craig-bampton substructured models: Review and improvement. *Mechanical Systems and Signal Processing*, *114*, 579–603.

Kuether, R. J., Allen, M. S., & Hollkamp, J. J. (2017). Modal substructuring of geometrically nonlinear finite element models with interface reduction. *AIAA Journal*.

Lindberg, E., & Höransson, P. (2013). Component mode synthesis using undeformed interface coupling modes to connect soft and stiff substructures. *Shock and Vibration*, *20*, 157–170.

MacNeal, R. H. (1971). Special issue on structural dynamics a hybrid method of component mode synthesis. *Computers and Structures*, *1*(4), 581–601.

Moore, B. (1981). Principal component analysis in linear systems: Controllability, observability and model reduction. *IEEE Transaction on Automatic Control* **AC-26**, 17–32 (1981).

Noor, A. K. (1994). Recent advances and applications of reduction methods. *Applied Mechanics Reviews*, *47*(5), 125–146.

Qu, Z. Q. (2004). *Model order reduction techniques: with application in finite element analysis*. New York: Springer.

Rahrovani, S., Vakilzadeh, M. K., & Abrahamsson, T. J. S. (2014). Modal dominancy analysis based on modal contribution to frequency response function 2-norm. *Mechanical Systems and Signal Processing*, *48*(1–2), 213–231.

Rixen, D. (2002). High order static correction modes for component mode synthesis. In *Fifth World Congress on Computational Mechanics*. Vienna, Austria.

Rixen, D.J. (2009). Dual craig-bampton with enrichment to avoid spurious modes. In *IMAC-XXVII: International Modal Analysis Conference, Orlando, FL*. Bethel, CT: Society for Experimental Mechanics.

Rixen, D.J. (2011). Interface reduction in the dual craig-bampton method based on dual interface modes. In *IMAC-XXIX: International Modal Analysis Conference, Jacksonville, FL: International Modal Analysis Conference, Jacksonville, FL*. Bethel, CT: Society for Experimental Mechanics.

Rixen, D. (2004). A dual Craig-Bampton method for dynamic substructuring. *Journal of Computational and Applied Mathematics*, *168*(1–2), 383–391.

Rubin, S. (1975). Improved component-mode representation for structural dynamic analaysis. *AIAA Journal*, *13*(8), 995–1006.

Rubin, S. (1975). Improved component-mode representation for structural dynamic analysis. *AIAA Journal*, *13*(8), 995–1006.

Shyu, W. H., Ma, Z. D., & Hulbert, G. M. (1997). A new component mode synthesis method: Quasi-static mode compensation. *Finite elements in Analysis and Design*, *24*(4), 197–298.

Skelton, R. E. (1980). Cost decomposition of linear systems with application to model reduction. *International Journal of Control*, *32*(6), 1031–1055.

Skelton, R. E., & Yousuff, A. (1983). Component cost analysis of large scale systems. *International Journal of Control*, *37*(2), 285–304.

Stoev, J., & Möhring, J. (2010). Hybrid reduction of non-symmetric mimo systems. In P.S. et al. (Ed.) *ISMA*. K.U. Leuven.

Su, T. J., & Craig, R. (1991). Model reduction and control of flexible structures using krylov vectors. *Journal of Guidance, Control, and Dynamics*, *14*(2), 260–267.

Vakilzadeh, M. K., Rahrovani, S., & Abrahamsson, T. J. S. (2014). Modal reduction based on accurate input-output relation preservation. In R. Allemang, J. D. Clerck, C. Niezrecki, & A. Wicks (Eds.), *Topics in Modal Analysis* (Vol. 7, pp. 333–342). New York: Springer.

Vakilzadeh, M. K., Yaghoubi, V., McKelvey, T., Abrahamsson, T. J. & Ljung, L. (2015). Experiment design for improved frequency domain subspace system identification of continuous-time systems. *IFAC-PapersOnLine*, 886–891.

Varga, A. (1995). Enhanced modal approach for model reduction. *Mathematical Modelling of Systems*, *1*(2), 91–105. https://doi.org/10.1080/13873959508837010.

Voormeeren, S., Nortier, B., & Rixen, D. (2013). Error estimation and adaptive model reduction applied to offshore wind turbine modeling. In *IMAC-XXXI: International Modal Analysis Conference, Error Estimation and Adaptive Model Reduction Applied to Offshore Wind Turbine Modeling*. Bethel, CT: Society for Experimental Mechanics.

Voormeeren, S. N., van der Valk, P. L. C., & Rixen, D. J. (2011). Generalized methodology for assembly and reduction of component models for dynamic substructuring. *AIAA Journal*, *49*, 1010–1020.

Chapter 4
Experimental Substructuring

Abstract In this chapter, we outline techniques to use measured dynamic characteristics of components to build an assembled model. Although the basic theory is pretty straightforward, the difficulty in experimental substructuring lies in the limitations in what can be measured and the special techniques needed to alleviate errors in the measured signals.—*(Chapter Authors: Daniel Rixen, Maarten van der Seijs, Randall Mayes, Matt Allen & Thomas Abrahamsson)*

4.1 Introduction

The substructuring methods discussed in previous chapters are equally applicable when any of the data for a substructure is replaced with experimental data. However, established methods exist to measure some quantities, such as the free-free natural frequencies of a structure, while other quantities, such as the constraint modes in a Hurty/Craig–Bampton model, may be practically impossible to measure. In the theory so far, we have seen that a wide variety of substructuring methods are based on the following formulations (recalled here for the frequency domain):

$$\text{General form} \quad \boxed{\begin{cases} \mathbf{Z}\bar{\mathbf{u}} = \bar{\mathbf{f}} + \bar{\mathbf{g}} \\ \mathbf{B}\bar{\mathbf{u}} = \mathbf{0} \\ \mathbf{L}^T\bar{\mathbf{g}} = \mathbf{0} \end{cases}} \tag{2.33}$$

$$\text{Primal Assembly} \quad \boxed{\mathbf{Z}_g\mathbf{u}_g = \mathbf{f}_g \quad \text{with} \quad \mathbf{Z}_g = \mathbf{L}^T\mathbf{Z}\mathbf{L} \quad \text{and} \quad \mathbf{f}_g = \mathbf{L}^T\bar{\mathbf{f}}} \tag{2.39}$$

$$\text{Dual Assembly} \quad \boxed{\begin{bmatrix} \mathbf{Z} & \mathbf{B}^T \\ \mathbf{B} & \mathbf{0} \end{bmatrix} \begin{bmatrix} \bar{\mathbf{u}} \\ \boldsymbol{\lambda} \end{bmatrix} = \begin{bmatrix} \bar{\mathbf{f}} \\ \mathbf{0} \end{bmatrix}} \tag{2.45}$$

The key consideration now is, what domain should we use to describe $\mathbf{Z}$ (i.e., modal domain, time domain, etc.) and what measurements can we obtain? A few common options are reviewed briefly here:

M. S. Allen et al., *Substructuring in Engineering Dynamics*,
CISM International Centre for Mechanical Sciences 594,
https://doi.org/10.1007/978-3-030-25532-9_4

Frequency Response Function (FRF) Measurements

Recall that for frequency-domain measurements, $\mathbf{Z}$ is the dynamic stiffness of a structure, and while it cannot typically be measured, one can measure the inverse or frequency response $\mathbf{Y}(\omega) = \mathbf{Z}(\omega)^{-1}$.

$$\mathbf{Y}^{(s)}(\omega) = \mathbf{Z}^{(s)}(\omega)^{-1} = \left(-\omega^2\mathbf{M}^{(s)} + i\omega\mathbf{C}^{(s)} + \mathbf{K}^{(s)}\right)^{-1}. \tag{2.16}$$

If we apply a force (i.e., with a shaker, impact hammer, etc.) or input at point j and measure the response (i.e., with an accelerometer, laser vibrometer, etc.) or output at point i, then the element $\mathbf{Y}_{ij}^{(s)}(\omega)$ can be measured at the excitation frequency, ω. In practice, a force is applied over some range of frequencies, transient measurements are acquired in the time domain, and FFT techniques are used to obtain the FRF over a range of frequencies.

Typically, a set of measurements is obtained by including multiple response transducers so $i = 1...N_o$ where N_o is the number of output points and multiple inputs (or drive points) can also be used so that $j = 1...N_i$. Hence, one can eventually obtain a matrix of FRFs, $\mathbf{Y}^{(s)}(\omega)$ at enough points so that an adequate model of the substructure can be obtained and coupled to other substructures, as elaborated later. Measurements may be used to create a model for one of the substructures (e.g., for $s = 1$) and the others would presumably be modeled numerically.

Measuring Modal Parameters

In the physical domain, a system model can be described in terms of the mass, damping, and stiffness matrices, but as was the case with dynamic stiffness, these matrices cannot typically be measured.

$$\mathbf{M}^{(s)}\ddot{\mathbf{u}}^{(s)} + \mathbf{C}^{(s)}\dot{\mathbf{u}}^{(s)} + \mathbf{K}^{(s)}\mathbf{u}^{(s)} = \mathbf{f}^{(s)} + \mathbf{g}^{(s)}. \tag{2.2}$$

However, given a set of either time- or frequency-domain measurements with $N_i \geq 1$, one can extract a modal model of the structure of the following form:

$$\tilde{\mathbf{M}}^{(s)}\ddot{\mathbf{q}}^{(s)} + \tilde{\mathbf{C}}^{(s)}\dot{\mathbf{q}}^{(s)} + \tilde{\mathbf{K}}^{(s)}\mathbf{q}^{(s)} = \tilde{\mathbf{f}}^{(s)} + \tilde{\mathbf{g}}^{(s)}, \tag{2.9}$$

where $\mathbf{q} \in \mathbb{R}^{N_m}$ with N_m the number of vibration modes that are extracted from the measurements. Assuming that the modal masses have been mass normalized, the mass matrix is an identity, $\tilde{\mathbf{M}}^{(s)} = [\mathbf{I}]$, and $\tilde{\mathbf{C}}^{(s)} = \left[\backslash 2\zeta_r\omega_{r\backslash}\right]$, $\tilde{\mathbf{K}}^{(s)} = \left[\backslash \omega_{r\backslash}^2\right]$ are defined by the modal parameters, i.e., the modal natural frequency ω_r and modal damping ratio ζ_r for modes $r = 1...N_m$. The measured mode shapes at points $i = 1...N_o$ are used to populate the mass normalized mode shape matrix, $\boldsymbol{\phi}$, which appears in the force term $\tilde{\mathbf{f}}^{(s)} = \boldsymbol{\phi}^{(s)T}\mathbf{f}^{(s)}$. The resulting equation of motion is as follows:

$$[\mathbf{I}]^{(s)}\ddot{\mathbf{q}}^{(s)} + \left[\backslash 2\zeta_r\omega_{r\backslash}\right]^{(s)}\dot{\mathbf{q}}^{(s)} + \left[\backslash \omega_{r\backslash}^2\right]^{(s)}\mathbf{q}^{(s)} = \boldsymbol{\phi}^{(s)T}\mathbf{f}^{(s)} + \tilde{\mathbf{g}}^{(s)}. \tag{4.1}$$

This experimental and modal representation of the dynamics of substructures can also be used to build an assembled system as will be shown later, one challenge being to express the interface conditions between physical DOF when only modal amplitudes are used as unknowns.

Fitting a modal model for each component and performing the assembly using the modal information is a very attractive approach, especially when the measurements of the frequency responses are not good enough to estimate the impedances of the substructures. Although such an approach can lead to good results for some structures (as explained later), the local modal information could be insufficient to properly combine components. In addition, the modal approach is hardly applicable for components with a high modal density. Note that the so-called "structural modification" techniques as implemented in several commercial testing software can be seen as a special case of substructuring, where simple components (lumped masses, springs...) can be added to the measured component, often using modal models of the test object (see, e.g., Avitabile 2003).

4.2 Why Is Experimental Substructuring So Difficult?

To elucidate some of the issues that are encountered in experimental substructuring, consider two beam substructures to be connected in a lap joint configuration with ten bolts as shown in Fig. 4.1.

1. In classical frequency-based substructuring, one requires the drive point and transfer FRFs to be measured for every connection DOF. For this case, if one considers one connection node per bolt, and three translations and three rotations per node, then a matrix of 60×60 connection FRFs must be measured for each beam and there are two beams. Also, in order to measure the rotations, we would need a rotational accelerometer in three directions, mounted exactly at the connection point and a hammer or shaker that applies a pure moment (i.e., no translational force) as an input; hardware does not exist that can reliably meet these requirements. Hence, one may decide, for logistical reasons, to measure only the translational components of force and motion. However, this will eliminate $3/4$ of the FRF matrix, and in many cases one cannot expect that $3/4$ of the FRF matrix is negligible.

Fig. 4.1 Sample substructuring problem where two beams are joined by 10 bolts, represented by vertical lines

   a. How does one place a sensor or actuator at every DOF. Remember there is a "hole" at the idealized connection point. Should the bolt be left in and apply forces and mount sensors on that? How does one apply a perfect shear force, or measure a shear acceleration right at the idealized point of connection?
   b. If the two 60×60 FRF matrices are successfully measured, how does one account for the compliance and damping of the interface? Traditional methods "weld" the connection DOF together with no interface compliance or damping, but actual joints are known to have finite stiffness and often have non-negligible damping.
   c. Drive point FRFs are already ill-conditioned for the required inversion, and measurement errors are known to make the inversion even more ill-conditioned often producing spurious resonances.
   d. This example has idealized connection to 10 nodes, but in reality, connections between parts are continuous and span the area of the lap joint.

   Some of the difficulties enumerated above can be addressed by performing substructuring in the modal domain, i.e., CMS substructuring.

2. Consider the requirements associated with classical CMS substructuring.

   a. Here, there is not a requirement to apply a force to every connection DOF, but only to excite all the modes active in the required frequency band. With this approach, at least, the pesky requirement to apply moments is eliminated.
   b. Again it is difficult to make the translation and rotation measurements and accurately extract all the connection DOF mode shapes.
   c. For a free-mode basis set, extremely accurate rigid body mode shapes are required as well. (In fact, the rigid body mode shapes are often more important than any of the elastic mode shapes.) These are often not easy to measure. Often these analytical estimates or mass properties' measurement provide better estimates of these than low-frequency modal methods.
   d. The extracted mode shapes eliminate the random noise on FRFs but do not eliminate gage bias errors or bias errors introduced by the modal extraction.
   e. In general, fixed base modes are difficult to obtain by testing simply because not truly fixed base exists, so one extracts some sort of "free" set.
   f. The free-mode shapes of substructures in the bandwidth of interest are often not an adequate basis set to span the space of the connected motion. In the analytical realm, free modes are augmented with residual flexibilities and then they often form a suitable basis. Efforts have been made to measure these residuals, but this once again requires force and moment inputs at all of the connection DOF. Since the residuals represent the flexibility of out of band modes, one might think to instead extract several modes above the frequency bandwidth of interest, but this is typically difficult and often many, many modes are needed to obtain convergence.

► An example is shown in Fig. 4.2. A finite element model was used to represent the red cylindrical substructure and measurements were used to obtain the free modes of the flange with the beam attached (i.e., the plate-beam system in cyan and green). The residual flexibilities were not measured, and the CMS method was used to couple a free-mode model of the plate beam with the FEM of the cylinder and then the modes of that model were used to reconstruct the drive point FRF at the end of the beam. The substructuring prediction is shown in red, with the test data measured from an actual assembly in blue. The plate attaches to the cylinder by eight bolts and all rotations and translations are included in the connection DOF. 100 modes of the cylinder were used and all the modes of the plate and beam to 4000 Hz. The goal was to produce an accurate model up to 2000 Hz. This example clearly shows that free modes are not adequate to obtain a reasonable match for this drive point in the combined system. It should be noted that when two components are coupled together loosely, i.e., using rubber bushings, then a free-mode model often can prove effective. The discussion here focuses primarily on substructures that connect move rigidly.

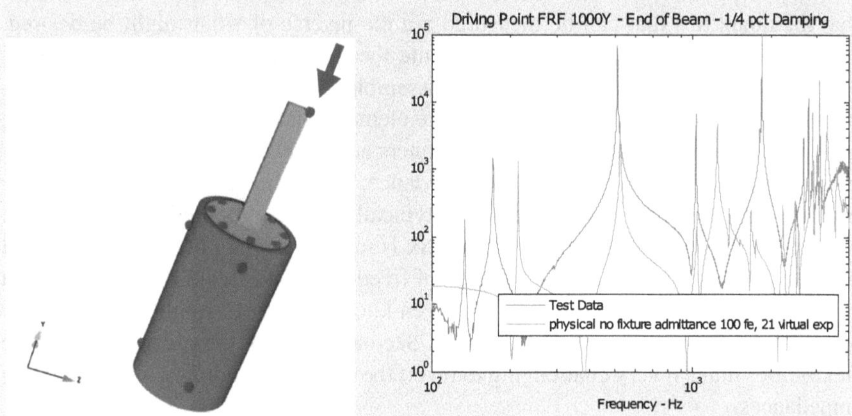

Fig. 4.2 Substructuring prediction of response at point 1000y (axial direction) for the cylinder-plate-beam system

◄

In summary, the CMS approach that is traditionally used in finite element analysis does not generally address the interface compliance, interface damping, or continuous connection issues that are problematic in experiments. Techniques to address these several problems in the physical DOF space will be addressed in this lecture note.

4.3 Basics of Frequency-Based Substructuring (FBS)

In this section, the Frequency-Based Substructuring method (FBS), as typically applied when the components are dynamically identified by measurements, will be explained. Although the substructuring method is mainly discussed for the frequency domain (since that is often the domain in which experimental data is considered), writing it for the time or the Laplace domain is straightforward. It will become clear why a dual approach to substructuring is commonly the method of choice in practical applications, although a primal assembly impedances would be mathematically equivalent.

The idea to build a dynamic model of an assembly using the measured dynamic properties of its components was already considered in the early 70s Klosterman (1971), Crowley et al. (1984) where the Structural Modification Using Response Functions (SMURF) method was developed. However, although not very challenging from a theoretical point of view, it became rapidly clear that the inaccuracy in the measurements and the number of measurements needed for a complex structure render the application of the idea in practice difficult. The main problem is the fact that the quantities that can be measured are the inverse of what might be desired. Computational models can readily compute the dynamic impedance (or dynamic stiffness), and it is then a trivial matter to assemble the stiffnesses from various components (as is done when assembling finite element stiffness matrices). In contrast, in experiments one can measure the component admittances or frequency responses, which are the inverse of the dynamic impedance. While the impedance is simply the inverse of the admittance, this inverse can typically not be computed for two reasons: First, the full impedance/admittance matrix is needed to compute the inverse, and one can never measure all of the degrees of freedom (there are infinitely many for a continuous structure) and one does not even know how many might be needed for an adequate finite element representation. Second, errors in the measured dynamic flexibilities make it very challenging to invert the component admittances and obtain impedances.

The topic was further investigated at the end of 80s Jetmundsen et al. (1988), Urgueira (1989) where different formalisms were proposed, but the basic problems of the method still remained. Around the same time, Imregun (1987), Otte et al. (1991), Lim and Steyer (1992), Martinez et al. (1984), Martinez et al. (1985) looked into techniques to"clean" the measurements, for instance, using truncated singular value decomposition techniques in order to regularize the inverse inherent to experimental substructuring. But applications remained only partly successful.

Motivated by the tremendous potential of experimental substructuring and the advent of affordable and high-quality measurement hardware and software, the topic received new attention in the years 2000 Rixen et al. (2006), de Klerk (2006), Sjövall (2006), Carne and Dohrmann (2006), Allen and Mayes (2007). Over the last 15 years, a general formalism has been developed for the method and several new ideas where proposed to properly perform the measurements of the components and to avoid situations in which measurement errors cause the assembled models to be meaningless. The most important techniques will be outlined later in this text.

4.3.1 Lagrange Multiplier FBS—the Dual Interface Problem

Let us consider again the dual form of an assembled system made of several components (see (2.45), Sect. 2.3), dropping for simplicity the bar superscript introduced earlier to indicate that the DOF is to be interpreted as complex amplitudes in the frequency domain:

$$\begin{bmatrix} \mathbf{Z} & \mathbf{B}^T \\ \mathbf{B} & \mathbf{0} \end{bmatrix} \begin{bmatrix} \mathbf{u} \\ \boldsymbol{\lambda} \end{bmatrix} = \begin{bmatrix} \mathbf{f} \\ \mathbf{0} \end{bmatrix}, \tag{4.2}$$

where

$$\mathbf{Z} = \begin{bmatrix} \mathbf{Z}^{(1)} & & \mathbf{0} \\ & \ddots & \\ \mathbf{0} & & \mathbf{Z}^{(N^{sub})} \end{bmatrix}$$

$$\mathbf{u} = \begin{bmatrix} \mathbf{u}^{(1)} \\ \vdots \\ \mathbf{u}^{(N^{sub})} \end{bmatrix} \quad \mathbf{f} = \begin{bmatrix} \mathbf{f}^{(1)} \\ \vdots \\ \mathbf{f}^{(N^{sub})} \end{bmatrix} \quad \boldsymbol{\lambda} = \begin{bmatrix} \lambda_1 \\ \vdots \\ \lambda_{n_\lambda} \end{bmatrix}$$

are the block diagonal matrices containing the impedances, DOF, and external forces of the unassembled substructures, where $\mathbf{B}$ is the signed Boolean matrix of suitable dimension, which expresses the interface compatibility constraints (see (2.29), (2.30)) and $\boldsymbol{\lambda}$ the associated Lagrange multipliers representing the internal forces in the interface between substructures. An interpretation of the $\mathbf{B}$ and the Lagrange multipliers is given Fig. 2.3.

The substructure displacements, $\mathbf{u}$, in this dual form can be solved to obtain the substructure displacements as functions of the forces, $\boldsymbol{\lambda}$, using the dynamic equilibrium:

$$\mathbf{u} = \mathbf{Y}(\mathbf{f} - \mathbf{B}^T\boldsymbol{\lambda}), \tag{4.3}$$

where

$$\mathbf{Y} = \begin{bmatrix} \mathbf{Y}^{(1)} & & \mathbf{0} \\ & \ddots & \\ \mathbf{0} & & \mathbf{Y}^{(N^{sub})} \end{bmatrix} = \begin{bmatrix} \mathbf{Z}^{(1)^{-1}} & & \mathbf{0} \\ & \ddots & \\ \mathbf{0} & & \mathbf{Z}^{(N^{sub})^{-1}} \end{bmatrix} = \mathbf{Z}^{-1} \tag{4.4}$$

is the block matrix containing the substructure admittances. Equation (4.3) contains the unassembled substructure dynamic equilibrium equations

$$\mathbf{u}^{(s)} = \mathbf{Y}^{(s)}(\mathbf{f}^{(s)} - \mathbf{B}^{(s)^T}\boldsymbol{\lambda}), \tag{4.5}$$

where the coupling is guaranteed by the Lagrange multipliers, unique between matching DOF across the interface, representing the interface forces and the amplitude of which is determined by the compatibility condition contained in the last line of (4.2). Substituting (4.3) into the compatibility condition yields

$$\mathbf{BY}(\mathbf{f} - \mathbf{B}^T\boldsymbol{\lambda}) = \mathbf{0}$$

that can be re-organized as

$$\boxed{(\mathbf{BYB}^T)\boldsymbol{\lambda} = \mathbf{BYf}} \tag{4.6}$$

or, substructure-wise,

$$\left(\sum_s^{N^{sub}} \mathbf{B}^{(s)}\mathbf{Y}^{(s)}\mathbf{B}^{(s)^T}\right)\boldsymbol{\lambda} = \sum_s^{N^{sub}} \mathbf{B}^{(s)}\mathbf{Y}^{(s)}\mathbf{f}^{(s)}. \tag{4.7}$$

Problem (4.6) (or its form (4.7)) is commonly referred to as the *Dual Interface Problem*: it contains n_λ compatibility equations for the unknown internal Lagrange multipliers on the interface. It can be interpreted as follows:

The external forces $\mathbf{f}^{(s)}$ *are applied to the substructures, creating substructure responses* $\mathbf{Y}^{(s)}\mathbf{f}^{(s)}$ *and generating interface incompatibilities* $\mathbf{BYf}$ *(Fig. 4.3b). In order to prevent those interface "gaps", internal forces must exist on the interface and have amplitudes such that the response* $\mathbf{YB}^{(s)^T}\boldsymbol{\lambda}$ *that they generate on each side of the interface closes the gap originating from the external forces (Fig. 4.3c).*

The dual interface problem for the amplitudes of the Lagrange multiplier is a global problem since a Lagrange multiplier between two substructures will affect the response of those substructures also on the interface with other substructures. The operator

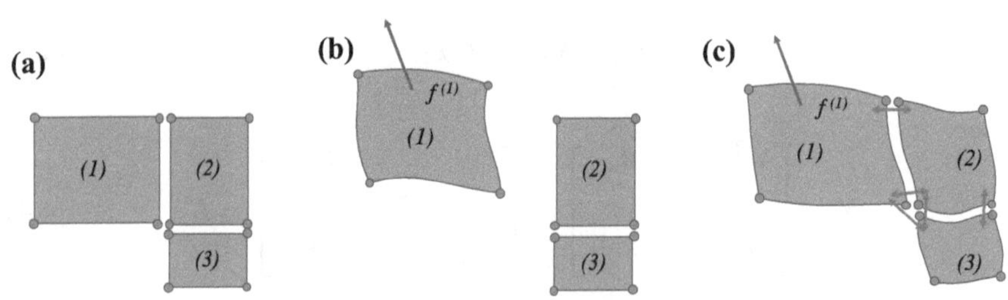

Fig. 4.3 Interpretation of the dual interface problem (4.7): **a** Substructured problem, **b** Effect of local force in $\Omega^{(1)}$, **c** Global effect of $\boldsymbol{\lambda}$ to enforce compatiblity

$$\mathbf{F}_I = (\mathbf{BYB}^T) = \left(\sum_s^{N^{sub}} \mathbf{B}^{(s)} \mathbf{Y}^{(s)} \mathbf{B}^{(s)^T} \right) \tag{4.8}$$

in the dual interface problem is often called the *dually assembled interface flexibility*.

4.3.2 Lagrange Multiplier FBS—the Dually Assembled Admittance

The dual interface problem (4.48) can be used to compute the interface Lagrange multipliers, namely,

$$\boldsymbol{\lambda} = (\mathbf{BYB}^T)^{-1}\mathbf{BYf}, \tag{4.9}$$

and substituting this result in the dynamic equation (4.3) for the substructure response finally yields

$$\mathbf{u} = \mathbf{Y}\left(\mathbf{f} - \mathbf{B}^T(\mathbf{BYB}^T)^{-1}\mathbf{BYf}\right) = \left(\mathbf{Y} - \mathbf{YB}^T(\mathbf{BYB}^T)^{-1}\mathbf{BY}\right)\mathbf{f}. \tag{4.10}$$

The different terms in this formula can easily be interpreted if one considers the way it was found:

$$\mathbf{u} = \Big(\overbrace{\mathbf{Y}}^{uncoupl.} \; \overbrace{-\mathbf{YB}^T \underbrace{(\mathbf{BYB}^T)^{-1} \underbrace{\mathbf{BY}}_{\Delta}}_{\lambda}}^{coupling} \Big)\mathbf{f}.$$

This will be illustrated in the simple example below.

Comparing now this result with the definition for the assembled admittance of the system

$$\mathbf{u} = \mathbf{Y}_{g,dual}\mathbf{f},$$

we can write

$$\boxed{\mathbf{Y}_{g,dual} = \left(\mathbf{Y} - \mathbf{YB}^T(\mathbf{BYB}^T)^{-1}\mathbf{BY}\right).} \tag{4.11}$$

This is the final formula for the dual assembly of substructure admittances.

Note: this admittance operates on the force vector $\mathbf{f}$, which contains the external forces defined on each uncoupled substructure, and returns the response, $\mathbf{u}$, for all DOF of the separated substructures. In other words, the dually assembled admittance is defined for all DOF on each side of an interface. Obviously, by construction, $\mathbf{Y}_{g,dual}$

is identical for all corresponding DOFs on the interface and thus the information in $\mathbf{Y}_{g,dual}$ is redundant. This redundancy can easily be eliminated by keeping in $\mathbf{Y}_{g,dual}$ only the row and line corresponding to one of the redundant DOFs on the interface.

► In order to illustrate the technique, let us consider a simple assembly problem of two bars as described in Fig. 4.4 for which the block matrices of the non-assembled problem can readily be seen to be

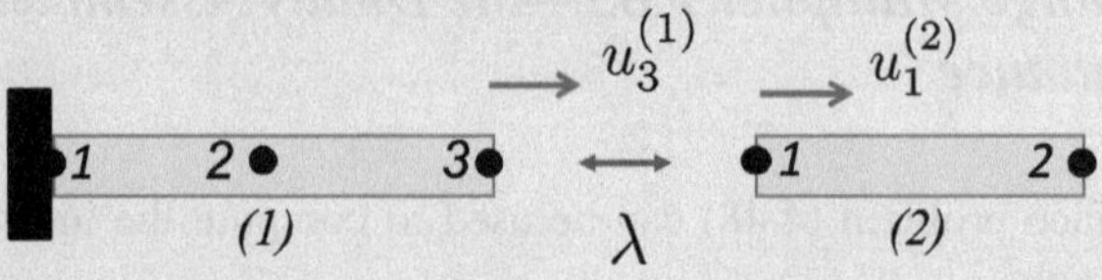

Fig. 4.4 A simple assembly problem of two bars

$$\mathbf{u} = \begin{bmatrix} u_2^{(1)} \\ u_3^{(1)} \\ u_1^{(2)} \\ u_2^{(2)} \end{bmatrix} \qquad \mathbf{f} = \begin{bmatrix} f_2^{(1)} \\ f_3^{(1)} \\ f_1^{(2)} \\ f_2^{(2)} \end{bmatrix} \qquad \mathbf{Y} = \begin{bmatrix} Y_{22}^{(1)} & Y_{23}^{(1)} & 0 & 0 \\ Y_{32}^{(1)} & Y_{33}^{(1)} & 0 & 0 \\ 0 & 0 & Y_{11}^{(2)} & Y_{12}^{(2)} \\ 0 & 0 & Y_{21}^{(2)} & Y_{22}^{(2)} \end{bmatrix}$$

$$\mathbf{B} = \begin{bmatrix} 0 & 1 & -1 & 0 \end{bmatrix}.$$

Applying the formula (4.11) to compute the assembled admittance, we find

$$\begin{bmatrix} u_2^{(1)} \\ u_3^{(1)} \\ u_1^{(2)} \\ u_2^{(2)} \end{bmatrix} = \mathbf{Y}_{assemb.} \begin{bmatrix} f_2^{(1)} \\ f_3^{(1)} \\ f_1^{(2)} \\ f_2^{(2)} \end{bmatrix},$$

where

$$\mathbf{Y}_{assemb.} = \begin{bmatrix} Y_{22}^{(1)} & Y_{23}^{(1)} & 0 & 0 \\ Y_{32}^{(1)} & Y_{33}^{(1)} & 0 & 0 \\ 0 & 0 & Y_{11}^{(2)} & Y_{12}^{(2)} \\ 0 & 0 & Y_{21}^{(2)} & Y_{22}^{(2)} \end{bmatrix} - \begin{bmatrix} Y_{23}^{(1)} \\ Y_{33}^{(1)} \\ -Y_{11}^{(2)} \\ -Y_{21}^{(2)} \end{bmatrix} \left(Y_{33}^{(1)} + Y_{11}^{(2)} \right)^{-1} \begin{bmatrix} Y_{32}^{(1)} & Y_{33}^{(1)} & -Y_{11}^{(2)} & -Y_{12}^{(2)} \end{bmatrix}.$$

This result can be interpreted as follows:

- When applied to the non-assembled (uncoupled) system, the external forces produce a response

$$\begin{bmatrix} u_2^{(1)} \\ u_3^{(1)} \\ u_1^{(2)} \\ u_2^{(2)} \end{bmatrix}_{uncoupl.} = \begin{bmatrix} Y_{22}^{(1)} & Y_{23}^{(1)} & 0 & 0 \\ Y_{32}^{(1)} & Y_{33}^{(1)} & 0 & 0 \\ 0 & 0 & Y_{11}^{(2)} & Y_{12}^{(2)} \\ 0 & 0 & Y_{21}^{(2)} & Y_{22}^{(2)} \end{bmatrix} \begin{bmatrix} f_2^{(1)} \\ f_3^{(1)} \\ f_1^{(2)} \\ f_2^{(2)} \end{bmatrix},$$

  which corresponds to the first part of the solution.

- This response does not account for interface coupling and would create an interface incompatibility (or "gap") equal to

$$\Delta = \mathbf{B}\mathbf{u}_{uncoupl.} = u_3^{(1)} - u_1^{(2)} = \begin{bmatrix} Y_{32}^{(1)} & Y_{33}^{(1)} & -Y_{11}^{(2)} & -Y_{12}^{(2)} \end{bmatrix} \begin{bmatrix} f_2^{(1)} \\ f_3^{(1)} \\ f_1^{(2)} \\ f_2^{(2)} \end{bmatrix}.$$

- In order to close this incompatibility gap on the interface, an interface force $\mathbf{g}$ of amplitude λ

$$\lambda = \left(Y_{33}^{(1)} + Y_{11}^{(2)} \right)^{-1} \Delta$$

  must exist, where $Y_{33}^{(1)} + Y_{11}^{(2)}$ is the flexibility of the interface similar to the flexibility of two springs in series.

- Since the interface force $\mathbf{g}$ is internal (actio-reactio):

$$\begin{bmatrix} g_2^{(1)} \\ g_3^{(1)} \\ g_1^{(2)} \\ g_2^{(2)} \end{bmatrix} = -\mathbf{B}^T\lambda = \begin{bmatrix} 0 \\ -\lambda \\ \lambda \\ 0 \end{bmatrix}$$

  and it will produce a response in the substructure equal to

$$\begin{bmatrix} u_2^{(1)} \\ u_3^{(1)} \\ u_1^{(2)} \\ u_2^{(2)} \end{bmatrix}_{coupling} = \begin{bmatrix} Y_{22}^{(1)} & Y_{23}^{(1)} & 0 & 0 \\ Y_{32}^{(1)} & Y_{33}^{(1)} & 0 & 0 \\ 0 & 0 & Y_{11}^{(2)} & Y_{12}^{(2)} \\ 0 & 0 & Y_{21}^{(2)} & Y_{22}^{(2)} \end{bmatrix} \begin{bmatrix} 0 \\ -\lambda \\ \lambda \\ 0 \end{bmatrix} = - \begin{bmatrix} Y_{23}^{(1)} \\ Y_{33}^{(1)} \\ -Y_{11}^{(2)} \\ -Y_{21}^{(2)} \end{bmatrix} \lambda.$$

So, the dually assembled response (4.12) can be interpreted as the combined response to the external forces on the unassembled system and to the internal interface forces necessary to keep the substructure together when excited by the external forces:

$$\overbrace{\begin{bmatrix} Y_{22}^{(1)} & Y_{23}^{(1)} & 0 & 0 \\ Y_{32}^{(1)} & Y_{33}^{(1)} & 0 & 0 \\ 0 & 0 & Y_{11}^{(2)} & Y_{12}^{(2)} \\ 0 & 0 & Y_{21}^{(2)} & Y_{22}^{(2)} \end{bmatrix}}^{uncoupl.} \overbrace{- \begin{bmatrix} Y_{23}^{(1)} \\ Y_{33}^{(1)} \\ -Y_{11}^{(2)} \\ -Y_{21}^{(2)} \end{bmatrix} \underbrace{\left(Y_{33}^{(1)} + Y_{11}^{(2)}\right)^{-1} \underbrace{\left[Y_{32}^{(1)}\; Y_{33}^{(1)}\; -Y_{11}^{(2)}\; -Y_{12}^{(2)}\right]}_{\Delta}}_{\lambda}}^{coupling}$$

Since, after coupling, $u_3^{(1)} = u_1^{(2)}$, the second and third rows of the so-obtained assembled admittance $\mathbf{Y}_{assemb.}$ are identical. Also, since it makes no difference how the external force on the assembled interface is split in[a] $f_3^{(1)}$ and $f_1^{(2)}$, the second and third columns of $\mathbf{Y}_{assemb.}$ are identical. Therefore, one could delete, for instance, the third row and column and write the nonredundant relations

$$\begin{bmatrix} u_2^{(1)} \\ u_3^{(1)} = u_1^{(2)} \\ u_2^{(2)} \end{bmatrix} = \mathbf{Y}_g \begin{bmatrix} f_2^{(1)} \\ f_3^{(1)} + f_1^{(2)} \\ f_2^{(2)} \end{bmatrix},$$

and using the notation for the global primal variables as in (2.39) one can write this also as

$$\mathbf{u}_g = \mathbf{Y}_g \mathbf{f}_g = \mathbf{Z}_g^{-1} \mathbf{f}_g.$$

◀

[a]Different splittings of the external force on the interface will produce the same assembled response, but will produce different amplitudes λ.

4.3.3 On the Usefulness of Dual Assembly of Admittance in Experimental Substructuring

The LM-FBS discussed in the previous sections is based on dual assembly (also sometimes called *admittance-based substructuring*) and was explained in frequency domain. Dual assembly approaches can also, without further difficulty, be written in the time domain. The dual assembly in the time domain (so in essence the inverse Fourier transform of the LM-FBS) was proposed in Gordis (1995), and the general formulation was later written in Rixen (2010), Rixen and van der Valk (2013). The method is based on the impulse responses at the interface of the substructures and was first developed as a simulation approach (also for multibody dynamics Géradin and Rixen 2017). First attempts to use it as an experimental technique were only partly successful Rixen (2010).

At this point, now that we have shown the basic algebraic techniques to find the assembled admittance in a dual approach, one might wonder why the primal approach (which is more commonly known by engineers and seemingly more straightforward) is not the method of choice in experimental dynamics.

As explained in Sect. 2.3, primal and dual assembly are mathematically equivalent. Since the properties of the components are obtained during testing as admittances (forces applied as input, responses measured as output), the dual approach seems more natural, although it would be not less natural to compute the component impedances (inverting the measurements) and applying a primal assembly. The result would be identical. So the real reason must be elsewhere.

The main reason to apply dual assembly in experimental dynamics comes from the fact that measurements are never perfect, whereas numerical models are. (That is, numerical models are at least consistent and physically realizable representations of some dynamic system, even if they don't capture the structure of interest as accurately as might be desired.) Consistency will be further discussed in Sect. 4.9.2, but it is easy to understand that, since measurement errors can be of any kind, there is no reason why the measured admittance represents exactly a structural system. For instance, it is very common that the eigenfrequencies or modal dampings as seen in the signals of different outputs or as obtained from different inputs are not coinciding (e.g., additional mass effect of sensors or influence of shaker links). When using such measurements of components, one typically observes so-called "spurious peaks" in the FRFs of the assembled system (whether the assembly was performed in a primal or in a dual manner, since the methods are mathematically equivalent). An example of spurious peaks in the assembled FRF due to inconsistencies in the local admittances is shown in Fig. 4.5. Spurious peaks were already considered in Imregun (1987) where a smoothing procedure was proposed and a discussion of the appearance of spurious peaks can be found for instance in Rixen (2008), Carne and Dohrmann (2006), Nicgorski and Avitabile (2010). An analysis of error and uncertainty propagation in the LM-FBS was presented in Voormeeren et al. (2010). It allows giving a confidence interval to the assembled FRFs when the uncertainty of the substructure admittance

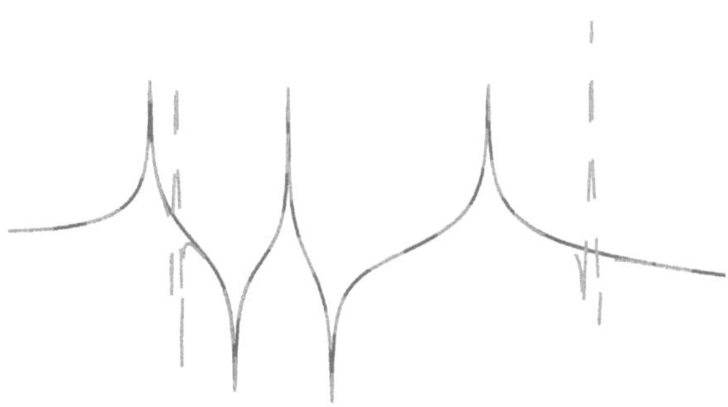

Fig. 4.5 Example of spurious peaks in an assembled FRF Rixen (2008)

is known (for instance, from the measurements). This will, however, not be further discussed here.

In dual assembly techniques, the compatibility condition is explicitly considered next to the substructure equilibrium equations, the interface equilibrium being ensured by the very nature of the Lagrange multipliers. Therefore, as will be explained in the next section (and further developed in other parts), **it is possible to weaken the interface compatibility and thereby mitigate the effect of the errors in the measurements**.

Mathematically, the effects of the measurement errors manifest themselves when the dual interface flexibility (4.8), needed to compute the interface Lagrange multipliers, is inverted in the dually assembled admittance (4.11):

$$\mathbf{Y}_{g,dual} = \left(\mathbf{Y} - \mathbf{Y}\mathbf{B}^T\mathbf{F}_I^{-1}\mathbf{B}\mathbf{Y}\right) \quad \mathbf{F}_I = (\mathbf{B}\mathbf{Y}\mathbf{B}^T) = \left(\sum_s^{N^{sub}} \mathbf{B}^{(s)}\mathbf{Y}^{(s)}\mathbf{B}^{(s)T}\right). \tag{4.12}$$

Errors in the substructure admittances will be dramatically amplified, especially if the conditioning of the admittance matrices is bad: if a structure is lightly damped, then the admittance matrices will be nearly rank one in the vicinity of each resonance. As a result, small measurement errors will generate large errors in the estimation of the stiffness coefficients on the interface (that is, the stiffness of a single DOF when all other DOF are fixed). The same type of error amplification would be found in a primal assembly like (2.39) since the each local measured admittance would need to be inverted to compute the local impedances.

The options to mitigate the effect of measurements errors can be classified into three categories:

1. "clean" the measured admittances of the substructures, for instance, by applying symmetrization on the measurement (e.g., Otte et al. 1991), by fitting a modal model (this can remove noise but also useful information not well identified, and it is applicable only if the modal density is not too high), or/and by applying further corrections to impose consistency (see Sect. 2.3).
2. Apply (Tikhonov) regularization to compute the inverse by removing the components corresponding to small singular values in $\mathbf{F}_I$ (see, for instance, Otte et al. 1991, Gialamas et al. 2001, Moorhouse 2003).
3. The last option consists in weakening the interface compatibility in order to enforce the interface to match only for smooth interface displacements, leaving unassembled some deformation components related to high stiffness on either side of the interface.

Although some successful applications were shown with the options 1 and 2, strategy 3 can be based on mechanical interpretation and engineering insight can be used. This will be briefly discussed in the following section.

4.3.4 Weakening the Interface Compatibility

As explained above, weakening the interface compatibility requirements between substructures can mitigate the measurement errors in the substructure admittances that are responsible for very high errors in the computation (through inversion) of the impedances (in the solution of the dual interface dual problem for the dual approach, or in the computation of local impedances for the primal approach).

Let us consider once more the dually assembled Problem (4.2)

$$\begin{bmatrix} \mathbf{Z} & \mathbf{B}^T \\ \mathbf{B} & \mathbf{0} \end{bmatrix} \begin{bmatrix} \mathbf{u} \\ \boldsymbol{\lambda} \end{bmatrix} = \begin{bmatrix} \mathbf{f} \\ \mathbf{0} \end{bmatrix}, \tag{4.2}$$

where we have assumed strong compatibility of the interface, meaning that we have imposed all matching interface DOFs to be exactly equal (second set of equation in (4.2)). If one, however, has a good idea of how the interface displacement will look in the assembled system, one can weaken the compatibility requirement by only requiring compatibility for the components of the interface displacements that are important for the assembled solution.

Mathematically speaking, let us assume that we know a good representation space $\mathbf{V}_{\Gamma^{rs}}$ for the displacements in the assembled system around an interface Γ^{rs} between two substructures s and r. The matrix $\mathbf{V}_{\Gamma^{rs}}$ contains, in its columns, a set of representation vectors that can well approximate the motions of the DOF around the assembled interface. These vectors can be set up from engineering know-how (assuming, for instance, a rigid behavior around the interface, i.e., the virtual point idea explained later) or could be obtained from an approximate finite element model of the interface region. In that subspace, the interface DOF of the substructures can be expressed as

$$\mathbf{u}_b^{(s)} \simeq \mathbf{V}_{\Gamma_b^{rs}} \boldsymbol{\beta}^{(s)} \quad \text{and} \quad \mathbf{u}_b^{(r)} \simeq \mathbf{V}_{\Gamma_b^{rs}} \boldsymbol{\beta}^{(r)}, \tag{4.13}$$

where $\mathbf{V}_{\Gamma_b^{rs}}$ is the restriction to the interface of the common interface modes and where $\boldsymbol{\beta}^{(r)}$ and $\boldsymbol{\beta}^{(s)}$ can be seen as the corresponding displacement amplitudes on each side of the interface. Here, we assume that the number of representation modes in $\mathbf{V}_{\Gamma^{rs}}$ is smaller than the numbers of interface DOF.

The compatibility condition[1]

$$\mathbf{u}_b^{(s)} - \mathbf{u}_b^{(r)} = 0 \tag{4.14}$$

can be replaced by

$$\mathbf{V}_{\Gamma_b^{rs}} \boldsymbol{\beta}^{(s)} - \mathbf{V}_{\Gamma_b^{rs}} \boldsymbol{\beta}^{(r)} = 0 \tag{4.15}$$

[1] Here, to simplify the presentation, we assume that the numbering on the DOF on each side of the interface is identical. If this would not be the case, the constraint matrix **B** should be considered as explained earlier.

or equivalently (if $\mathbf{V}_{\Gamma_b^{rs}}$ is full rank, i.e., if the vectors in the columns of $\mathbf{V}_{\Gamma_b^{rs}}$ are linearly independent)

$$\boldsymbol{\beta}^{(s)} - \boldsymbol{\beta}^{(r)} = 0, \tag{4.16}$$

hence requesting that only the part of the substructure displacements living in the subspace of $\mathbf{V}_{\Gamma_b^{rs}}$ be compatible. The remainder of the interface displacement, on each side of the interface, that is not representable in $\mathbf{V}_{\Gamma_b^{rs}}$ *is left free on the interface and does not participate to the coupling of the substructures. This weak compatibility is the key to include only the important behavior in the coupling (usually well captured by the measurements), leaving the remainder unassembled (since they may be strongly affected or even dominated by measurement errors).*

In order to write the weak compatibility as a condition on the DOF of the original problem, the coordinates in the reduced space $\mathbf{V}_{\Gamma_b^{rs}}$ can be computed by observing that

$$\mathbf{u}_b^{(s)} = \mathbf{V}_{\Gamma_b^{rs}} \boldsymbol{\beta}^{(s)} + \mathbf{r}^{(s)}, \tag{4.17}$$

where $\mathbf{r}$ is a remainder that will not be made compatible and that one can request to remain orthogonal to the interface representation space:

$$\mathbf{V}_{\Gamma_b^{rs}}^T \mathbf{W} \mathbf{r}^{(s)} = \mathbf{0} \tag{4.18}$$

$\mathbf{W}$ being any positive definite matrix defining the norm of the projection (usually taken as the identity matrix). This condition leads to

$$\boldsymbol{\beta}^{(s)} = (\mathbf{V}_{\Gamma_b^{rs}}^T \mathbf{W} \mathbf{V}_{\Gamma_b^{rs}})^{-1} \mathbf{V}_{\Gamma_b^{rs}}^T \mathbf{W} \mathbf{u}_b^{(s)} = \mathbf{V}_{\Gamma_b^{rs}}^+ \mathbf{u}^{(s)}, \tag{4.19}$$

since $(\mathbf{V}_{\Gamma_b^{rs}}^T \mathbf{W} \mathbf{V}_{\Gamma_b^{rs}})^{-1} \mathbf{V}_{\Gamma_b^{rs}}^T \mathbf{W}$ can be seen as a (weighted) pseudo-inverse of $\mathbf{V}_{\Gamma_b^{rs}}$. The pseudo-inverse can equivalently be computed using a (weighted) singular value decomposition. Similarly, we can write the following for the other side of the interface:

$$\boldsymbol{\beta}^{(r)} = \mathbf{V}_{\Gamma_b^{rs}}^+ \mathbf{u}_b^{(r)}, \tag{4.20}$$

so that the compatibility condition (4.16) becomes

$$\mathbf{V}_{\Gamma_b^{rs}}^+ \left(\mathbf{u}_b^{(s)} - \mathbf{u}_b^{(r)} \right) = 0. \tag{4.21}$$

The number of compatibility conditions in this form is equal to the number of representation vector in $\mathbf{V}_{\Gamma^{rs}}$, namely, less than the number of interface DOF, indicating once more that the compatibility condition is weakened. The pseudo-inverse can be seen as a (weighted) projection in the space of $\mathbf{V}_{\Gamma^{rs}}$ of the interface degrees of freedom.

To summarize, the weakening of the compatibility equation in general leads to a dual assembly problem of the form

$$\begin{bmatrix} \mathbf{Z} & \tilde{\mathbf{B}}^T \\ \tilde{\mathbf{B}} & \mathbf{0} \end{bmatrix} \begin{bmatrix} \mathbf{u} \\ \boldsymbol{\lambda} \end{bmatrix} = \begin{bmatrix} \mathbf{f} \\ \mathbf{0} \end{bmatrix} \qquad \tilde{\mathbf{B}} = \mathbf{BT}, \tag{4.22}$$

where $\mathbf{T}$ is a projection of the interface DOF of the substructures onto a common subspace (usually per local interface). The Lagrange multipliers now give the amplitude of the generalized forces associated with each deformation in the interface representation space. For example, for a simple case where two structures are coupled using a rigid interface with six DOFs (three translations and three rotations), the Lagrange multipliers would give the net force in each direction and the net moment in each direction on each interface.

Following the same procedure as earlier to find the dually assembled admittance, one finds

$$\boxed{\mathbf{Y}_{g,dual,weak} = \left(\mathbf{Y} - \mathbf{Y}\tilde{\mathbf{B}}^T(\tilde{\mathbf{B}}\mathbf{Y}\tilde{\mathbf{B}}^T)^{-1}\tilde{\mathbf{B}}\mathbf{Y}\right)} \qquad \tilde{\mathbf{B}} = \mathbf{BT}. \tag{4.23}$$

Remark 1: A different interpretation of compatibility weakening

It is interesting to note that the weakened interface compatibility (4.21) can be transformed by premultiplying it with $(\mathbf{V}_{\Gamma_b^{rs}}^T \mathbf{W} \mathbf{V}_{\Gamma_b^{rs}})$, resulting in the equivalent weak condition

$$\mathbf{V}_{\Gamma_b^{rs}}^T \mathbf{W}(\mathbf{u}_b^{(s)} - \mathbf{u}_b^{(r)}) = \mathbf{0}. \tag{4.24}$$

This shows that the weak compatibility can also be interpreted as satisfying the compatibility only when projected on a subspace, which in general can be expressed by the weakening

$$\mathbf{Bu} = 0 \quad \rightarrow \quad \mathbf{V}^T\mathbf{WBu} = 0.$$

This is strongly related to the so-called coarse space problem in dual assembly theory (e.g., in the FETI parallel solution techniques).

Remark 2: The virtual point constraint

In order to illustrate the weakening idea, let us discuss two weak compatibility conditions often used in practice. One possible choice for the representation of the interface behavior consists in assuming that a part of the interface behaves rigidly (for instance, a local bolted connection that can be considered as a point connection as illustrated in Fig. 4.25). In that case, the representation space for the signal measured around the connections can be taken as

$$\mathbf{V}_{\Gamma^{rs}} = \mathbf{R}_{\Gamma^{rs}},$$

where $\mathbf{R}_{\Gamma^{rs}}$ contains in its columns the six rigid body modes (three translations, three rotations) around a virtual point on the interface. With those rigid modes, the DOF on each side of the connection point where admittances were acquired can be described.

The weak coupling described by (4.21)–(4.23) then represents the assembly at the interface of a virtual point having six DOFs, i.e., the coordinates $\boldsymbol{\beta}^{(s)}$, $\boldsymbol{\beta}^{(r)}$ that are made equal on the interface represent the rigid body DOF of the interface. The remainder of the interface behavior that cannot be described by the local rigid body motion around the point will not be assembled, thereby avoiding the computation of the high impedance related to deformation modes around the virtual point that is usually badly captured in the measurement and that is often not necessary to represent the interface behavior in the frequency range considered.

The virtual point method will be explained in more details in Sect. 4.6.

Remark 3: Interface extrapolation

Often, the interface response cannot be measured directly on the interface itself, for instance, because sensors and actuators cannot be easily installed on the true interface but rather in its neighborhood. This is true also for the case where the interface is to behave rigidly (see the discussion about the virtual point in Remark 2) and where the rigid behavior of the interface is measured away from the physical interface (see for instance Fig. 4.25).

In that case, the interface modes $\mathbf{V}_{\Gamma^{rs}}$ can still be used to define generalized interface DOF $\boldsymbol{\beta}^{(s)}$ and write the interface compatibility as before in (4.16). To that effect, taking the value of the modes $\mathbf{V}_{\Gamma^{rs}}$ at the measurement points m (that can thus be different from the true interface location), one can write

$$\boldsymbol{\beta}^{(s)} - \boldsymbol{\beta}^{(r)} = 0 \qquad \boldsymbol{\beta}^{(s)} = \mathbf{V}^{+}_{\underset{ms}{\Gamma^{rs}}} \mathbf{u}^{(s)} \qquad \boldsymbol{\beta}^{(r)} = \mathbf{V}^{+}_{\underset{mr}{\Gamma^{rs}}} \mathbf{u}^{(r)}, \tag{4.25}$$

where $\mathbf{V}_{\underset{ms}{\Gamma^{rs}}}$ and $\mathbf{V}_{\underset{mr}{\Gamma^{rs}}}$ are the values of the global mode in the vicinity of the interface for the measured DOFs in r and in s.

The dually weakly assembled forms of the system dynamics and of the admittance is then written like in (4.22) and (4.23), the difference being now that the substructure matrices $\mathbf{Y}$ and $\mathbf{Z}$ refer no longer to interface DOFs but to other measurement DOFs in the interface vicinity and that the projection matrix T contains pseudo-inverses for $\mathbf{V}_{\Gamma^{rs}}$ evaluated at the measurement points.

This technique can be interpreted as using common interface modes $\mathbf{V}_{\Gamma^{rs}}$ to extrapolate measured dynamics to the interface while at the same time imposing a weak compatibility.

4.3.5 Dual Assembly in the Modal Domain

Weakening of the interface compatibility is very useful when assembling substructures represented by three local free interface modes. Assume that the measured admittance $\mathbf{Y}^{(s)}$ of two components can be well represented by a modal fit. Assuming small damping and well-separated eigenfrequencies for each substructure, the modal fit can be used to write the admittances as (see, for instance, Géradin and Rixen 2015)

$$\mathbf{Y}^{(s)}(\omega) = \boldsymbol{\Phi}^{(s)} \left(\boldsymbol{\Omega}^{(s)^2} + 2i\omega\boldsymbol{\Omega}^{(s)}\boldsymbol{\varepsilon}^{(s)} - \omega^2\mathbf{I} \right)^{-1} \boldsymbol{\Phi}^{(s)^T}, \tag{4.26}$$

where $\boldsymbol{\Phi}^{(s)}$ contains in its column the identified modes of substructure s (assumed mass normalized), and where $\boldsymbol{\Omega}^{(s)}$ and $\boldsymbol{\varepsilon}^{(s)}$ are diagonal matrices containing the corresponding identified eigenfrequencies and modal damping ratios.

This modal fit has the advantage that it provides a fully consistent expression of the admittance, removing inconsistencies and noise from the measurement.

The admittances of the substructures can now easily be obtained as

$$\mathbf{Z}^{(s)}(\omega) = \boldsymbol{\Phi}^{(s)^{-T}} \left(\boldsymbol{\Omega}^{(s)^2} + 2i\omega\boldsymbol{\Omega}^{(s)}\boldsymbol{\varepsilon}^{(s)} - \omega^2\mathbf{I} \right) \boldsymbol{\Phi}^{(s)^{-1}} \tag{4.27}$$

so that the dually assembled form (4.2) (here discussed for the case of two substructures for simplicity) is written as

$$\begin{bmatrix} \ddots & \mathbf{0} & & \vdots \\ \mathbf{0} & \mathbf{Z}^{(s)}(\omega) & & \mathbf{B}^{(s)^T} \\ & & \ddots & \vdots \\ \cdots & \mathbf{B}^{(s)} & \cdots & \mathbf{0} \end{bmatrix} \begin{bmatrix} \vdots \\ \mathbf{u}^{(s)} \\ \vdots \\ \boldsymbol{\lambda} \end{bmatrix} = \begin{bmatrix} \vdots \\ \mathbf{f}^{(s)} \\ \vdots \\ \mathbf{0} \end{bmatrix}. \tag{4.28}$$

Considering that the substructure DOF can be expressed as a modal superposition, namely,

$$\mathbf{u}^{(s)} = \boldsymbol{\Phi}^{(s)}\eta^{(s)},$$

where $\eta^{(s)}$ are the modal coordinates in substructure s, and multiplying the local dynamic equations in (4.28) by $\boldsymbol{\Phi}^{(s)^T}$, the dual assembly now is written as

$$\begin{bmatrix} \ddots & \mathbf{0} & & \vdots \\ \mathbf{0} & \mathbf{Z}_m^{(s)}(\omega) & & \boldsymbol{\Phi}^{(s)^T}\mathbf{B}^{(s)^T} \\ & & \ddots & \vdots \\ \cdots & \mathbf{B}^{(s)}\boldsymbol{\Phi}^{(s)} & \cdots & \mathbf{0} \end{bmatrix} \begin{bmatrix} \vdots \\ \eta^{(s)} \\ \vdots \\ \boldsymbol{\lambda} \end{bmatrix} = \begin{bmatrix} \vdots \\ \boldsymbol{\Phi}^{(s)^T}\mathbf{f}^{(s)} \\ \vdots \\ \mathbf{0} \end{bmatrix}, \tag{4.29}$$

where

$$\mathbf{Z}_m^{(s)}(\omega) = \left(\boldsymbol{\Omega}^{(s)^2} + 2i\omega\boldsymbol{\Omega}^{(s)}\boldsymbol{\varepsilon}^{(s)} - \omega^2\mathbf{I} \right).$$

This form can be interpreted as follows: the local dynamics of the substructures are described in the modal space using the modal eigenfrequencies and dampings identified from the measured FRFs. Since the compatibility still needs to be satisfied by the physical DOF of the interfaces, the constraint matrices $\mathbf{B}^{(s)}$ are now replaced by $\mathbf{B}^{(s)}\boldsymbol{\Phi}^{(s)}$, namely, the traces of the modes on the interface.[2]

The modes of the substructures are free interface modes and exhibit in general very different displacements on the interface. Therefore, requiring strong compatibility significantly limits the possible combination of modes allowed in the substructures. Often, the number n_ϕ^s of modes $\boldsymbol{\Phi}^{(s)}$ considered for a substructure is smaller, then the number of physical DOF $\mathbf{u}_b^{(s)}$ on its interface. In that case, the strong interface compatibility in (4.29) will significantly limit the number of possible combination of the local modes. In extreme cases where the number of interface compatibility conditions n_λ is higher than the number of substructures modes $\sum n_\phi^s$, the only possible solution satisfying the strong constraint is $\boldsymbol{\eta}^{(s)} = 0$, leading to a full looking of the solution space.

Even if the number of modes accounted for in the substructures is high, the motion remaining possible for the assembled system after the strong compatibility has been satisfied is often limited. Further, one should keep in mind that the local modes obtained by the experimental identification are free interface modes and are in general not well suited to represent the behavior of the substructure when assembled to its number (i.e., when the interface is not free).[3]

[2]It is interesting to note that this form of the dual assembly can deliver a model that is not only valid in the frequency domain. Indeed, (4.29) can also be written as

$$\left(\begin{bmatrix} \ddots & \mathbf{0} & \vdots \\ \mathbf{0} & \boldsymbol{\Omega}^{(s)^2} & \boldsymbol{\Phi}^{(s)^T}\mathbf{B}^{(s)^T} \\ & \ddots & \vdots \\ \dots & \mathbf{B}^{(s)}\boldsymbol{\Phi}^{(s)} \dots & \mathbf{0} \end{bmatrix} + i\omega \begin{bmatrix} \ddots & \mathbf{0} & \vdots \\ \mathbf{0} & 2\boldsymbol{\Omega}^{(s)}\boldsymbol{\varepsilon}^{(s)} & \mathbf{0} \\ & & \ddots \vdots \\ \dots & \mathbf{0} & \dots \mathbf{0} \end{bmatrix} \dots \right.$$

$$\left. -\omega^2 \begin{bmatrix} \ddots & \mathbf{0} & \vdots \\ \mathbf{0} & \mathbf{I} & \mathbf{0} \\ & & \ddots \vdots \\ \dots & \mathbf{0} \dots & \mathbf{0} \end{bmatrix} \right) \begin{bmatrix} \vdots \\ \boldsymbol{\eta}^{(s)} \\ \vdots \\ \boldsymbol{\lambda} \end{bmatrix} = \begin{bmatrix} \vdots \\ \boldsymbol{\Phi}^{(s)^T}\mathbf{f}^{(s)} \\ \vdots \\ \mathbf{0} \end{bmatrix},$$

indicating that the dual assembly of the substructure modal model yield the stiffness, damping, and mass matrix of the assembled system for the substructure modal coordinates as DOF. These matrices can be used, for instance, to directly write the dynamic equation in the time domain.

[3]Later, in Sect. 4.4.3, we will discuss a technique to measure the substructure modes when the interface is interacting with substructures in a way similar to what it will experience in the assembly. This so-called transmission simulator technique will significantly enhance the basic modal-based substructuring explained here.

From the discussion above, it is clear that it is desirable to relax the compatibility on the interface in order to allow the free interface modes of the substructures to properly represent the local dynamics. To that effect, the compatibility relaxation explained in the previous section can be used as follows.

Using the technique outlined in the previous section, we can weaken the interface compatibility by replacing the Boolean matrices $\mathbf{B}$ by $\tilde{\mathbf{B}} = \mathbf{BT}$, where $\mathbf{T}$ contains a projection operator on interface modes. For the case of two subdomains, calling $\mathbf{V}_{\Gamma_b^{12}}$ the representative modes in the vicinity of the interface, the weak compatibility would then be written as

$$\mathbf{V}^+_{\Gamma_b^{12}} \boldsymbol{\Phi}_b^{(1)} \boldsymbol{\eta}^{(1)} - \mathbf{V}^+_{\Gamma_b^{12}} \boldsymbol{\Phi}_b^{(2)} \boldsymbol{\eta}^{(2)} = 0. \tag{4.30}$$

In case the measurement points are not directly on the interface but in its vicinity (see Remark 3 in the previous section), the weak compatibility conditions would be written as

$$\mathbf{V}^+_{\Gamma_{m_s}^{12}} \boldsymbol{\Phi}_{m_s}^{(1)} \boldsymbol{\eta}^{(1)} - \mathbf{V}^+_{\Gamma_{m_r}^{12}} \boldsymbol{\Phi}_{m_r}^{(2)} \boldsymbol{\eta}^{(2)} = 0. \tag{4.31}$$

The modal-based substructuring will be further explained and used in Sect. 4.4.3.

4.3.6 A Special Case: Substructures Coupled Through Bushings

It is very common in engineering applications to couple substructures through flexible bushings (e.g., rubber bushings between the power train and the chassis of a car, or isolation supports of power units installed on ships). Such connecting elements are usually small and soft, their function being primarily to introduce a low impedance coupling and thereby minimize vibration transmission. Hence, bushings are often considered as connecting elements characterized by a dynamic stiffness of the form (in the frequency domain)

$$\mathbf{Z}^{(bush)}(\omega) = \mathbf{K}^{(bush)} - i\omega\mathbf{C}^{(bush)}, \tag{4.32}$$

assuming that the main contribution of the bushing in the assembly is determined by its static stiffness and its damping, the small mass of the bushing and its internal dynamics being neglected.

In the primal assembly setting, the dynamic equation of an assembly (see (2.39)) of two components connected by a bushing can easily be expressed as (see Fig. 4.6)

$$\mathbf{Z}_g = \begin{bmatrix} \mathbf{Z}_{ii}^{(1)} & \mathbf{Z}_{ib}^{(1)} & \mathbf{0} & \mathbf{0} \\ \mathbf{Z}_{bi}^{(1)} & \mathbf{Z}_{bb}^{(1)} + \mathbf{Z}_{b^1,b^1}^{(bush)} & \mathbf{Z}_{b^1,b^2}^{(bush)} & \mathbf{0} \\ \mathbf{0} & \mathbf{Z}_{b^2,b^1}^{(bush)} & \mathbf{Z}_{bb}^{(2)} + \mathbf{Z}_{b^2,b^2}^{(bush)} & \mathbf{Z}_{bi}^{(2)} \\ \mathbf{0} & \mathbf{0} & \mathbf{Z}_{bi}^{(2)} & \mathbf{Z}_{ii}^{(2)} \end{bmatrix} \begin{bmatrix} \mathbf{u}_i^{(1)} \\ \mathbf{u}_{b^1} \\ \mathbf{u}_{b^2} \\ \mathbf{u}_i^{(2)} \end{bmatrix} = \begin{bmatrix} \mathbf{f}_i^{(1)} \\ \mathbf{f}_{b^1} \\ \mathbf{f}_{b^2} \\ \mathbf{f}_i^{(2)} \end{bmatrix}, \tag{4.33}$$

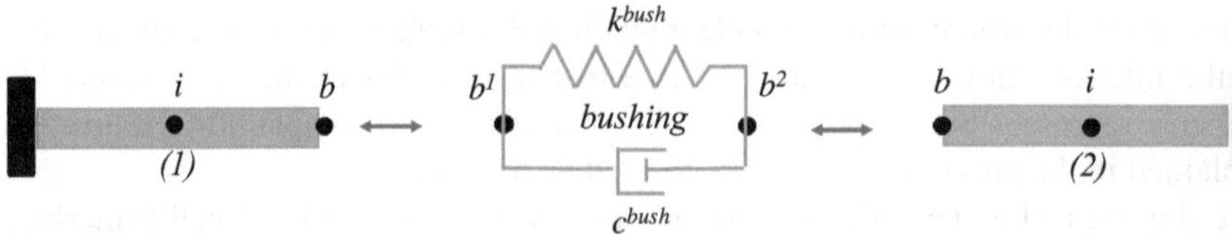

Fig. 4.6 Assembly over a bushing element

where we assumed the numbering of the degrees of freedom of the bushing to fit the numbering of the interfaces of the substructures it connects, and where the subscript b and i denote boundary DOF on the interface of the bushing and internal DOF, respectively.

In the experimental substructuring framework, considering a bushing element can be challenging since, before being assembled in the system, it is a free floating components and its impedance matrix is singular due to the fact that a rigid motion (translation or rotation) can be given to the bushing without force (its inertia being neglected):

$$\mathbf{Z}^{(bush)}\mathbf{R}^{(bush)} = \mathbf{0}, \tag{4.34}$$

where $\mathbf{R}^{(bush)}$ is a vector containing in its column a set of independent rigid body modes of the bushing (usually 6).

► If we consider the example of Fig. 4.6 as a one-dimensional bushing, its dynamic stiffness matrix would be

$$\mathbf{Z}^{(bush)} = \begin{bmatrix} z^{bush} & -z^{bush} \\ -z^{bush} & z^{bush} \end{bmatrix} \qquad z^{bush} = k^{bush} - i\omega c^{bush}$$

which is clearly singular. There is one rigid body mode in this case

$$\mathbf{R}^{(bush)} = \begin{bmatrix} 1 \\ 1 \end{bmatrix}.$$

◄

Hence, the inverse of $\mathbf{Z}^{(bush)}$ does not exist, meaning that no solution can be computed for a set of forces applied to the bushing unless those forces are self-equilibrated (zero resulting force and moment).[4]

There are several ways to introduce a bushing in the general framework of LM-FBS, as will be shortly discussed next. First, we will consider the bushing as a connecting element between the interfaces of two substructures that introduces an interface incompatibility, thereby slightly modifying the dual interface problem. In

[4]For a more detailed discussion on singular matrices in structural dynamics, see for instance Géradin and Rixen (2015).

a second approach, the bushing is treated as a substructure similar to connected parts, and the singularity of its impedance will be handled either by applying a regularization or by using a pseudo-inverse technique.

Bushings as Compatibility Relaxation

In this first approach, we consider the system schematically depicted in Fig. 4.6 as an assembly of two substructures for which the interface compatibility is relaxed by the flexibility of the bushing. In other words, the Boolean matrix defining the interface compatibility on either side of the bushing is of the form $\mathbf{B} = [\mathbf{B}^{(1)}\ \mathbf{B}^{(2)}]$ and the interface condition for the displacements now is written as

$$\mathbf{Bu} = \boldsymbol{\delta} \neq \mathbf{0}, \tag{4.35}$$

where $\boldsymbol{\delta}$ is the interface gap allowed by the bushing. This gap introduces in the bushing internal forces $\boldsymbol{\lambda}$ proportional to the dynamic stiffness of the bushing. Calling $\boldsymbol{\Upsilon}$ the matrix describing the dynamic flexibility of the bushing between its interface DOF, one can write

$$\mathbf{Bu} = \boldsymbol{\delta} = \boldsymbol{\Upsilon}\boldsymbol{\lambda}. \tag{4.36}$$

Note that $\boldsymbol{\Upsilon}$ is not equal to $\mathbf{Z}^{(bush)^{-1}}$ since it expresses the flexibility related to interface gaps (hence in the space of $\boldsymbol{\lambda}$) and not directly the flexibility between interface DOFs.

► If one considers the example depicted in Fig. 4.7 as a one-dimensional problem, one would have for this approach

$$\mathbf{B} = \begin{bmatrix} 0 & 1 & -1 & 0 \end{bmatrix} \qquad \delta = u_b^{(1)} - u_b^{(2)} \qquad \delta = \frac{1}{k^{bush} - i\omega c^{bush}}\lambda,$$

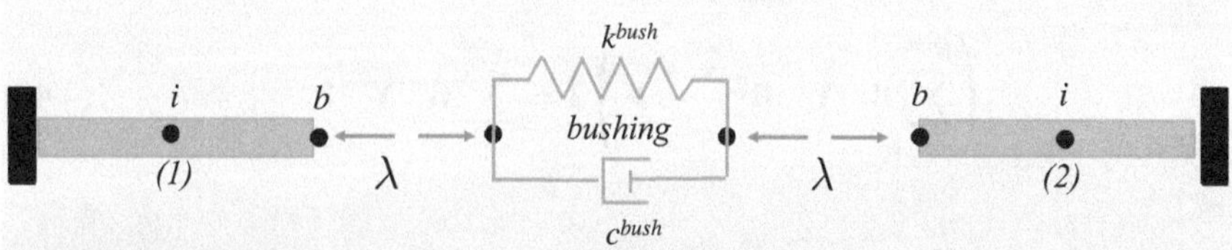

Fig. 4.7 Bushings as compatibility relaxation

where $k^{bush} - i\omega c^{bush}$ is the dynamic stiffness of the rubber between the interfaces. In general, this bushing stiffness can be different for different gaps defined across an interface (for instance, in different translational or rotational directions).

◄

The internal forces $\boldsymbol{\lambda}$ in the bushing are equal in amplitude and opposite on either side of the bushing (since the bushing has no inertia): they correspond to the Lagrange multipliers defined in the dual assembly of structures, and the dual assembly form

(4.2) in the presence of a bushing is now written as (here as example for a system with two substructures):

$$\begin{bmatrix} \mathbf{Z}^{(1)} & \mathbf{0} & \mathbf{B}^{(1)^T} \\ \mathbf{0} & \mathbf{Z}^{(2)} & \mathbf{B}^{(2)^T} \\ \mathbf{B}^{(1)} & \mathbf{B}^{(2)} & -\boldsymbol{\Upsilon} \end{bmatrix} \begin{bmatrix} \mathbf{u}^{(1)} \\ \mathbf{u}^{(2)} \\ \boldsymbol{\lambda} \end{bmatrix} = \begin{bmatrix} \mathbf{f}^{(1)} \\ \mathbf{f}^{(2)} \\ \mathbf{0} \end{bmatrix}. \tag{4.37}$$

So the difference compared to the usual dual assembly form consists in the fact that the lower block diagonal term in the dual system matrix is no longer null but includes the bushing stiffness $\boldsymbol{\Upsilon}$. From a mathematical point of view, $\boldsymbol{\Upsilon}$ can also be seen as a perturbation of the compatibility constraints.[5]

The dual assembly (4.37) can directly be written for the general case of N^{sub} substructures even if bushings are present only at some interfaces: for the interfaces where no bushings are present, the corresponding entries in $\boldsymbol{\Upsilon}$ are set to zero (i.e., no interface flexibility). In block matrix form, one thus generalizes (4.2) to

$$\begin{bmatrix} \mathbf{Z} & \mathbf{B}^T \\ \mathbf{B} & -\boldsymbol{\Upsilon} \end{bmatrix} \begin{bmatrix} \mathbf{u} \\ \boldsymbol{\lambda} \end{bmatrix} = \begin{bmatrix} \mathbf{f} \\ \mathbf{0} \end{bmatrix}. \tag{4.38}$$

Using the first part of (4.38) to eliminate the substructure DOF (identical to (4.3))

$$\mathbf{u} = \mathbf{Y}(\mathbf{f} - \mathbf{B}^T\boldsymbol{\lambda}) \tag{4.3}$$

and substituting in the second part of (4.38), one finds that dual interface problem

$$\boxed{(\mathbf{B}\mathbf{Y}\mathbf{B}^T + \boldsymbol{\Upsilon})\boldsymbol{\lambda} = \mathbf{B}\mathbf{Y}\mathbf{f}} \tag{4.39}$$

or, substructure-wise,

$$\left(\sum_s^{N^{sub}} \mathbf{B}^{(s)}\mathbf{Y}^{(s)}\mathbf{B}^{(s)^T} + \boldsymbol{\Upsilon}\right)\boldsymbol{\lambda} = \sum_s^{N^{sub}} \mathbf{B}^{(s)}\mathbf{Y}^{(s)}\mathbf{f}^{(s)}. \tag{4.40}$$

---

[5]One can transform the form (4.37) in a primal assembled problem by eliminating $\boldsymbol{\lambda}$ to find

$$\boldsymbol{\lambda} = \boldsymbol{\Upsilon}^{-1}\begin{bmatrix} \mathbf{B}^{(1)} & \mathbf{B}^{(2)} \end{bmatrix}\begin{bmatrix} \mathbf{u}^{(1)} \\ \mathbf{u}^{(2)} \end{bmatrix}$$

$$\left(\begin{bmatrix} \mathbf{Z}^{(1)} & \mathbf{0} \\ \mathbf{0} & \mathbf{Z}^{(2)} \end{bmatrix} + \begin{bmatrix} \mathbf{B}^{(1)^T} \\ \mathbf{B}^{(2)^T} \end{bmatrix}\boldsymbol{\Upsilon}^{-1}\begin{bmatrix} \mathbf{B}^{(1)} & \mathbf{B}^{(2)} \end{bmatrix}\right)\begin{bmatrix} \mathbf{u}^{(1)} \\ \mathbf{u}^{(2)} \end{bmatrix} = \begin{bmatrix} \mathbf{f}^{(1)} \\ \mathbf{f}^{(2)} \end{bmatrix}.$$

In this form, the interface flexibility can be seen as an augmentation method for the compatibility constraints. Comparing with (4.33), one observes that the bushing dynamic stiffness matrix has a specific form induced by the connectivity between the interfaces of the bushing and the bushing intrinsic stiffness.

Finally, the dually assembled admittance expressed by (4.11) in the absence of bushing is now found to be solving the dual interface problem (4.39) for $\boldsymbol{\lambda}$ and replacing in the substructure equilibrium equations:

$$\boxed{\mathbf{Y}_{g,dual} = \left(\mathbf{Y} - \mathbf{Y}\mathbf{B}^T(\mathbf{B}\mathbf{Y}\mathbf{B}^T + \boldsymbol{\Upsilon})^{-1}\mathbf{B}\mathbf{Y}\right).} \tag{4.41}$$

Bushings as Substructures

The approach explained above considers the bushing as an interface where the compatibility is relaxed. Comparing the dually assembled admittance obtained by the LM-FBS with and without bushing, we observe that the presence of bushing flexibilities in (4.41) modifies the way the dual assembly is organized. If one would like to treat the presence of bushings without changing the LM-FBS formulation, one should apply the standard form (4.11) and consider bushings as substructures of their own. Note that in that case the connectivity matrices are different from the previous approach. For instance, if one considers the bushing of the simple case of Fig. 4.6 as a third substructure,

$$\mathbf{B} = \left[\mathbf{B}^{(1)}\ \mathbf{B}^{(bush)}\ \mathbf{B}^{(2)}\right] = \begin{bmatrix} 0 & 1 & -1 & 0 & 0 & 0 \\ 0 & 0 & 0 & 1 & -1 & 0 \end{bmatrix} \qquad \mathbf{u} = \begin{bmatrix} \mathbf{u}^{(1)} \\ \mathbf{u}^{(bush)} \\ \mathbf{u}^{(2)} \end{bmatrix}.$$

In order to apply the classical LM-FBS approach and write the assembled admittance as (4.11), one needs to find a manner to write the admittance of the bushing substructure, which is not easy since its dynamic impedance is usually singular (see (4.34)).

A straightforward manner to build an admittance matrix for the bushing consists in regularizing its impedance by adding virtual lumped masses to its ends, namely,

$$\mathbf{Y}^{(bush)} \simeq \left(\mathbf{Z}^{(bush)} - \omega^2\mathbf{M}^{(bush)}\right)^{-1} \tag{4.42}$$

In order for this modification not to affect (too much) the dynamic assembly of the assembly, three strategies can be considered:

- Choose $\mathbf{M}^{(bush)}$ as a diagonal matrix with small entries. For instance, one can choose the mass perturbation such that the local nonzero eigenfrequencies of the bushing are 10 times higher than the highest frequency of interest for the assembly.
- Choose any $\mathbf{M}^{(bush)}$ and subtract its effect by assembling an additional (virtual) substructure with the corresponding negative mass, namely, with admittance $\omega^2\mathbf{M}^{(bush)}$ (see also the concept of decoupling in Sect. 4.4.1).
- Choose a diagonal $\mathbf{M}^{(bush)}$ and remove the corresponding lumped masses from the neighboring substructure admittances (for instance by decoupling, see Sect. 4.4.1).

Obviously, the first strategy is the easiest one and can be readily applied in practice.

Bushings as Substructures via Pseudo-inverses

The last method to introduce a bushing in the LM-FBS consists in considering the bushing as a substructure (as in the previous method), but instead of regularizing its impedance by adding inertia terms like in (4.42), one can compute an admittance matrix using a pseudo-inverse, as explained next.

The fact that $\mathbf{Z}^{(bush)}$ is singular means that the substructure dynamics

$$\mathbf{Z}^{(bush)}\mathbf{u}^{(bush)} = \mathbf{f}^{(bush)} - \mathbf{B}^{(bush)^T}\boldsymbol{\lambda}$$

has a solution only if the applied forces are self-equilibrated or, in other words, if the applied forces have zero resulting forces and moments. Mathematically (see for instance Géradin and Rixen 2015), this is expressed by the fact that the applied forces must not produce any work on the rigid body modes:

$$\mathbf{R}^{(bush)^T}\left(\mathbf{f}^{(bush)} + \mathbf{B}^{(bush)^T}\boldsymbol{\lambda}\right) = \mathbf{0}. \tag{4.43}$$

Under this condition, there exist a family of solution that can be written as

$$\mathbf{u}^{(bush)} = \mathbf{Y}^{(bush)}\left(\mathbf{f}^{(bush)} - \mathbf{B}^{(bush)^T}\boldsymbol{\lambda}\right) + \mathbf{R}^{(bush)}\boldsymbol{\alpha}^{(bush)} \qquad \mathbf{Y}^{(bush)} = \mathbf{Z}^{(bush)^+}, \tag{4.44}$$

where $\mathbf{Z}^{(bush)^+}$ is a pseudo-inverse of $\mathbf{Z}^{(bush)}$ that can be computed either by singular value decomposition or, more efficiently, by a incomplete factorization (see, for instance, Géradin and Rixen (2015) or the simple example below). The amplitudes $\boldsymbol{\alpha}^{(bush)}$ of the contribution of the rigid body mode to the solution are additional unknowns that will be determined, thanks to the self-equilibrium condition (4.43). In other words, the overall position of the bushing needs to be such that the generated interface forces are in global static equilibrium for the bushing.

► For the simple example of Fig. 4.6, we can compute a pseudo-inverse in the form of

$$\mathbf{Y}^{(bush)} = \mathbf{Z}^{(bush)^+} = \begin{bmatrix} z^{bush} & -z^{bush} \\ -z^{bush} & z^{bush} \end{bmatrix}^+ = \begin{bmatrix} (k^{bush} - i\omega c^{bush})^{-1} & 0 \\ 0 & 0 \end{bmatrix}$$

and the general solution takes the form

$$\mathbf{u}^{(bush)} = \begin{bmatrix} (k^{bush} - i\omega c^{bush})^{-1} & 0 \\ 0 & 0 \end{bmatrix}\left(\mathbf{f}^{(bush)} - \mathbf{B}^{(bush)^T}\boldsymbol{\lambda}\right) + \begin{bmatrix} 1 \\ 1 \end{bmatrix}\alpha^{(bush)}$$

under the condition

$$[1\ 1]\left(\mathbf{f}^{(bush)} - \mathbf{B}^{(bush)^T}\boldsymbol{\lambda}\right) = 0.$$

The general solution can be understood as the solution obtained when the bushing is fixed at the right end and when an arbitrary rigid body motion is added. Note that fixing the bushing at one end does not introduce any additional force to the system (since all the external forces are self-equilibrated) but merely determines one specific solution.

◄

The local solutions of the substructure dynamics are thus either in the general form (4.44) for a bushing or in the regular form (4.5) for the other substructures. This can be written in a block form similar to (4.3) as

$$\mathbf{u} = \mathbf{Y}(\mathbf{f} - \mathbf{B}^T\boldsymbol{\lambda}) + \mathbf{R}\boldsymbol{\alpha}, \tag{4.45}$$

if we define the block matrix of the rigid body modes for the bushings and their corresponding amplitudes

$$\mathbf{R} = \begin{bmatrix} \mathbf{0} & \cdots & \mathbf{0} \\ \vdots & & \\ \mathbf{R}^{(bush1)} & \mathbf{0} & \vdots \\ \vdots & \mathbf{R}^{(bush2)} & \vdots \\ \vdots & \vdots & \vdots \\ \mathbf{0} & \cdots & \mathbf{0} \end{bmatrix} \qquad \boldsymbol{\alpha} = \begin{bmatrix} \boldsymbol{\alpha}^{(bush1)} \\ \boldsymbol{\alpha}^{(bush2)} \\ \vdots \end{bmatrix}. \tag{4.46}$$

The dually assembled problem can now be found as before: if we consider for the bushing the local solution (4.45) and substitute in the compatibility condition, one finds

$$\mathbf{B}\mathbf{u} = \mathbf{B}\left(\mathbf{Y}(\mathbf{f} - \mathbf{B}^T\boldsymbol{\lambda}) + \mathbf{R}\boldsymbol{\alpha}\right) = \mathbf{0}. \tag{4.47}$$

Rearranging this equation together with the self-equilibrium condition (4.43) leads to the dual interface problem (which is an extension of the standard form (4.48)):

$$\begin{bmatrix} \mathbf{BYB}^T & \mathbf{BR} \\ \mathbf{R}^T\mathbf{B}^T & \mathbf{0} \end{bmatrix} \begin{bmatrix} \boldsymbol{\lambda} \\ \boldsymbol{\alpha} \end{bmatrix} = \begin{bmatrix} \mathbf{BYf} \\ \mathbf{R}^T\mathbf{f} \end{bmatrix}. \tag{4.48}$$

This extended dual problem can be solved for $\boldsymbol{\lambda}$ and $\boldsymbol{\alpha}$. Substituting the solution into the governing equation then returns the dually assembled admittance. It can be written as an extended form of (4.11) but will not be shown in detail here (see Mahmoudi et al. 2019 for further details and examples).

4.4 Substructure Decoupling

In the literature, two types of disassembling problems are treated. In the first type of problem, one assumes that the overall system is made of several substructures, and the question to be answered is *to find the dynamical properties of one substructure, the dynamics of the rest of the structure being known.* This problem will be referred to as *Substructure Decoupling*.

A similar, yet different, problem arises when considering structures assembled through bushings, i.e., structural elements that are characterized only by stiffness and damping and having negligible mass (see Sect. 4.3.6). In this case, one knows a priori that the forces on each side of the bushing are in equilibrium (since the internal dynamics of the bushing are negligible) and therefore it is possible (as will be shown later) to extract the dynamic properties of the separated bushing and substructures, *without having any pre-knowledge of the dynamics of any component.* This problem is commonly referred to as *inverse substructuring* in the literature.

We will first outline the basics of substructure decoupling, and then those of inverse substructuring. In the last part, substructure decoupling will be applied in the modal domain and the technique of transmission simulator will be outlined.

4.4.1 Theory of Decoupling

The problem of substructure decoupling has been tackled in the recent years by several others (e.g., Ind and Ewins 2003, Sjövall 2007, Groult 2008, D'Ambrogio 2009, Voormeeren and Rixen 2009, Cloutier and Avitabile 2011, D'Ambrogio and Fregalent 2014, D'Ambrogio and Fregolent 2015, Höller and Gibbs 2015) including methods for simple nonlinear systems Kalaycıoğlu and Özgüven (2018), Tol and Özgüven (2015). The presentation here is based on Voormeeren and Rixen (2012).

Let us consider a system made of two substructures called A and B in this section. The coupling problem as discussed in the previous section consists of building the dynamics of the assembly AB using the dynamics of the components, Fig. 4.8. The *substructure decoupling* problem consists in finding the dynamic properties of substructure B when the dynamics of the assembly AB and of component A is known, Fig. 4.8.

This problem is often encountered in practice, for instance, when a component cannot be disassembled to measure its FRF (an early application of the idea was used for determining the dynamics of microsystems attached to a support in Epp et al. 2004).

The problem of decoupling can be solved by considering the interpretation described in Fig. 4.9: if the part B in the assembly AB is excited by a force $\mathbf{f}_i^{(B)}$ in B, the response at a DOF $\mathbf{u}_i^{(B)}$ in B when B would be disconnected from A can be obtained if, in addition to the external force $\mathbf{f}_i^{(B)}$, one applies to the assembly AB the opposite of the internal force experienced by B in the assembly AB. This interface force can be obtained by imposing on the interface of A the same displacements as the one experienced by the assembly and by calculating the corresponding interface forces.

Mathematically speaking, this means that the response of the uncoupled substructure B obeying

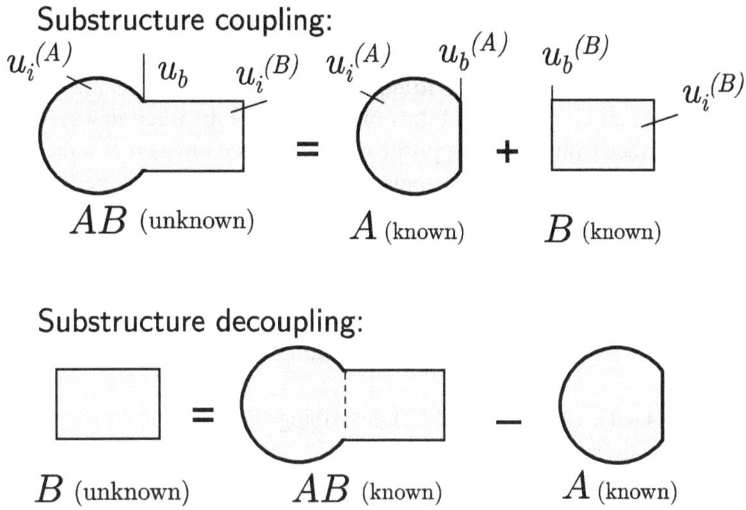

Fig. 4.8 The concepts of substructure coupling and decoupling

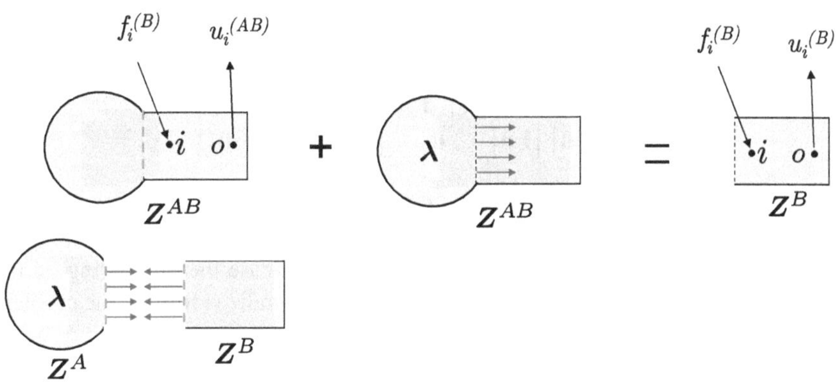

Fig. 4.9 Interpretation of the decoupling problem

$$\mathbf{Z}^B \mathbf{u}^B = \mathbf{f}^B \tag{4.49}$$

can be computed from

$$\mathbf{Z}^{AB} \mathbf{u}^{AB} = \mathbf{f}^{AB} + \mathbf{g}^{AB} \tag{4.50}$$

or

$$\begin{bmatrix} \mathbf{Z}^{AB}_{i_A,i_A} & \mathbf{Z}^{AB}_{i_A,b} & \mathbf{0} \\ \mathbf{Z}^{AB}_{b,i_A} & \mathbf{Z}^{AB}_{b,b} & \mathbf{Z}^{AB}_{b,i_B} \\ \mathbf{0} & \mathbf{Z}^{AB}_{i_B,b} & \mathbf{Z}^{AB}_{i_B,i_B} \end{bmatrix} \begin{bmatrix} \mathbf{u}^{AB}_{i_A} \\ \mathbf{u}^{AB}_{b} \\ \mathbf{u}^{AB}_{i_B} \end{bmatrix} = \begin{bmatrix} \mathbf{0} \\ \mathbf{f}^{AB}_b = \mathbf{f}^B_b \\ \mathbf{f}^{AB}_{i_B} = \mathbf{f}^B_i \end{bmatrix} + \begin{bmatrix} \mathbf{0} \\ \boldsymbol{\lambda} \\ \mathbf{0} \end{bmatrix},$$

where the subscripts i_A and i_B denote the internal DOFs in the A and B part of AB, and where we assumed a special ordering of the DOFs.

The interface forces $\boldsymbol{\lambda}$ that one has to apply as external forces to AB must annihilate the effect of the coupling on B and thus must be minus the force that AB receives through its interface from A. The opposite of the interface force on B is the interface force on A (actio-reactio): it can be computed as the force that A experiences when $\mathbf{u}_b^{AB}$ is imposed on it, namely (see Fig. 4.8),

$$\boldsymbol{\lambda} \quad \text{such that} \quad \mathbf{Z}^A \begin{bmatrix} \mathbf{u}_i^A \\ \mathbf{u}_b^A \end{bmatrix} = \begin{bmatrix} \mathbf{0} \\ \boldsymbol{\lambda} \end{bmatrix} \tag{4.51}$$

$$\mathbf{u}_b^A = \mathbf{u}_b^{AB}. \tag{4.52}$$

Equations (4.4.1), (4.51), and (4.52) describing the dynamic of the uncoupled substructure B can be summarized in the form

$$\begin{bmatrix} \mathbf{Z}^{AB} & \mathbf{0} & \begin{bmatrix} \mathbf{0} \\ -\mathbf{I} \\ \mathbf{0} \end{bmatrix} \\ \mathbf{0} & \mathbf{Z}^A & -\begin{bmatrix} \mathbf{0} \\ \mathbf{I} \end{bmatrix} \\ \begin{bmatrix} -\mathbf{I} & \mathbf{0} & \mathbf{0} \end{bmatrix} & \begin{bmatrix} \mathbf{I} & \mathbf{0} \end{bmatrix} & \mathbf{0} \end{bmatrix} \begin{bmatrix} \mathbf{u}_{i_A}^{AB} \\ \mathbf{u}_b^{AB} \\ \mathbf{u}_{i_B}^{AB} \\ \mathbf{u}_i^A \\ \mathbf{u}_b^A \\ \boldsymbol{\lambda} \end{bmatrix} = \begin{bmatrix} \mathbf{0} \\ \mathbf{f}_b^B \\ \mathbf{f}_i^B \\ \mathbf{0} \\ \mathbf{0} \\ \mathbf{0} \end{bmatrix}. \tag{4.53}$$

This relation shows that the dynamics of the isolated substructure B can be obtained by "assembling" AB and A, enforcing the compatibility on the interface, but applying the interface force in the opposite direction on AB. In case the numbering on the interface of AB and of B are not corresponding, the generalized form of the problem is written as

$$\begin{bmatrix} \mathbf{Z}^{AB} & \mathbf{0} & \mathbf{B}^{AB} \\ \mathbf{0} & \mathbf{Z}^A & -\mathbf{B}^A \\ \mathbf{B}^{AB} & \mathbf{B}^A & \mathbf{0} \end{bmatrix} \begin{bmatrix} \mathbf{u}^{AB} \\ \mathbf{u}^A \\ \boldsymbol{\lambda} \end{bmatrix} = \begin{bmatrix} \mathbf{0} \\ \mathbf{f}^B \\ \mathbf{0} \end{bmatrix}, \tag{4.54}$$

where $\mathbf{B}^{AB}$ and $\mathbf{B}^A$ are the compatibility constraint matrices as used for the assembly.

This relation can be equivalently written as

$$\boxed{\begin{bmatrix} \mathbf{Z}^{AB} & \mathbf{0} & \mathbf{B}^{AB} \\ \mathbf{0} & -\mathbf{Z}^A & \mathbf{B}^A \\ \mathbf{B}^{AB} & \mathbf{B}^A & \mathbf{0} \end{bmatrix} \begin{bmatrix} \mathbf{u}^{AB} \\ \mathbf{u}^A \\ \boldsymbol{\lambda} \end{bmatrix} = \begin{bmatrix} \mathbf{0} \\ \mathbf{f}^B \\ \mathbf{0} \end{bmatrix},} \tag{4.55}$$

showing that *the decoupling problem is equivalent to a coupling problem with A having a negative impedance*. Hence, one can apply the formula (4.11) of the frequency-based substructuring to find the admittance of the uncoupled system if one considers substructure A as having a negative impedance.

Assuming again that the numbering of the interface DOFs in AB and A are matching, one can easily work out the impedance of B by solving (4.53), resulting in Voormeeren and Rixen (2012)

$$\mathbf{Y}^{(B)} = \begin{bmatrix} \mathbf{Y}^{B}_{b,b} & \mathbf{Y}^{B}_{b,i} \\ \mathbf{Y}^{B}_{i,b} & \mathbf{Y}^{B}_{i,i} \end{bmatrix} = \begin{bmatrix} \mathbf{Y}^{AB}_{b,b} & \mathbf{Y}^{AB}_{b,i_B} \\ \mathbf{Y}^{AB}_{i_B,b} & \mathbf{Y}^{AB}_{i_b,i_B} \end{bmatrix} - \begin{bmatrix} \mathbf{Y}^{AB}_{b,b} \\ \mathbf{Y}^{AB}_{i_B,b} \end{bmatrix} \left(\mathbf{Y}^{AB}_{b,b} - \mathbf{Y}^{A}_{b,b}\right)^{-1} \begin{bmatrix} \mathbf{Y}^{AB}_{b,b} & \mathbf{Y}^{AB}_{b,i_B} \end{bmatrix}. \tag{4.56}$$

Remark 1: Post-processing of decoupled dynamics

Note that, due to the errors inherent to the fact that the method is based on measurement data, the FRFs of the uncoupled system B are usually not fully accurate and might exhibit nonphysical properties such as non-passivity or negative mass. Several authors have proposed different cures to improve the FRFs obtained by decoupling Carne and Dohrmann (2006), Mayes and Ross (2012). Other techniques to enhance the consistency of the FRFs in state space are discussed in Sect. 4.9.2.

Remark 2: Decoupling with overdetermined compatibilities

The decoupling method formalized in (4.55), exactly like for substructure coupling, is very sensitive to small errors in the interface admittances of the components and, therefore, it is desirable to weaken the interface compatibility using one of the techniques described in Sect. 4.3.4.

However, for decoupling, there is an additional manner to relax the strong compatibility (4.52) required between the assembly AB and the substructure A to be removed. Indeed, instead of requiring that $\boldsymbol{\lambda}$ applied to AB should be such that only the interface DOFs of A and AB are equal (see (4.52)), one could require that all DOFs measured in A and those measured in the part A of AB should be equal when they experience the same $\boldsymbol{\lambda}$ on the interface.

$$\begin{bmatrix} \mathbf{u}^{A}_{b} \\ \mathbf{u}^{A}_{i} \end{bmatrix} = \begin{bmatrix} \mathbf{u}^{AB}_{b} \\ \mathbf{u}^{AB}_{i_B} \end{bmatrix}. \tag{4.57}$$

Requiring that also the DOFs inside A must be identical in A and AB is, in theory, unnecessary since if all interface DOFs in AB and A are identical, the internal ones will automatically be identical. However, in practice, measurements are not perfect and thus the data used for $\mathbf{Y}^{A}$ might not be fully consistent with the information about A contained in AB. Hence, requiring also explicitly the compatibility inside A helps mitigating inaccuracies when computing the interface forces $\boldsymbol{\lambda}$.

With the extended compatibility requirement (4.57), the decoupling problem (4.58) is written as

$$\begin{bmatrix} \mathbf{Z}^{AB} & \mathbf{0} & \begin{bmatrix} \mathbf{0} \\ -\mathbf{I} \\ \mathbf{0} \end{bmatrix} \\ \mathbf{0} & \mathbf{Z}^{A} & -\begin{bmatrix} \mathbf{0} \\ \mathbf{I} \end{bmatrix} \\ \begin{bmatrix} -\mathbf{I} & \mathbf{0} & \mathbf{0} \\ \mathbf{0} & -\mathbf{I} & \mathbf{0} \end{bmatrix} \begin{bmatrix} \mathbf{I} & \mathbf{0} \\ \mathbf{0} & \mathbf{I} \end{bmatrix} & & \mathbf{0} \end{bmatrix} \begin{bmatrix} \mathbf{u}_{i_A}^{AB} \\ \mathbf{u}_b^{AB} \\ \mathbf{u}_{i_B}^{AB} \\ \mathbf{u}_i^{A} \\ \mathbf{u}_b^{A} \\ \boldsymbol{\lambda} \end{bmatrix} = \begin{bmatrix} \mathbf{0} \\ \mathbf{f}_b^B \\ \mathbf{f}_i^B \\ \mathbf{0} \\ \mathbf{0} \\ \mathbf{0} \\ \mathbf{0} \end{bmatrix}. \tag{4.58}$$

Note that, although compatibility is required for all DOFs of A, only interface forces $\boldsymbol{\lambda}$ are considered. Therefore, the interface problem will be *overdetermined* and will be solved using a pseudo-inverse, meaning that one computes the interface forces that guarantees the best compatibility for all DOFs on A in a least square sense. The admittance of B, given by (4.56) for the basic decoupling form, now becomes Voormeeren and Rixen (2012)

$$\mathbf{Y}^{(B)} = \begin{bmatrix} \mathbf{Y}_{b,b}^{B} & \mathbf{Y}_{b,i}^{B} \\ \mathbf{Y}_{i,b}^{B} & \mathbf{Y}_{i,i}^{B} \end{bmatrix} = \begin{bmatrix} \mathbf{Y}_{b,b}^{AB} & \mathbf{Y}_{b,i_B}^{AB} \\ \mathbf{Y}_{i_B,b}^{AB} & \mathbf{Y}_{i_B,i_B}^{AB} \end{bmatrix} - \begin{bmatrix} \mathbf{Y}_{b,b}^{AB} \\ \mathbf{Y}_{i_B,b}^{AB} \end{bmatrix} \left(\begin{bmatrix} \mathbf{Y}_{i_A,b}^{AB} \\ \mathbf{Y}_{b,b}^{AB} \end{bmatrix} - \begin{bmatrix} \mathbf{Y}_{i,b}^{A} \\ \mathbf{Y}_{b,b}^{A} \end{bmatrix} \right)^{+} \begin{bmatrix} \mathbf{Y}_{i_A,b}^{AB} & \mathbf{Y}_{i_A,i_B}^{AB} \\ \mathbf{Y}_{b,b}^{AB} & \mathbf{Y}_{b,i_B}^{AB} \end{bmatrix}. \tag{4.59}$$

In that way, inaccuracies in $\mathbf{Y}^{AB}$ and $\mathbf{Y}^{A}$ will be mitigated. This idea was already proposed in Sjövall (2006), although in a different form.

Here, only one variant of the different possibilities to impose compatibility and equilibrium for a decoupling problem was outlined. Other approaches modifying the basic decoupling form (4.55) are discussed in Voormeeren and Rixen (2012), where also application on an analytical example and on a hardware example can be found.

Remark 3: Transmission simulator and interface substitute

The substructure decoupling technique is very useful also when applying the FBS methods to assemble measured components. Indeed, the results obtained from the FBS techniques are significantly improved when the components are not measured with free boundary conditions on the interface, but when the measurement is performed with a so-called *transmission simulator* introduced in Allen et al. (2010); Mayes and Arviso (2010), Mayes and Ross (2012) for the modal domain coupling (see Sect. 4.4.3). The methodology is also sometimes called *interface substitute* van der Seijs (2013) when used in the physical domain. The idea is illustrated in Fig. 4.10. The coupling substitute must be removed from the assembled system using a decoupling method as discussed in this section, assuming that the substitute is a well-defined system for which a very good (numerical or measured) model exists.

Such a procedure seems useless since in theory adding a coupling substitute to obtain the component dynamics, then removing it in the assembly does not affect the final result. In practice, however, this can make a very significant difference and dramatically improve the accuracy of the assembled system obtained by FBS, the two main reasons being the following:

- the dynamics of the component measured with a substitute on its interface exhibits, during the component measurement, dynamics that are similar to the dynamics it

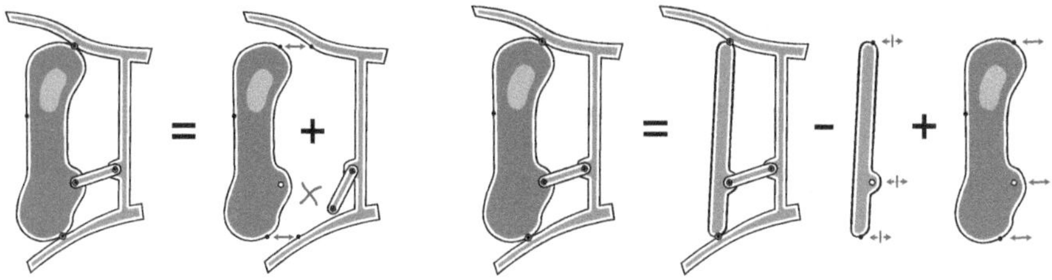

Fig. 4.10 The idea of the transmission simulator (coupling substitute) van der Seijs (2013)

will exhibit in the assembly later. Therefore, when inverting the interface flexibility to compute the assembled dynamics (see 4.11), the local admittance has much better accuracy for the problem to be solved.

- the model of the component, when the substitute has been decoupled, contains in practice not only the dynamics of the isolated component but includes also the dynamics of the interface as it was present in the assembly with the substitute. Hence, this interface dynamics (stiffness and damping) will also be included in the model of the final assembly. This is essential, for instance, for assemblies where a significant part of the damping comes from joints.

In practice, the substitute should be designed such that it exercises the interface of the component in a manner similar to what the component will experience in the real assembly, i.e., the substitute should have similar dynamics as the components connected in the final assembly and it should be attached through the same physical connections (e.g., bolted, riveted..).

4.4.2 *Inverse Substructuring*

The *inverse substructuring* technique can be applied to problems where components are assembled through bushings (see Sect. 4.3.6) and when one assumes that the bushings have no significant inertia. The method then allows to find the admittances of the bushings and of the individual components, based only on the admittances measured on the assembly (hence, contrary to the substructure decoupling techniques, not requiring any pre-knowledge of the dynamics of any of the substructures).

The inverse substructuring techniques were developed several years ago and the formulation improved over the years (see, for instance, Zhen et al. 2004, Zhen et al. 2004, Wang and Wang 2011, Wang et al. 2018). Many of those publications are rather difficult to follow due to the cumbersome algebraic developments and the ad hoc approach of the problem. A more structured manner to derive the equations was proposed in Keersmaekers et al. (2015). Recently, the inverse substructuring technique was written in a very clear and intuitive way Moorhouse et al. (2013);

Meggitt et al. (2015), Meggitt et al. (2018), Meggitt and Moorhouse (2018). We shortly outline the method based on the approach of those latter publications.

Referring to the schematic representation of Fig. 4.6, let us consider the primal form (4.33) of the dynamic problem where two substructures (called 1 and 2) are connected by a bushing, recalled here for convenience:

$$\mathbf{Z}^{(1+bush+2)}\mathbf{u}^{(1+bush+2)} = \begin{bmatrix} \mathbf{Z}_{ii}^{(1)} & \mathbf{Z}_{ib}^{(1)} & \mathbf{0} & \mathbf{0} \\ \mathbf{Z}_{bi}^{(1)} & \mathbf{Z}_{bb}^{(1)} + \mathbf{Z}_{b^1,b^1}^{(bush)} & \mathbf{Z}_{b^1,b^2}^{(bush)} & \mathbf{0} \\ \mathbf{0} & \mathbf{Z}_{b^2,b^1}^{(bush)} & \mathbf{Z}_{bb}^{(2)} + \mathbf{Z}_{b^2,b^2}^{(bush)} & \mathbf{Z}_{bi}^{(2)} \\ \mathbf{0} & \mathbf{0} & \mathbf{Z}_{bi}^{(2)} & \mathbf{Z}_{ii}^{(2)} \end{bmatrix} \begin{bmatrix} \mathbf{u}_i^{(1)} \\ \mathbf{u}_{b^1} \\ \mathbf{u}_{b^2} \\ \mathbf{u}_i^{(2)} \end{bmatrix} = \begin{bmatrix} \mathbf{f}_i^{(1)} \\ \mathbf{f}_{b^1} \\ \mathbf{f}_{b^2} \\ \mathbf{f}_i^{(2)} \end{bmatrix}. \tag{4.60}$$

If a substructure decoupling approach as explained in Sect. 4.4.2 would be used, one would consider the bushings as a third component, measure $\mathbf{Y}^{(1+bush+2)}$ and the admittances $\mathbf{Y}^{(1)}$ and $\mathbf{Y}^{(2)}$. Then one would decouple the admittances of substructure 1 and 2 by assembling to $\mathbf{Y}^{(1+bush+3)}$ the negative admittances $-\mathbf{Y}^{(1)}$ and $-\mathbf{Y}^{(2)}$.

In the inverse substructuring, only the measurement of the admittance $\mathbf{Y}^{(1+bush+2)}$ of the assembly is used and inverted to obtain

$$\mathbf{Z}^{(1+bush+2)} = \left(\mathbf{Y}^{(1+bush+2)}\right)^{-1}. \tag{4.61}$$

Inverse substructuring then uses the fact that the off-diagonal terms in the measured impedance (4.61) (comparing with the theoretical form (4.60)) are properties of the bushing alone, namely,

$$\mathbf{Z}_{b^1,b^2}^{(1+bush+2)} = \mathbf{Z}_{b^1,b^2}^{(bush)}. \tag{4.62}$$

Let us assume now that the DOFs at the two sides of the bushing are connected with the topology shown in Fig. 4.11, i.e., each DOF on one side of the interface is only coupled to one DOF on the other side and there are no cross couplings with any other DOF. Additionally, it is assumed that the joint element (i.e., the rubber) has negligible mass. If these assumptions hold, then the stiffness matrix of the rubber isolator has the following properties:

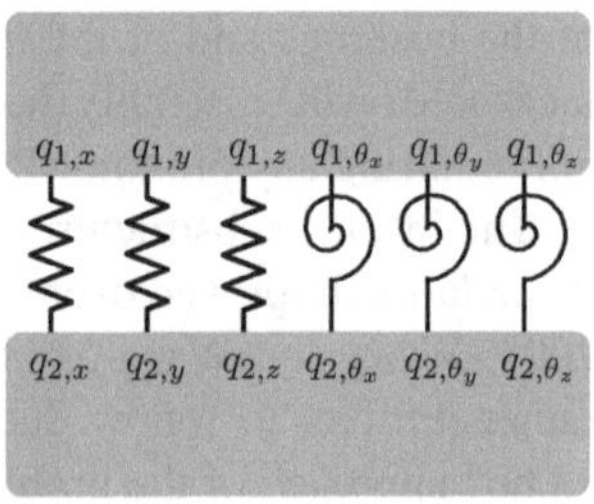

Fig. 4.11 Schematic representation of the inverse substructuring assumption for a bushing

$$\mathbf{Z}_{b^1,b^1}^{(bush)} = \mathbf{Z}_{b^2,b^2}^{(bush)} = -\mathbf{Z}_{b^1,b^2}^{(bush)} = -\mathbf{Z}_{b^2,b^1}^{(bush)}, \tag{4.63}$$

where $\mathbf{Z}_{b^1,b^1}^{(bush)}$ is actually a diagonal matrix. Thus, by simply inverting the measured FRF matrix of the assembly, the whole rubber isolator can be identified from just taking the off-diagonal blocks of the matrix (provided that the above-stated assumptions hold). Hence, the name "inverse substructuring". In this case, it is straightforward to derive the dynamic stiffnesses of the two separate substructures 1 and 2 from the measurements:

$$\begin{bmatrix} \mathbf{Z}_{ii}^{(1)} & \mathbf{Z}_{ib}^{(1)} & \mathbf{0} & \mathbf{0} \\ \mathbf{Z}_{bi}^{(1)} & \mathbf{Z}_{bb}^{(1)} & \mathbf{0} & \mathbf{0} \\ \mathbf{0} & \mathbf{0} & \mathbf{Z}_{bb}^{(2)} & \mathbf{Z}_{bi}^{(2)} \\ \mathbf{0} & \mathbf{0} & \mathbf{Z}_{bi}^{(2)} & \mathbf{Z}_{ii}^{(2)} \end{bmatrix} = \left(\mathbf{Y}^{(1+bush+2)}\right)^{-1} - \begin{bmatrix} \mathbf{0} & \mathbf{0} & \mathbf{0} & \mathbf{0} \\ \mathbf{0} & \mathbf{Z}_{b^1,b^1}^{(bush)} & \mathbf{Z}_{b^1,b^2}^{(bush)} & \mathbf{0} \\ \mathbf{0} & \mathbf{Z}_{b^2,b^1}^{(bush)} & \mathbf{Z}_{b^2,b^2}^{(bush)} & \mathbf{0} \\ \mathbf{0} & \mathbf{0} & \mathbf{0} & \mathbf{0} \end{bmatrix}. \tag{4.64}$$

This comes with the great practical advantage that it is not necessary to know anything about the dynamic properties of the two structures A and B which are joint by the isolator. Thus, one is able to identify the dynamic properties of all involved substructures 1, 2 and the bushing separately from only one set of measurements performed on the assembly.

► *Application to the identification of a rubber bushing*
To illustrate the inverse substructuring and the substructure decoupling techniques, we summarize the results obtained for a rubber bushing as reported in Haeussler et al. (2018). In that work, the dynamic stiffness of the rubber bushing depicted in Fig. 4.12 is looked for.

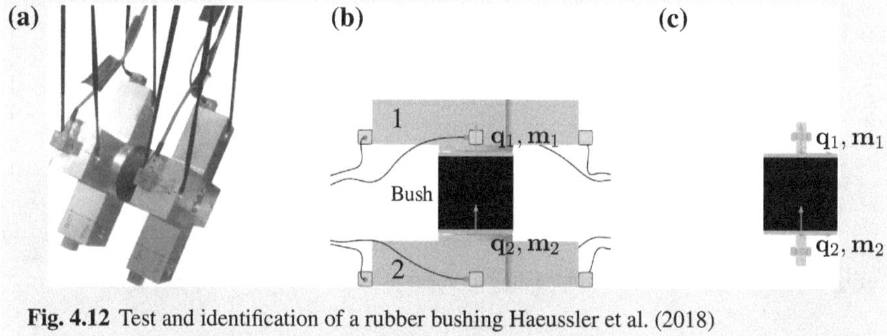

Fig. 4.12 Test and identification of a rubber bushing Haeussler et al. (2018)

The idea is to attach two well-defined and modeled rigid crosses on each side of the rubber and consider them as substructures 1 and 2 connected by the bushing. The crosses can be considered as very rigid for the frequency range considered and, since the masses and inertias of the crosses are accurately known (from a CAD model), their admittances are accurately known so that the substructure decoupling techniques can be applied (using the rigid body

behavior of the crosses to define virtual points and assemble the model of the system, see Sect. 4.6). The results obtained from the substructure decoupling method for the dynamic properties of the bushing are then compared to the ones obtained by applying the inverse substructuring technique.
The experiment is further described in Fig. 4.12 which shows (a) the actual measurement setup of the assembly: rubber isolator between upper and lower fixture hung up with rubber bands, (b) the schematics of the free-free measurement of bushing between fixtures 1 and 2, and (c) the substructure of the rubber isolator which is to be determined. The DOFs $\mathbf{q}_1$ and $\mathbf{q}_2$ represent the rigid body DOF on the top and bottom of the rubber isolator, respectively, $\mathbf{m}_1$ and $\mathbf{m}_2$, containing the translational and rotational resulting forces on each side.
The results of the substructure decoupling and inverse decoupling are illustrated in Fig. 4.13. It shows the impedance values obtained for the axial DOF of the rubber. One clearly sees that both methods give very similar results up to 100 Hz, but differ significantly for higher frequencies. In fact, the substructure decoupling yields the dynamic properties of the rubber including the bolts attached to its ends (see Fig. 4.12c), whereas the inverse substructuring approach assumes that the entire bushing is massless (Fig. 4.13).

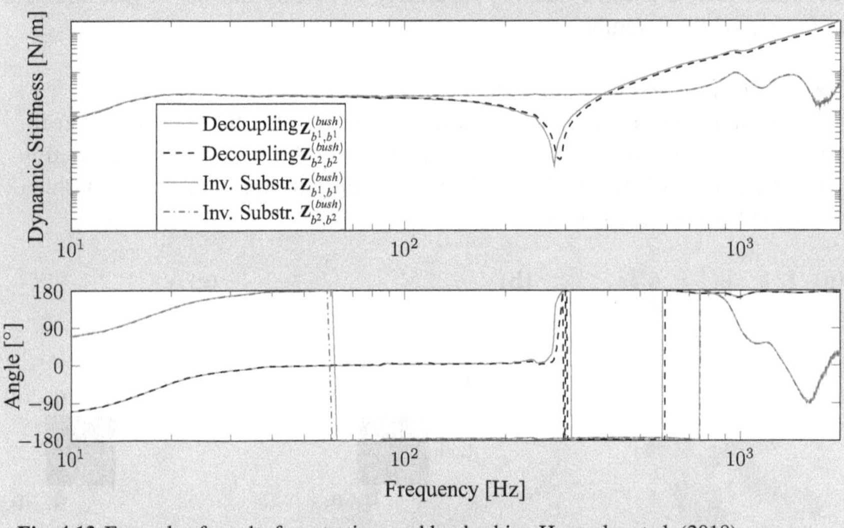

Fig. 4.13 Example of results from testing a rubber bushing Haeussler et al. (2018)

◀

4.4.3 The Transmission Simulator Method for Substructuring and Substructure Decoupling

(*The material in this section was taken from Paper 161 at the 2012 IMAC.*)

The modal constraint for fixture and subsystem (MCFS) method was introduced at IMAC in 2007 for the component mode synthesis approach Allen and Mayes (2007) and the frequency-based substructuring approach Mayes and Stasiunas (2007). It uses a fixture in the experimental dynamic substructure called the transmission simulator, so is also called the transmission simulator method. An analytical model of the transmission simulator is generally required with this method (or a very accurate modal model).

Physically, the transmission simulator is mounted to the experimental substructure with exactly the same joint geometry and material as that to which the experimental substructure will ultimately be connected. Because of this, the resulting experimental substructure inherently includes the linearized stiffness and damping in the joint, which classical methods neglect, to their peril. The fixture is also designed to provide accessible locations to mount response sensors and to apply input forces. Often the actual connection points are poor locations to mount sensors. For example, the connection point may be at a bolt or a threaded screw interface. Special features can also be included to provide good driving point measurements, which are extremely important to obtain accurate modal mass for scaling the mode shapes.

The analytical model of the transmission simulator is used in multiple ways. It is generally mounted with an assumed welded connection to the analytical substructure (often a finite element model) to which the experimental substructure is to be connected. By constraining the transmission simulator on the experimental substructure to have the same motion as the transmission simulator on the analytical substructure, the systems are joined. Then the mass and stiffness of the analytical and experimental transmission simulators are subtracted from the assembled system. The analytical model of the transmission simulator can also be thought of as an aid to interpolate from the measurement sensor locations back to the actual connection degrees of freedom (DOFs). A truncated set of the mode shapes of the transmission simulator is used that spans the frequency bandwidth of interest. The sensor set can be chosen to provide a set of sensor locations for which all chosen transmission shapes are linearly independent. These sensors can all be translational—no rotations are required. The rotations at the connection points are inherently carried out in the modal coordinates of the transmission simulator. The connections can actually be continuous, not just discrete, as long as the transmission simulator mode shape set spans the space of the connection motion reasonably well. In addition, the transmission of simulator mass stresses the joint. This stress across the joint provides a much better Ritz vector shape basis than simple free modes where there is no stress at the joint. It provides enough improvement in the basis vectors that residuals, which are difficult to measure, do not need to be added to the basis set.

Although this method requires fabrication of a fixture and generation of its associated analytical model, it provides extensive benefits for the investment. One does not have to measure exactly at the connection points. One does not have to measure rotations because they are inherently included in the analytical model modal coordinates (so they are not just being neglected). The stiffness and damping of the joint are inherently included. No residual measurements are required.

One problem that can result in this method is that the mass matrix can be indefinite when the transmission simulator is subtracted. However, methods to correct the mass matrix have been developed recently Mayes et al. (2012). This has cleared the way so that in practice, refinements can be made without theoretical roadblocks.

An example from Mayes et al. (2012) shows conceptually how the method was implemented for one actual hardware case in Fig. 4.14. The transmission simulator hardware, which is a ring with tabs, is mounted to the experimental substructure with eight bolts, just as it will be attached in the real system to the cylinder substructure, so it contains the joint. An analytical model of the transmission simulator is welded into the flange of the cylinder analytical model. The transmission simulator and the cylinder flange actually occupy the same space, which can be done with an analytical model. Then transmission simulators for each substructure are forced to have the same motion, which connects them. Finally, the stiffness and mass of the two transmission simulators are analytically subtracted.

The free-free experimental modal test setup is shown in Fig. 4.14 with 12 triaxial accels on the transmission simulator and 2 triaxial accels at points of interest on the beam. Twenty-five rigid body and elastic modes were extracted out to 4 kHz. This structure had eight discrete bolted attachments, which would require 48 discrete constraints for the three rotations and three translations at each connection. Twelve

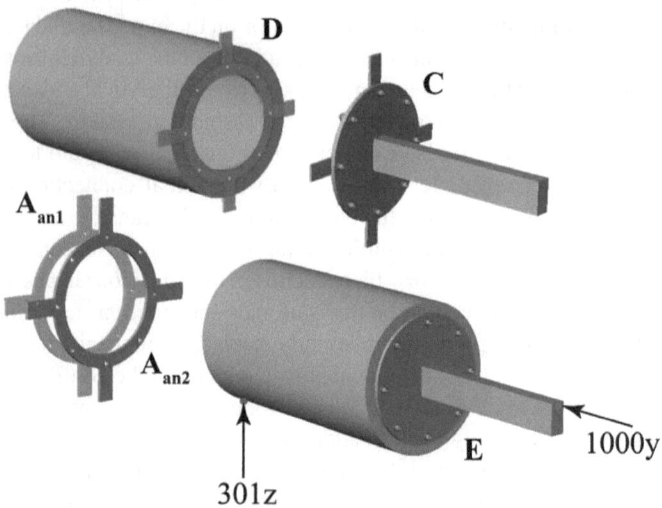

Fig. 4.14 Coupling of experimental substructure C with analytical substructure D to generate full system E after transmission simulators (A_{an1} and A_{an2}) are subtracted. Reproduced from Allen et al. (2010)

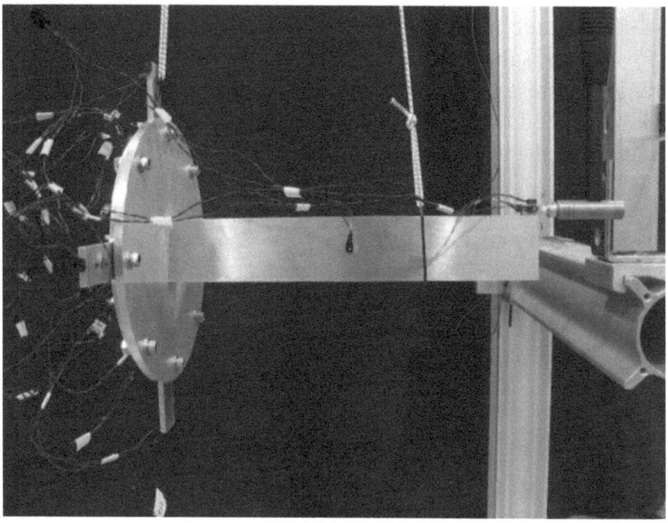

Fig. 4.15 Experimental substructure free modal test setup with 36 sensors on the transmission simulator ring and a few sensors at points of interest on the substructure. Reproduced from Allen et al. (2010)

triaxial accelerometers were placed on the TS, as shown in Fig. 4.14 none of which were at the eight connection points. The sensor locations are shown in Allen et al. (2010). Then, 18 modes of the transmission simulator analytical model were used to span the connection space motion, using 36 measurement points to describe the mode shapes. The method thus reduces the number of constraints down from the classic 48 connections to 18. The 18 modes covered a frequency bandwidth of 2 kHz (Fig. 4.15).

Component Mode Synthesis Theory Using Primal Formulation In A Generalized Framework

Following the notation outlined previously and in Klerk et al. (2008), and assuming that each substructure has been approximated with a reduced model, whether experimental or analytical, the displacements are approximated with the modal substitution as

$$\mathbf{u} \cong \mathbf{R}\eta, \tag{4.65}$$

where $\mathbf{u}$ is the vector of physical displacements, η is the vector of generalized coordinates from a modal test or eigenvector analysis, and R is the truncated mode shape matrix relating the generalized coordinates to the physical coordinates. Then, the equations of motion for the substructure can be written as

$$\mathbf{M}^{(s)}\mathbf{R}\ddot{\eta}^{(s)} + \mathbf{C}^{(s)}\mathbf{R}\dot{\eta}^{(s)} + \mathbf{K}^{(s)}\mathbf{R}\eta^{(s)} = \mathbf{f}^{(s)} + \mathbf{g}^{(s)} + \mathbf{r}^{(s)}, \tag{4.66}$$

where **M**, **C**, and **K** are mass stiffness and damping matrices, the superscript *(s)* denotes the particular substructure, $f^{(s)}$ denotes the external forces applied to the substructure, $\mathbf{g}^{(s)}$ represents the equilibrium joining forces from another substructure that will be applied to the substructure, and $\mathbf{r}^{(s)}$ is the residual force due to the mismatch caused by the approximation of the displacements in Eq. (4.65). Premultiplying by $\mathbf{R}^T$ yields

$$\mathbf{R}^T\mathbf{M}^{(s)}\mathbf{R}\ddot{\eta}^{(s)} + \mathbf{R}^T\mathbf{C}^{(s)}\mathbf{R}\dot{\eta}^{(s)} + \mathbf{R}^T\mathbf{K}^{(s)}\mathbf{R}\eta^{(s)} = \mathbf{R}^T\mathbf{f}^{(s)} + \mathbf{R}^T\mathbf{g}^{(s)} + \mathbf{R}^T\mathbf{r}^{(s)}. \tag{4.67}$$

It can be shown that, because of orthogonality of the mode shapes with respect to **M**, **C**, and **K**,

$$\mathbf{R}^T\mathbf{r}^{(s)} = \mathbf{0}, \tag{4.68}$$

which leaves

$$\mathbf{R}^T\mathbf{M}^{(s)}\mathbf{R}\ddot{\eta}^{(s)} + \mathbf{R}^T\mathbf{C}^{(s)}\mathbf{R}\dot{\eta}^{(s)} + \mathbf{R}^T\mathbf{K}^{(s)}\mathbf{R}\eta^{(s)} = \mathbf{R}^T\mathbf{f}^{(s)} + \mathbf{R}^T\mathbf{g}^{(s)}, \tag{4.69}$$

or in a renamed form

$$\mathbf{M}_m^{(s)}\ddot{\eta}^{(s)} + \mathbf{C}_m^{(s)}\dot{\eta}^{(s)} + \mathbf{K}_m^{(s)}\eta^{(s)} = \mathbf{f}_m^{(s)} + \mathbf{g}_m^{(s)}, \tag{4.70}$$

where subscript m denotes modal quantities. The matrices are, if the mode shapes are mass normalized,

$$\begin{aligned}
\mathbf{M}_m^{(s)} &= \mathbf{R}^T\mathbf{M}^{(s)}\mathbf{R} = \mathbf{I}\\
\mathbf{C}_m^{(s)} &= \mathbf{R}^T\mathbf{C}^{(s)}\mathbf{R} = \left[\backslash 2\zeta_r\omega_r\backslash\right]\\
\mathbf{K}_m^{(s)} &= \mathbf{R}^T\mathbf{K}^{(s)}\mathbf{R} = \left[\backslash\omega_r^2\backslash\right]\\
\mathbf{f}_m^{(s)} &= \mathbf{R}^T\mathbf{f}^{(s)}\\
\mathbf{g}_m^{(s)} &= \mathbf{R}^T\mathbf{g}^{(s)}.
\end{aligned} \tag{4.71}$$

At this point, let us concatenate the various substructures together in the uncoupled form as

$$\mathbf{M}_m\ddot{\eta} + \mathbf{C}_m\dot{\eta} + \mathbf{K}_m\eta = \mathbf{f}_m + \mathbf{g}_m. \tag{4.72}$$

Compatibility is now enforced with a constraint equation from

$$\mathbf{B}\mathbf{u}_c = \mathbf{0}, \tag{4.73}$$

where **B** is a Boolean matrix of ones, zeros, and negative ones, and for convenience we will include only connection DOF in the displacement vector. Take the partition of Eq. (4.65) for only the connection DOF and again make the modal substitution

$$\mathbf{B}\mathbf{R}_c\eta \cong \mathbf{0}, \tag{4.74}$$

where the subscript c indicates taking only the partition of **R** necessary for the connection DOF.

Up to this point, we have followed the generalized framework rather strictly, but at this point, the transmission simulator method affects the rest of the development. It is assumed that there is an analytical model of the transmission simulator; one uses a truncated set of its mode shapes, $\boldsymbol{\Psi}_c$, as a basis to span the space of the connection motion for each substructure. Now we use the pseudo-inverse (denoted with a superscript +) to project the constraint on the space of the transmission simulator vector space by premultiplying both sides by the block diagonal pseudo-inverse as

$$\boldsymbol{\Psi}_c^{BD+}\mathbf{B}\mathbf{R}_c\boldsymbol{\eta} \cong \boldsymbol{\Psi}_c^{BD+}\mathbf{0}, \tag{4.75}$$

where

$$\boldsymbol{\Psi}_c^{BD+} = \begin{bmatrix} \boldsymbol{\Psi}_c^+ & \mathbf{0} & \mathbf{0} \\ \mathbf{0} & \dots & \mathbf{0} \\ \mathbf{0} & \mathbf{0} & \boldsymbol{\Psi}_c^+ \end{bmatrix} \tag{4.76}$$

will have as many block rows as there are substructures. The right-hand side of Eq. (4.75) is still a vector of zeros, although the number of constraints (rows) is reduced since the matrix $\boldsymbol{\Psi}_c$ is selected so that it always has more DOF than modes. The final constraint is

$$\tilde{\mathbf{B}}\boldsymbol{\eta} = \mathbf{0}, \tag{4.77}$$

where

$$\tilde{\mathbf{B}} = \boldsymbol{\Psi}_c^{BD+}\mathbf{B}\mathbf{R}_c. \tag{4.78}$$

There are multiple reasons for premultiplying by $\boldsymbol{\Psi}_c^{BD+}$. First, it softens the constraint (reduces the number of constraints). The advantage of this is that it gives a least squares fit through the measured motions of the transmission simulator at the DOF to which the constraints will be applied. The mode shapes of the transmission simulator provide a smoothing effect through the measured motions, which always have experimental error. The modified matrix greatly improves the conditioning of the problem. Also, one does not HAVE to use motions measured directly at the attachment points and also does not HAVE to measure rotations, which are inherently carried along in the generalized DOF of the transmission simulator. The generalized modal DOF of the transmission simulator is $\boldsymbol{\gamma}$ in the following

$$\boldsymbol{\Psi}_c\boldsymbol{\gamma} \cong \mathbf{R}_c^{(s)}\boldsymbol{\eta}^{(s)}, \tag{4.79}$$

but can be expressed as

$$\boldsymbol{\gamma} \cong \boldsymbol{\Psi}_c^+\mathbf{R}_c^{(s)}\boldsymbol{\eta}^{(s)}, \tag{4.80}$$

which can be seen in the left-hand side of Eq. (4.75) as converting the constraint to the transmission simulator modal coordinates.

Now continue the development in the general framework from Eq. (4.77). We perform another modal-like substitution with

$$\eta = \tilde{\mathbf{L}}\mathbf{q}. \tag{4.81}$$

Substitute Eq. (4.81) into Eq. (4.77) to give

$$\tilde{\mathbf{B}}\tilde{\mathbf{L}}\mathbf{q} = \mathbf{0}. \tag{4.82}$$

If one chooses $\tilde{\mathbf{L}}$ such that it is in the null space of $\tilde{\mathbf{B}}$, then Eq. (4.82) is guaranteed to be satisfied because

$$\tilde{\mathbf{B}}\tilde{\mathbf{L}} = \mathbf{0}. \tag{4.83}$$

All the rows of $\tilde{\mathbf{B}}$ are orthogonal to all the columns of $\tilde{\mathbf{L}}$. Since $\tilde{\mathbf{B}}$ is known, a one line command in MATLAB can provide $\tilde{\mathbf{L}}$. Substituting Eq. (4.81) back into the uncoupled equations of motion in Eq. (4.72) and premultiplying by $\tilde{\mathbf{L}}^T$ gives

$$\tilde{\mathbf{L}}^T\mathbf{M}_m\tilde{\mathbf{L}}\ddot{\mathbf{q}} + \tilde{\mathbf{L}}^T\mathbf{C}_m\tilde{\mathbf{L}}\dot{\mathbf{q}} + \tilde{\mathbf{L}}^T\mathbf{K}_m\tilde{\mathbf{L}}\mathbf{q} = \tilde{\mathbf{L}}^T\mathbf{f}_m + \tilde{\mathbf{L}}^T\mathbf{g}_m, \tag{4.84}$$

which couples the equations of motion, reducing the number of rows in Eq. (4.72) by the number of constraints (rows) in Eq. (4.77). This leads to the primal coupling formulation in this framework. In this formulation, $\tilde{\mathbf{L}}^T\mathbf{g}_m = \tilde{\mathbf{L}}^T\mathbf{R}^T\mathbf{g} = \mathbf{0}$, since the rows of $\tilde{\mathbf{L}}^T$ are orthogonal to a linear combination of the columns of $\mathbf{R}^T$, leaving

$$\tilde{\mathbf{M}}_m\ddot{\mathbf{q}} + \tilde{\mathbf{C}}_m\dot{\mathbf{q}} + \tilde{\mathbf{K}}_m\mathbf{q} = \tilde{\mathbf{f}}_m, \tag{4.85}$$

where

$$\begin{aligned} \tilde{\mathbf{M}}_m &= \tilde{\mathbf{L}}^T\mathbf{M}_m\tilde{\mathbf{L}} \\ \tilde{\mathbf{C}}_m &= \tilde{\mathbf{L}}^T\mathbf{C}_m\tilde{\mathbf{L}} \\ \tilde{\mathbf{K}}_m &= \tilde{\mathbf{L}}^T\mathbf{K}_m\tilde{\mathbf{L}} \\ \tilde{\mathbf{f}}_m &= \tilde{\mathbf{L}}^T\mathbf{f}_m \end{aligned}. \tag{4.86}$$

Frequency-Based Substructuring Dual Formulation In The General Framework

Again following Klerk et al. (2008), for the physical DOF, the uncoupled equations of motion, the compatibility and the equilibrium are written as

$$\mathbf{M}\ddot{\mathbf{u}} + \mathbf{C}\dot{\mathbf{u}} + \mathbf{K}\mathbf{u} = \mathbf{f} + \mathbf{g} \tag{4.87}$$

$$\mathbf{B}\mathbf{u} = \mathbf{0} \tag{4.88}$$

$$\mathbf{L}^T\mathbf{g} = \mathbf{0}, \tag{4.89}$$

where **M**, **C**, and **K** are block diagonals with as many blocks as substructures. The constraint forces **g** between the substructures can be written as

$$\mathbf{g} = -\mathbf{B}^T\boldsymbol{\lambda} \tag{4.90}$$

where $\boldsymbol{\lambda}$ are Lagrange multipliers corresponding physically to the interface forces. Equations (4.87) and (4.88) can now be written in matrix form as

$$\begin{bmatrix} \mathbf{M} & \mathbf{0} \\ \mathbf{0} & \mathbf{0} \end{bmatrix} \begin{bmatrix} \ddot{\mathbf{u}} \\ \boldsymbol{\lambda} \end{bmatrix} + \begin{bmatrix} \mathbf{C} & \mathbf{0} \\ \mathbf{0} & \mathbf{0} \end{bmatrix} \begin{bmatrix} \dot{\mathbf{u}} \\ \boldsymbol{\lambda} \end{bmatrix} + \begin{bmatrix} \mathbf{K} & \mathbf{B}^T \\ \mathbf{B} & \mathbf{0} \end{bmatrix} \begin{bmatrix} \mathbf{u} \\ \boldsymbol{\lambda} \end{bmatrix} = \begin{bmatrix} \mathbf{f} \\ \mathbf{0} \end{bmatrix}. \tag{4.91}$$

Taking the Fourier transform to bring this into the frequency domain, where each quantity is now a function of frequency, gives

$$\begin{bmatrix} \mathbf{Z} & \mathbf{B}^T \\ \mathbf{B} & \mathbf{0} \end{bmatrix} \begin{bmatrix} \bar{\mathbf{u}} \\ \bar{\boldsymbol{\lambda}} \end{bmatrix} = \begin{bmatrix} \bar{\mathbf{f}} \\ \mathbf{0} \end{bmatrix}, \tag{4.92}$$

where $\mathbf{Z}$ is the block diagonal impedance matrix resulting from the mass, stiffness, and damping. The frequency response function matrix, $\mathbf{H}$, which is often measured experimentally is the inverse of $\mathbf{Z}$. The dual formulation coupled formulation in terms of $\mathbf{H}$ is derived from Eq. (4.92) by eliminating $\bar{\boldsymbol{\lambda}}$, yielding

$$\bar{\mathbf{u}} = \mathbf{H}\bar{\mathbf{f}} - \mathbf{H}\mathbf{B}^T(\mathbf{B}\mathbf{H}\mathbf{B}^T)^{-1}\mathbf{B}\mathbf{H}\bar{\mathbf{f}}. \tag{4.93}$$

The transmission simulator method modifies the $\mathbf{B}$ matrix in Eq. (4.88) as

$$\boldsymbol{\Psi}^{BD+}\mathbf{B}\bar{\mathbf{u}} \cong \mathbf{0} \tag{4.94}$$

So one has

$$\hat{\mathbf{B}}\bar{\mathbf{u}} = \mathbf{0}, \tag{4.95}$$

where

$$\hat{\mathbf{B}} = \boldsymbol{\Psi}^{BD+}\mathbf{B}. \tag{4.96}$$

Now simply substitute Eq. (4.96) into Eq. (4.93) to give the frequency-based transmission simulator equation as

$$\bar{\mathbf{u}} = \mathbf{H}\bar{\mathbf{f}} - \mathbf{H}\hat{\mathbf{B}}^T(\hat{\mathbf{B}}\mathbf{H}\hat{\mathbf{B}}^T)^{-1}\hat{\mathbf{B}}\mathbf{H}\bar{\mathbf{f}}. \tag{4.97}$$

The $\hat{\mathbf{B}}$ matrix transforms the physical connection DOF frequency response functions (FRFs) into generalized DOF FRFs cast on the space of the transmission simulator mode shapes. This collapses the size of the physical connection DOF FRFs down to the size of the number of modes of the transmission simulator, providing some least squares smoothing, and makes the matrix inversion in Eq. (4.97) much better conditioned.

► *Example Combining Experimental Plate-Beam Substructure with Analytical Cylinder Substructure Using the CMS approach*

The example is based on the hardware and analytical finite element models depicted in the first three figures. In the equations below, the finite element (FE) substructure is considered to have the analytical model of the transmission simulator attached, and the experimental (EXP) substructure has the physical transmission simulator attached. Two transmission simulators must be subtracted. In this example, 100 modes were utilized from the finite element substructure, 25 modes from the experimental substructure, and 18 modes from the transmission simulator giving 143 uncoupled equations of motion. Consider only the eigenvalue equations of motion, and then the final coupled modal parameters can be used to analytically form any desired full system response FRF. Damping will be predicted in a simplified method later. The undamped frequency-domain equations of motion using modal coordinates when forces are removed are

$$\begin{bmatrix} \omega_{FE}^2 & \mathbf{0} & \mathbf{0} \\ \mathbf{0} & \omega_{EXP}^2 & \mathbf{0} \\ \mathbf{0} & \mathbf{0} & -2\omega_{TS}^2 \end{bmatrix} \begin{Bmatrix} \eta_{FE} \\ \eta_{EXP} \\ \eta_{TS} \end{Bmatrix} - \omega^2 \begin{bmatrix} \mathbf{I}_{FE} & \mathbf{0} & \mathbf{0} \\ \mathbf{0} & \mathbf{I}_{EXP} & \mathbf{0} \\ \mathbf{0} & \mathbf{0} & -2\mathbf{I}_{TS} \end{bmatrix} \begin{Bmatrix} \ddot{\eta}_{FE} \\ \ddot{\eta}_{EXP} \\ \ddot{\eta}_{TS} \end{Bmatrix} = \mathbf{0} \tag{4.98}$$

and the physical displacements, $\mathbf{y}$, on each substructure are

$$\begin{Bmatrix} \mathbf{y}_{FE} \\ \mathbf{y}_{EXP} \\ \mathbf{y}_{TS} \end{Bmatrix} = \begin{bmatrix} \mathbf{R}_{FE} & \mathbf{0} & \mathbf{0} \\ \mathbf{0} & \mathbf{R}_{EXP} & \mathbf{0} \\ \mathbf{0} & \mathbf{0} & \mathbf{R}_{TS} \end{bmatrix} \begin{Bmatrix} \eta_{FE} \\ \eta_{EXP} \\ \eta_{TS}, \end{Bmatrix} \tag{4.99}$$

where $\mathbf{R}$ are a truncated set of the mass normalized mode shapes of each substructure coming from experiment or analysis. The motion of the transmission simulator in all three substructures should be the same when all the substructures are coupled, so two sets of physical constraints can be written

$$\mathbf{y}_{FEmeas} = \mathbf{y}_{EXPmeas} \text{ and } \mathbf{y}_{EXPmeas} = \mathbf{y}_{TSmeas}. \tag{4.100}$$

Just consider the first of these constraints, invoke the modal substitution from Eq. (4.99), and premultiply by the pseudo-inverse of transmission simulator mode shapes, $\mathbf{R}_{TS}^{+}$to give

$$\mathbf{R}_{TS}^{+}\mathbf{R}_{FEmeas}\eta_{FE} = \mathbf{R}_{TS}^{+}\mathbf{R}_{EXPmeas}\eta_{EXP}. \tag{4.101}$$

Or moving everything to the left-hand side gives

$$\mathbf{R}_{TS}^{+}\mathbf{R}_{FEmeas}\eta_{FE} - \mathbf{R}_{TS}^{+}\mathbf{R}_{EXPmeas}\eta_{EXP} = \mathbf{0}. \tag{4.102}$$

A similar process can be applied to the second constraint of Eq. (4.100). The constraints can now be written in the form given in Eqs. (4.76)–(4.78) as

$$\begin{bmatrix} \mathbf{R}_{TS}^{+} & \mathbf{0} \\ \mathbf{0} & \mathbf{R}_{TS}^{+} \end{bmatrix} \begin{bmatrix} \mathbf{I} & -\mathbf{I} & \mathbf{0} \\ \mathbf{0} & \mathbf{I} & -\mathbf{I} \end{bmatrix} \begin{bmatrix} \mathbf{R}_{FEmeas} & \mathbf{0} & \mathbf{0} \\ \mathbf{0} & \mathbf{R}_{EXPmeas} & \mathbf{0} \\ \mathbf{0} & \mathbf{0} & \mathbf{R}_{TS} \end{bmatrix} \begin{Bmatrix} \eta_{FE} \\ \eta_{EXP} \\ \eta_{TS} \end{Bmatrix} = \begin{Bmatrix} \mathbf{0} \\ \mathbf{0}, \end{Bmatrix} \tag{4.103}$$

where $\boldsymbol{\Psi}_c^{BD+}$ in Eq. (4.75) is $\mathbf{R}_c^{BD+}$ and $\mathbf{B}$ is the Boolean matrix here, thus

$$\tilde{\mathbf{B}} = \begin{bmatrix} \mathbf{R}_{TS}^{+} & \mathbf{0} \\ \mathbf{0} & \mathbf{R}_{TS}^{+} \end{bmatrix} \begin{bmatrix} \mathbf{I} & -\mathbf{I} & \mathbf{0} \\ \mathbf{0} & \mathbf{I} & -\mathbf{I} \end{bmatrix} \begin{bmatrix} \mathbf{R}_{FEmeas} & \mathbf{0} & \mathbf{0} \\ \mathbf{0} & \mathbf{R}_{EXPmeas} & \mathbf{0} \\ \mathbf{0} & \mathbf{0} & \mathbf{R}_{TS} \end{bmatrix}. \tag{4.104}$$

Repeating Eq. (4.81) and (4.82) here for clarity gives

$$\eta = \tilde{\mathbf{L}}\mathbf{q} \tag{4.81}$$

and

$$\tilde{\mathbf{B}}\tilde{\mathbf{L}}\mathbf{q} = \mathbf{0} \tag{4.82}$$

and $\tilde{\mathbf{L}}$ is the null space of $\tilde{\mathbf{B}}$, which is known. Substitute Eq. (4.81) into Eq. (4.98) and premultiply by $\tilde{\mathbf{L}}^T$ to give

$$\mathbf{L}^T \begin{bmatrix} \omega_{FE}^2 & 0 & 0 \\ 0 & \omega_{EXP}^2 & 0 \\ 0 & 0 & -2\omega_{TS}^2 \end{bmatrix} \mathbf{L}\mathbf{q} - \omega^2 \mathbf{L}^T \begin{bmatrix} \mathbf{I}_{FE} & \mathbf{0} & \mathbf{0} \\ \mathbf{0} & \mathbf{I}_{EXP} & \mathbf{0} \\ \mathbf{0} & \mathbf{0} & -2\mathbf{I}_{TS} \end{bmatrix} \mathbf{L}\ddot{\mathbf{q}} = \mathbf{0}, \tag{4.105}$$

providing the coupled equations from which the eigenvalue problem can be solved yielding. The solution will provide a set of eigenvectors, $\boldsymbol{\Phi}$, frequencies, $\boldsymbol{\omega}^2$, and modal coordinates, $\boldsymbol{\beta}$. Now the coupled displacements will be

$$\mathbf{y} = \mathbf{R}\mathbf{L}\boldsymbol{\Phi}\boldsymbol{\beta}. \tag{4.106}$$

The resulting new damping matrix is formed by

$$\boldsymbol{\Phi}^T \mathbf{L}^T \begin{bmatrix} 2\zeta_{FE}\omega_{FE} & 0 & 0 \\ 0 & 2\zeta_{EXP}\omega_{EXP} & 0 \\ 0 & 0 & -4\zeta_{TS}\omega_{TS} \end{bmatrix} \mathbf{L}\boldsymbol{\Phi}, \tag{4.107}$$

from which we usually just take the diagonal values to give $2\zeta_{new}\omega_{new}$ (Fig. 4.16).

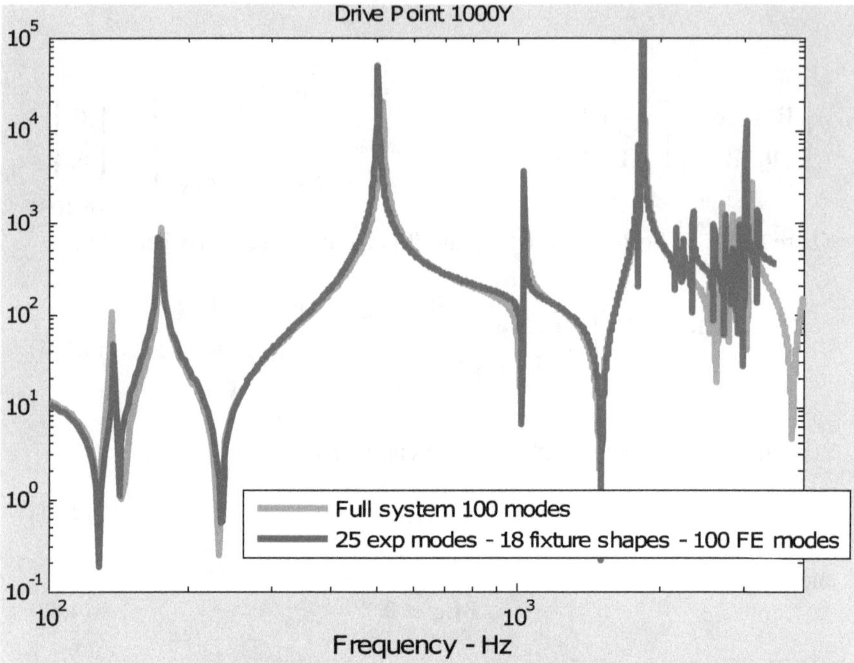

Fig. 4.16 Transmission simulator method FRF of coupled system (red) versus truth model (blue)—Driving point response at tip of beam in Fig. 4.14

In Fig. 4.16, one can see the axial FRF at the tip of the beam (see Fig. 4.14). The red FRF was constructed from the modal parameters of the coupled system mode shapes, frequencies, and damping using the transmission simulator method with the CMS approach. The blue FRF represents the truth data constructed from the modal parameters of a highly validated finite element model of the full system.

◄

► *Example Combining Experimental Plate/Beam Substructure with Analytical Cylinder Substructure Using the FBS approach*

Generally, the author does not execute Eq. (4.97) in a single step, since it makes the matrices very large and the resulting inversions are computationally too intensive. For this example, assume we have one step where systems C and D in the figure are too be joined. (Another step can be taken to subtract the transmission simulators, which will not be done here). Define **HC** and **HD** as the frequency response function matrices for substructures C and D, respectively. **HT** is the FRF matrix of the total system after C and D are coupled. Each substructure has a two-dimensional FRF matrix for every frequency line of the FRFs. The first subscript on any of these matrices represents the output response DOF and the second subscript represents the input force DOF. Perhaps, the two most useful equations from partitions of the classical method in Eq. (4.93) are

$$\mathbf{HT}_{ri} = \mathbf{HD}_{rc} \left(\mathbf{HD}_{cc} + \mathbf{HC}_{cc}\right)^{-1} \mathbf{HC}_{ci}, \tag{4.108}$$

where the force input is on substructure C and the response output is on substructure D, and

$$\mathbf{HT}_{ri} = \mathbf{HC}_{ri} - \mathbf{HC}_{rc} \left(\mathbf{HD}_{cc} + \mathbf{HC}_{cc}\right)^{-1} \mathbf{HC}_{ci}, \tag{4.109}$$

where the force input is on substructure C and the response output is also on substructure C. Here, the subscript r represents the output response and the subscript i represents the input force, and the subscript c represents the connection DOF between the two substructures. With the transmission simulator method, the Boolean matrix $\mathbf{B}$ is replaced with $\hat{\mathbf{B}}$ to convert all the connection DOF to modal DOF of the transmission simulator in Eq. (4.97), so that

$$\mathbf{HD}_{rc} = \mathbf{HD}_{rp}\mathbf{R}_{TS}^{T+} \tag{4.110}$$

$$\mathbf{HD}_{cc} = \mathbf{R}_{TS}^{+}\mathbf{HD}_{pp}\mathbf{R}_{TS}^{T+} \tag{4.111}$$

$$\mathbf{HC}_{cc} = \mathbf{R}_{TS}^{+}\mathbf{HC}_{pp}\mathbf{R}_{TS}^{T+} \tag{4.112}$$

$$\mathbf{HC}_{ci} = \mathbf{R}_{TS}^{+}\mathbf{HC}_{pi} \tag{4.113}$$

$$\mathbf{HC}_{rc} = \mathbf{HC}_{rp}\mathbf{R}_{TS}^{T+} \tag{4.114}$$

in Eqs. (4.113)–(4.114), the subscript p represents the DOF at the transmission simulator measurement locations on either substructure. The pseudo-inverse of the mode shape matrix $\mathbf{R}_{TS}$ of the transmission simulator reduces the size of the physical measurement FRF matrices down to the number of modes kept for the transmission simulator at all the connection DOFs.

◄

4.5 Measurement Methods for Substructuring

When developing experimental substructures, the measurements can "make or break" the experimental model. The requirements for the measurements are much more exacting than those for a modal test being used for model validation for which mode shapes may be acceptable with a bias error in the modal scaling. In the case of substructuring, such errors have a more dramatic effect when coupled to another structure. The FRF measurement importance can be appreciated when one considers FBS with FRFs. Basically, at any frequency line, one is coming up with a weighted sum of the substructure FRFs. Two great principles for the case of real modes arise from this. If a frequency line is supposed to have a resonant response in the coupled system, then the real part of the coupled FRF has to go to zero, and the imaginary part of the FRF will be a drastic amplification (on the order of the quality of the resonance, $Q = 1/(2 * \zeta)$). Restating this, if one is to get the coupled natural frequency correctly, the real part of the connection FRFs must be accurate. If one is to get the coupled amplitude of the resonant FRF correctly, the imaginary part of all the FRFs of concern must be accurate. Consider the real part of a FRF obtained from an impact test. On first consideration in the left figure below, it look good. Let us assume that when the substructures are coupled, the first resonance will drop down to about 100 Hz. When one zooms in on the real part between 0 and 300 Hz, one sees that there is a significant amount of uncertainty in the real part of the FRF. This could affect the substructured FRF so that it crosses zero at the wrong place, or even multiple places, yielding the resonant response at wrong frequencies (Figs. 4.17 and 4.18).

Consider the imaginary portion of the drive point FRF below. The response at left looks good. However, if a new mode is destined for the region just above 3000 Hz, look at the zoomed response in that bandwidth on the right. In theory, a drive point FRF should never have an imaginary part less than zero. Clearly, there are many frequencies with response below zero. This could not only give inaccurate estimate

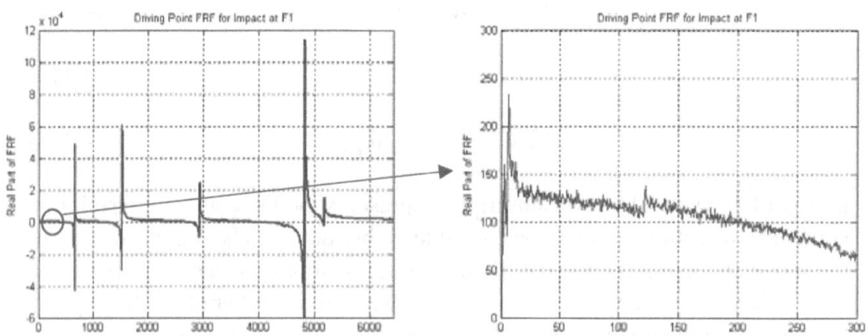

Fig. 4.17 Sample FRF showing contamination of the real parts

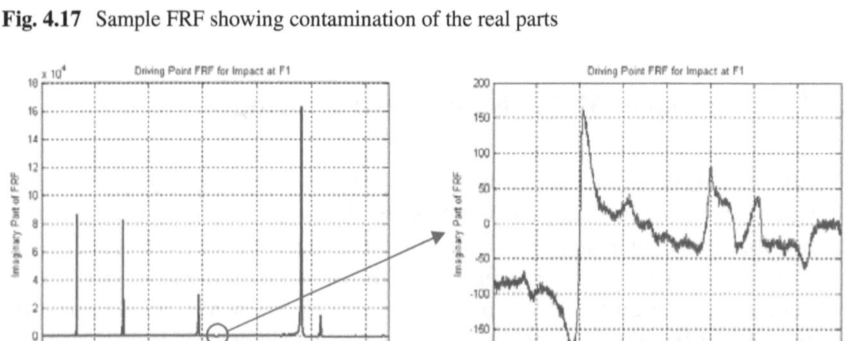

Fig. 4.18 Sample FRF showing contamination of the imaginary parts

of the amplitude of the imaginary portion of the FRF, but even the wrong sign! For this reason, structures with higher damping are typically easier to perform FBS than lightly damped structures as shown here.

1. **Response Sensor Measurement Errors and their Sources**
   Here is a list of common sources of measurement error, approximately ranked for probability of occurrence. Also included are some recommendations for correcting these.

   a. **Bookkeeping errors**
      (Wrong sign, wrong DOF, crossed wires, wrong sensitivity/gain, local coordinate system errors) Administrative double checks by multiple people on the mounting, channel table, label checks, and coordinate system definitions can help eliminate these. For "free" configuration tests, a great engineering check to identify such errors is to compare the low-frequency real part response of each FRF with the analytical response for that DOF using just the rigid body modes. Usually, they will be within about 10% of each other.
   b. **Choosing too wide a bandwidth**
      In general, the difficulty of developing an accurate experimental model increases with frequency. Minimize the bandwidth as much as possible to meet the requirement. Push back when customers ask for "the moon."
   c. **Undetected overloads, especially in drive point sensors**
      Undetected overloads where the sensor response, amplifier response, or the band edge of the DAQ hardware is beyond, the rated range can ruin or pollute the data, but the digital anti-aliasing filter can disguise it so that it looks "fine". Set the DAQ and filter bandwidths wide open, and collect at-level data in the time domain to verify that the sensors or signal conditioning equipment is not being driven beyond its rated voltage range (a typical integrated electronics sensor maximum is 5 volts)
   d. **Cross-axis sensitivity**
      In very lightly damped systems, the out-of-axis response to a strong resonance will have a peak in the real part and the imaginary part will cross zero, out of phase with a normal acceleration response. This is not easily corrected.
   e. **Base strain sensitivity**
      Base strain sensitivity allows strain in the surface of the structure to which the accelerometer is mounted to produce false acceleration voltage output. It can sometimes be observed as a large uncharacteristic low-frequency response sloping downward as frequency increases. This can be remedied by putting the accelerometer on a block, or (less preferably) on tape with an adhesive layer.
   f. **Mounting or mounting block resonances**
      At some high frequency, tape-mounted accelerometers will see phase shift from the true response and block-mounted accelerometers will have a mode of the block mass oscillating locally on the structure. If possible, testing an accelerometer mounted without tape and comparing the FRF mounted on

the tape will provide observation of this error. Mounting block resonances will change frequency if the size of the mounting block is changed. Beware of mounting accelerometers on a screw head or a nut as one may measure the fastener response instead of the desired structure response.

g. **Sensors with wrong amplitude sensitivity or broken sensors**
Sensors that are not responding can often be diagnosed by observing their time- or frequency-domain response, especially in relation to the noise floor with no stimulus applied. Sensors that have wrong sensitivity, for example, from a cracked crystal that is still partially functional, are difficult to identify by simple observation of the response. The check in the Bookkeeping errors section above is of value here for free systems. Another similar check for free systems is to excite the structure at a low frequency below the elastic modes in each rigid body direction and extract the real part of the FRF at that frequency as a shape. The shape should be an appropriate rigid body response shape.

2. **The Most Critical Sensor Locations for Substructuring**
By far the most critical sensors for experimental substructures are the connection DOF sensors. Those are the ones used in computing inverses or pseudo-inverses and ultimately determine the coupled modal frequencies as well as having large bearing on the amplitude of the coupled responses.

a. **Drive point sensors are the most critical**
In FBS, all the connected DOFs require drive point responses. In CMS substructuring, the drive point FRFs determine the scaling of the mode shapes, so they are very important. A low-frequency check of each connection DOF response as previously mentioned above in bookkeeping errors is a great first check of these FRFs for free substructures. If one performs the substructuring and the rigid body modes are appropriately coupled, one should obtain the correct coupled low-frequency response. There is no approximation with coupling of rigid body modes and no singularities if mass properties are appropriate, so this should always work. As mentioned before, if there are errors in this coupling, it may help identify bookkeeping or some other error in the setup.

b. **Neglect connection DOF (including rotations) to your peril**
As previously mentioned, 3/4 of the FRF matrices contain rotation terms. In many cases, rotations cannot be neglected without significant degradation of the results. Approximation of the connection DOF is better than elimination. The transmission simulator method inherently includes rotation estimates. The virtual point transformation is another approach to include the estimate of rotations.

c. **For CMS substructuring check synthesized FRFs against measured data for each substucture**
Once the measured FRFs are obtained, the modal parameters must be extracted for CMS substructuring. Errors in the extraction will cause detriment to the substructured results. A good first check is to compare the ana-

lytical Complex Mode Indicator Function (CMIF) with the CMIF from the measured data. They should compare "well". One can compare each analytical FRF with its measured counterpart as well.
In particular, the drive point FRF from which certain modes are extracted is very important. The drive point mode shape and modal mass are estimated from the drive point FRF. If the modal mass is not estimated correctly, the drive point mode shape will be estimated inversely proportional in error. The same is true if the damping is estimated incorrectly. If the drive point shape is in error, generally all the cross point mode shapes will be inversely proportional in error to the drive point mode shape error. This can be observed by considering the fit to the FRF with an error considered in each of the various terms singly (modal mass, damping, drive point shape, cross point shape). See the standard real modes FRF equation below:

$$H_{ij}(\omega) = \sum_{r=1}^{N} \frac{-\omega^2 \phi_i \phi_j}{m_r(\omega_r^2 - \omega^2 + 2j\zeta_r\omega\omega_r)}.$$

3. **Placing sensors to minimize theoretical errors with the transmission simulator**
The transmission simulator method is very amenable to analysis using the free modes of the FE model of the transmission simulator. One chooses at least 1.5 times the number of sensors as the transmission simulator has modes in the desired bandwidth. The mode shape matrix of a large number (perhaps three to ten times the number of modes) of candidate DOF is obtained. Then one optimizes down to the desired number of sensors with an algorithm to minimize the condition number of the mode shape matrix. This can be done iteratively. Condition numbers around 3–4 have been used successfully with the final sensor set. Since the pseudo-inverse of the mode shape matrix is required, the condition number is a good metric to ensure the matrix is easily invertible. Remember that these shapes may all be translations as the modal DOF that are inferred from these calculations will inherently contain the rotations as well.
4. **Force Sensor Measurement Errors and their Sources**
Drive point measurements are the ones from which modal mass is determined from the ratio of force/acceleration. Therefore, the force must be measured accurately.

    a. **The force is assumed to be co-located with the drive point response accelerometer.**
    Often this can be approximated with a force that is co-linear with the accelerometer. Often fixtures can greatly increase the probability that the force is co-linear with the accelerometer. For example, force caps over the accelerometer can be drilled to receive the force gage (for shaker testing) or with a normal surface and markings for hammer impact. Stubs or blocks can be incorporated in the transmission simulator fixture to allow for applying non-normal forces.

b. **Use of shaker versus hammer for force measurements**
Shakers and impact hammers each have advantages and disadvantages for different situations. Here are some of the advantages of each with the disadvantages of the other listed beneath.
   i. **A shaker can produce a more linear result and better signal-to-noise ratio for the same peak force on a nonlinear system**
   A hammer will produce an FRF that is less easily fit with modal parameters because of additional nonlinear distortion. In the CMIF below, one can see the chop just to the right of the 120 and 500 Hz resonances showing the nonlinear distortion of the typical FRFs due to hammer excitation.
   ii. **A shaker has more precise input direction, location and amplitude**
   A handheld hammer may miss the target, induce shear as well as normal force on the strike and has about three times more uncertainty in the calibration sensitivity. Pendulum fixtures may improve the impact location and direction.
   iii. **A hammer can provide much higher frequency input**
   The shaker begins to uncouple above the frequency of the armature resonance. The stinger also has lateral modes that uncouple the response and can also introduce false modes into the FRF.
   iv. **Fixtures associated with a shaker may provide unmeasured mass**
   The force gage associated with a shaker, with its additional fixturing, can also pollute the FRF by adding undesirable shear and rotation mass, which may be very significant for small test structures.
   v. **A hammer provides strong input at the resonances**
   The shaker force will tend to drop at the resonances, especially for lightly damped systems. For example, see Fig. 4.22a, where the circled regions are the force input at the resonant frequencies of the system. The autospectrum from a hammer does not drop out at resonance, as shown in Fig. 4.22b (Figs. 4.19, 4.20 and 4.21).

4.6 Virtual Point Transformation

4.6.1 Interface Modeling

Dynamic substructuring requires that compatibility is satisfied between the displacement of the interfaces of two components, and that the interconnecting forces are in equilibrium. This is easily established for a discrete dynamical system using the coupling equations as outlined in Chaps. 2 and 3, namely, by imposing the two coupling conditions on the respective coupling nodes of the substructures. In practice,

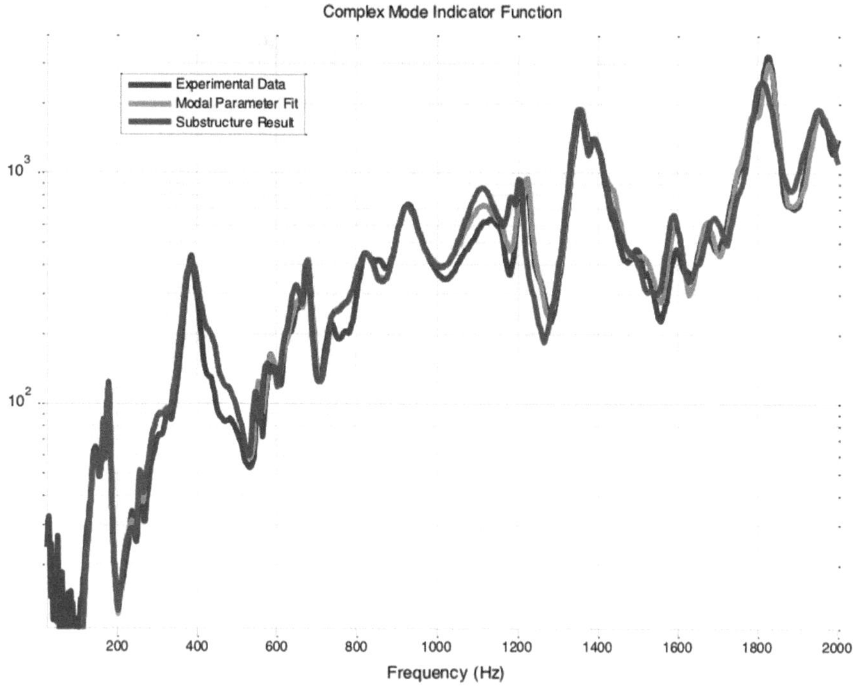

Fig. 4.19 Complex mode indicator function example. The figure compares the first singular value in a CMIF for the experimental data, the fit modal model, and a substructuring prediction

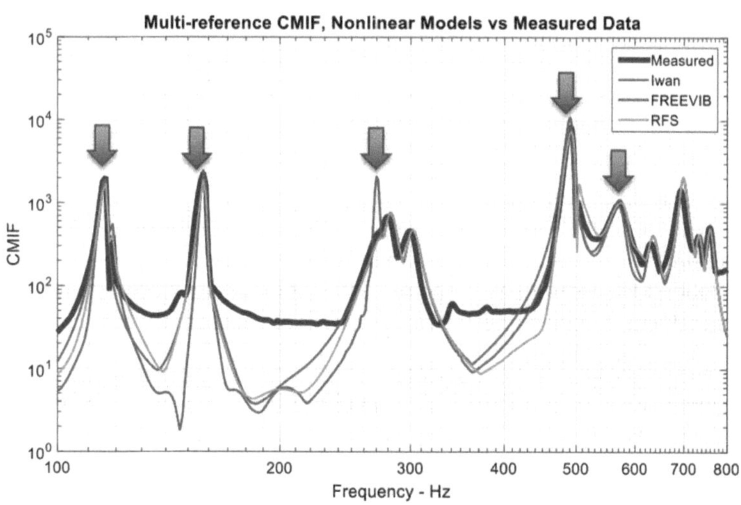

Fig. 4.20 Typical nonlinear distortions

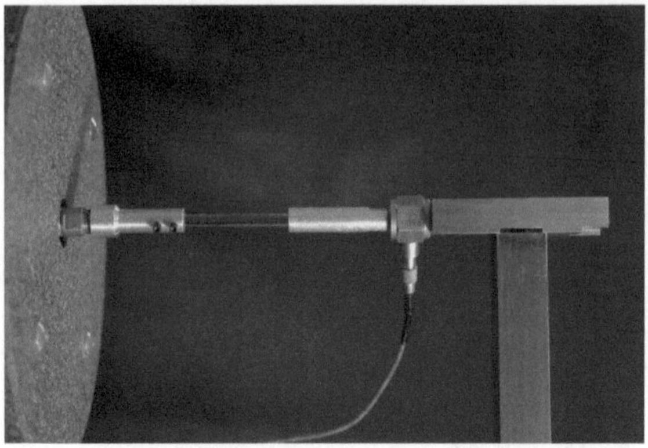

Fig. 4.21 Force gauges may induce a non-negligible shear force

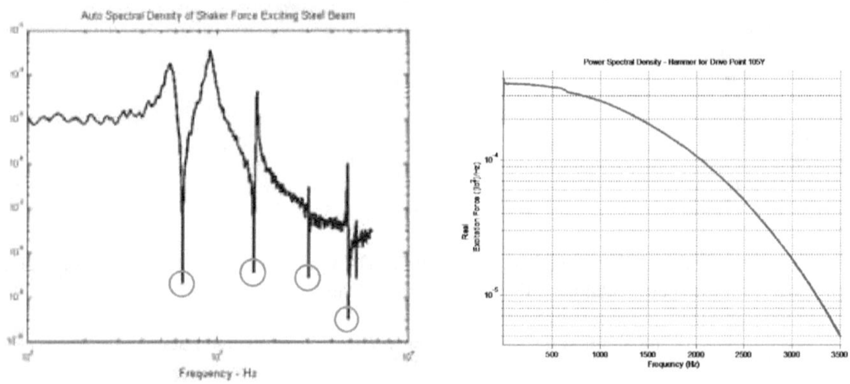

Fig. 4.22 Shaker autospectrum (left) versus hammer autospectrum (right)

however, components are often connected by bolts, welds, or adhesives, that physically show more resemblance to a line or surface connection than to a single point. Modeling such connectivity by experimental measurements requires more thoughtful contemplation of what is really going on at the interfaces between two structures.

In general, one can distinguish two categories of connectivity between substructures, of which one is normally the most appropriate for a given case:

1. The components are connected at a handful of points that can be regarded as discrete nodes (Fig. 4.23a). These points are permitted to displace with respect to each other, thereby allowing for significant deformations in the structures. The connection points themselves, however, behave fairly rigid in the area near the interfaces. Deformations in the connecting points are of minor order and not contributing to the coupled vibrations. Examples are truss structures in bridges and a steering system mounted in a vehicle subframe.

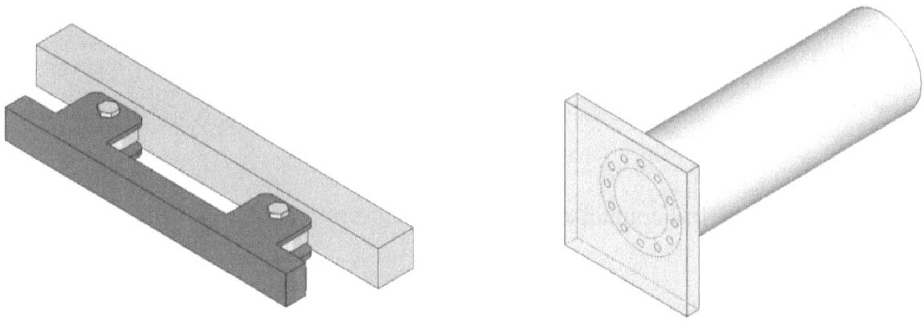

(a) Discrete interface points. (b) Continuous interface surface.

Fig. 4.23 Two examples of substructure connectivity

2. The interface resembles a continuous line or surface (Fig. 4.23b). The deformation of the interface is strongly coupled to the internal vibrations of the structure, and can therefore not be discretized to a small number of nodes. Examples include the coupling of a gearbox to an engine, glued carbon-fiber panels of a vehicle's bodywork and the partitions of an aircraft's fuselage.

This next section focuses on the discrete interface situation. Each connection point can be appointed linear and rotational DOFs for the nodal displacement and forces, which have a clear physical interpretation. In order to solve the second situation, one would instead define a set of deformation modes for the interface region and assign DOFs to their modal amplitudes and forces.

Discrete Interface Modeling

Let us consider a single bolted connection as illustrated in Fig. 4.24. The pretension in the bolt causes friction to occur at the interfaces, which prohibits the two structures to have relative motion with respect to each other. If both structures were represented by an FE model, the substructuring task amounts to coupling the displacements and interface forces of the coinciding nodes, as illustrated in Fig. 4.24b. By coupling a sufficient number of nodes over a larger area, any rotational coupling is implicitly accounted for.

Let us now consider the procedure of experimental modeling. An admittance measurement needs to be conducted on each substructure in order to obtain the respective FRF matrices. To facilitate substructure coupling, the simplest way would be to mount a single triaxial accelerometer on the connection point of each substructure and excite in X-, Y-, and Z-directions close to the sensor. This yields a 3-DOF description which is sometimes called a Single-Point Connection (SPC). In this way, one obtains a 3×3 FRF matrix for the substructure's interfaces, which can be used for substructure coupling.

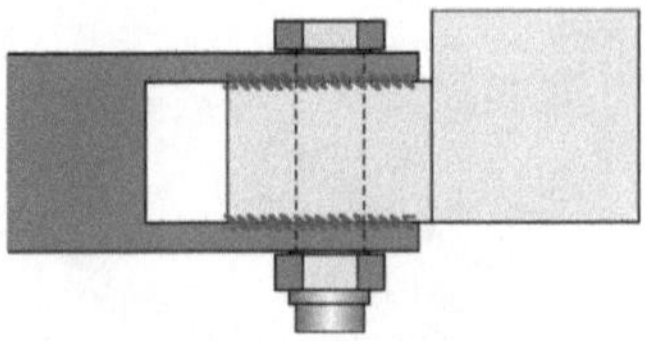

(a) Friction at the full contact area.

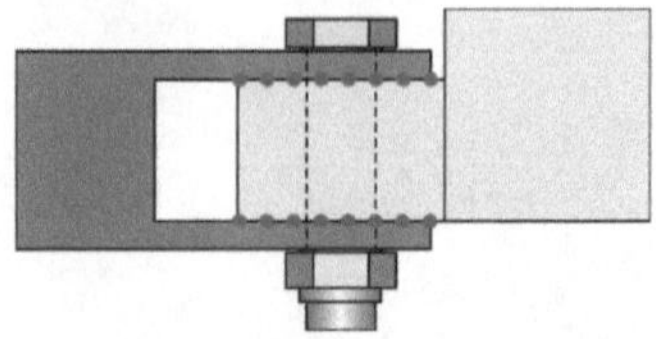

(b) Discrete coupling with coinciding nodes.

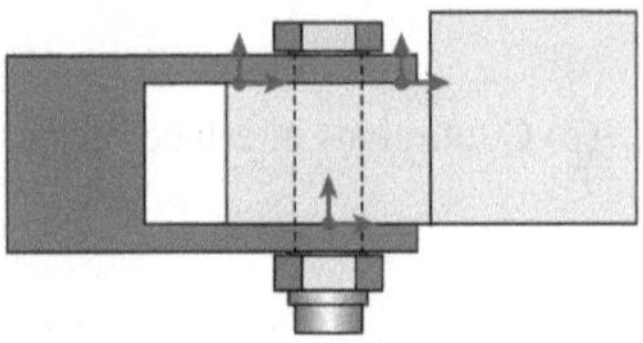

(c) EMPC: multiple points, only translational directions (9 DoFs).

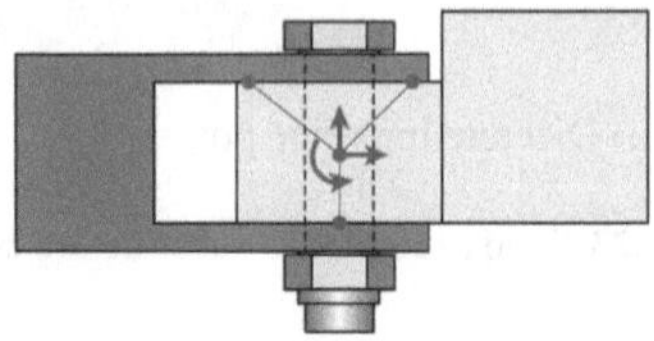

(d) Virtual point transformation: 6 collocated translations and rotations (9 $\rightarrow$ 6 DoFs).

Fig. 4.24 A bolted connection: four ways to consider the interface problem

The SPC method does not account for rotational coupling, which is known to be essential in many cases of substructuring Liu and Ewins (1999), Montalvao et al. (2004). Measurement and excitation (or in fact observation and controllability) of rotations has been an active topic of research among experimentalist and has led to several approaches:

1. *Direct measurement*. Rotation sensors and transducers can be thought of, but have not been the practical standard over the past. Some examples of unconventional hardware for rotation measurement or excitation can be found in Petersson (1987), Bello et al. (2003), Liu et al. (2013).
2. *Modal expansion*. If modal information of the structure is available, for instance, from static FEM analysis, rotations can be obtained from extrapolation of the measured data. This is the concept of the system equivalent reduction and expansion process (SEREP) O'Callahan et al. (1989).
3. *Finite differences*. Rotations can be approximated from translations measured at known distances from each other Duarte and Ewins (2000), Bregant and Sandersons (2000). Simple finite difference approaches such as $\theta = (u_0 - u_d)/d$ may suffice for simple plate or beam problems Sanderson and Fredö (1995), Elliott et al. (2012), but can be cumbersome for obtaining full 6-DOF kinematics.
4. *Multiple translations*. The equivalent multi-point connection (EMPC) method Pasteuning (2007), de Klerk et al. (2008) accounts for the rotations *implicitly*, by coupling translational FRFs of multiple points in the proximity of the interface. This is illustrated in Fig. 4.24c. In practice, a minimum of three triaxial accelerometers (not in line) is sufficient to couple all translational and rotational

directions,[6] introducing a total of NINE DOFs per coupling point. However, this condition can be too stringent: if the structure between the three connection points is very stiff, any discrepancy in motion (due to measurement errors) will be overcompensated for in the coupling equations. This unwanted "stiffening" of the interface results in spurious peaks in the coupled FRFs Rixen (2008).

5. *IDM filtering.* By defining six rigid interface displacement modes[7] (IDMs) per connection point and projecting the 9-DOF (or more) admittance matrix onto this subspace, one only retains the dynamics that load the interface area in a rigid manner Helderweirt et al. (2001), de Klerk et al. (2008). If substructuring is now performed with this "filtered" receptance, one only imposes compatibility and equilibrium on the motion of those nine or more DOFs that obey local rigid behavior, while the residual flexible motion is left uncoupled. The interface problem is thereby "weakened" and, due to the least square reduction step, measurement errors are averaged out.

A common problem of abovementioned methods is the practical difficulty to collocate applied forces with measured displacements/accelerations. Indeed, one can never excite exactly at the point where a sensor is located, which is required to measure a true "driving point" FRF. This already poses a problem for the class of academic structures, as reported, for instance, in Nicgorski (2008), Nicgorski and Avitabile (2010), Nicgorski and Avitabile (2010), Harvie and Avitabile (2013), making real-life structures even more formidable. In addition, it is difficult to define measurement points in such way that they appear identically on both structures *and* can be measured and excited in all required directions.

Concluding, it seems justified to not question if, but rather how much uncertainty is introduced in the experimental determination of substructure FRFs Ewins and Inman (2001), de Klerk and Voormeeren (2008), Allen et al. (2010). One assurance is that the effects of experimental errors become more apparent at higher frequencies.[8] Hence, it stands to reason to first "message" the raw FRF data of a measurement into a format that is better suited for substructuring.

Toward Collocated Virtual Point Dynamics

Let us look at the previously discussed IDM filtering method from a modal reduction point of view. By defining the IDMs for a single connection point, one confines the dynamics to 6-DOF-per-node kinematics instead of the nine or more DOFs of the measured FRFs (rows of the matrix). In particular, if the IDMs are defined to represent the three translations (X, Y, Z) and three rotations $(\theta_X, \theta_Y, \theta_Z)$ with respect to a single point in a global coordinate system, this kinematic description is similar to a node in an FE model. In that case, the IDMs can be used to transform the

[6]This is a similar reasoning as a chair being stable on three legs.

[7]The original definition of IDM is interface *deformation* mode de Klerk et al. (2008). However, as we are primarily interested in the modes for which the interface behaves rigid, the wording interface *displacement* seems more appropriate.

[8]This applies to most sources of errors. An example of a frequency-independent error is an offset in sensor sensitivity due to bad calibration.

measured translational displacements to 6-DOF motion of a single point. A similar approach can be followed to extract concentrated 6-DOF forces and moments from the excitation forces of the FRF measurement (columns of the matrix).

The concept is illustrated in Fig. 4.24d. The resulting 6×6 FRF matrices describe the dynamic responses of so-called *virtual point motion* to *virtual point loads*, or in other words the admittance of an interface concentrated in a *virtual point*. The point is said to be virtual, because no actual measurements need to be performed on the point itself. In fact, it can be chosen anywhere in the proximity of the interface, allowing to locate the virtual points *identically* for two substructures to couple.

Interesting is now how to set up an experiment that leads to a complete and reciprocal 6×6 virtual point FRF matrix, including proper driving point admittance on the diagonal. The next two sections propose a methodology to achieve this, based on de Klerk et al. (2008), van der Seijs et al. (2013). Section 4.6.4 introduces useful quality assurance criteria and Sect. 4.6.5 addresses practicalities such as sensor placement and impact locations.

4.6.2 *Virtual Point Transformation*

For the derivation of the virtual point transformation, let us consider the coupling problem of two experimental substructures A and B as illustrated in Fig. 4.25. The substructures have non-collocated interface DOFs by nature: neither sensors nor excitation points appear identically. The measured FRF matrices are denoted by $\mathbf{Y}^{\mathrm{A}}(\omega)$ and $\mathbf{Y}^{\mathrm{B}}(\omega)$. The governing equation of the uncoupled dynamics is of the admittance FRF form:

$$\mathbf{u} = \mathbf{Y}(\mathbf{f} + \mathbf{g}). \tag{4.115}$$

We shall concentrate on the interface DOFs, respectively, denoted by $\mathbf{Y}^{\mathrm{A}}_{22}$ and $\mathbf{Y}^{\mathrm{B}}_{22}$. As the interface DOFs are non-collocated, they cannot be equated directly by definition of a Boolean matrix. In other words, there is no direct relation between $\mathbf{u}^{\mathrm{A}}_2 \nLeftrightarrow \mathbf{u}^{\mathrm{B}}_2$ and

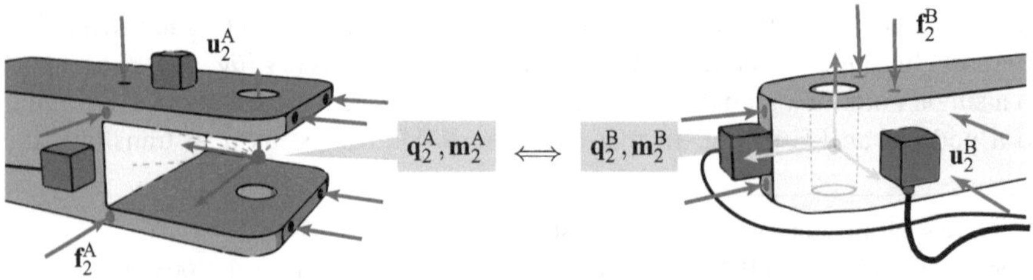

Fig. 4.25 The virtual point transformation workflow for coupling of substructure A (left) and B (right). Each substructure is instrumented by three acceleration sensors (blue, only four of the six sensors are visible here). The excitations of the FRF measurements are depicted by red arrows

$\mathbf{g}_2^{\mathrm{A}} \Leftrightarrow \mathbf{g}_2^{\mathrm{B}}$. In what follows, the assumption is used that the structures behave rigidly in the area close to the coupling point. The derivation focuses on the interface DOFs of a single substructure, denoted for simplicity by $\mathbf{u} \in \mathbb{R}^n$.

Interface Displacement Reduction

Let us express the n interface displacements $\mathbf{u}$ by $m < n$ interface deformation modes (IDMs) $\mathbf{q}$. The IDMs are contained in the columns of the $n \times m$ matrix $\mathbf{R}$, which is a frequency-independent mode shape matrix (the actual construction of the IDMs is discussed in Sect. 4.6.3). As the number of IDMs is smaller than the number of interface DOFs, a residual on the displacements is added, denoted by $\boldsymbol{\mu}$. This residual captures all displacements not in the subspace of the IDMs, which is normally the "flexible" motion.

$$\mathbf{u} = \mathbf{R}\mathbf{q} + \boldsymbol{\mu} \qquad \mathbf{q} \in \mathbb{R}^m. \tag{4.116}$$

To find $\mathbf{q}$ in a minimal-quadratic sense, one could apply the standard Moore–Penrose pseudo-inverse of $\mathbf{R}$, minimizing the norm of the residuals on the displacements and enforce $\mathbf{R}^T\boldsymbol{\mu} = \mathbf{0}$. To gain more flexibility, we introduce a symmetric weighting matrix $\mathbf{W}$ acting on the measured set of DOFs. The residual displacement shall then satisfy:

$$\mathbf{R}^T\mathbf{W}\boldsymbol{\mu} = \mathbf{0} \qquad \mathbf{W} \in \mathbb{R}^{n\times n}. \tag{4.117}$$

The coordinate transformation $\mathbf{u} \rightarrow \mathbf{q}$ follows from standard application of a weighted least square procedure, namely, by premultiplying (4.116) by $\mathbf{R}^T\mathbf{W}$ and solving for $\mathbf{q}$. In addition to the IDM coordinates $\mathbf{q}$, the "filtered" displacement can be retrieved by substituting $\mathbf{q}$ back into (4.116). The resulting displacement is denoted by $\tilde{\mathbf{u}}$. The residual, finally, is simply the remainder of $\mathbf{u}$ minus $\tilde{\mathbf{u}}$:

$$\left\{\begin{aligned} \mathbf{q} &= \left(\mathbf{R}^T\mathbf{W}\mathbf{R}\right)^{-1}\mathbf{R}^T\mathbf{W}\mathbf{u} && (4.118\text{a})\\ \tilde{\mathbf{u}} &= \mathbf{R}\left(\mathbf{R}^T\mathbf{W}\mathbf{R}\right)^{-1}\mathbf{R}^T\mathbf{W}\mathbf{u} && (4.118\text{b})\\ \boldsymbol{\mu} &= \left[\mathbf{I} - \mathbf{R}\left(\mathbf{R}^T\mathbf{W}\mathbf{R}\right)^{-1}\mathbf{R}^T\mathbf{W}\right]\mathbf{u} && (4.118\text{c}) \end{aligned}\right.$$

To simplify notation, let us introduce the following two substitutions:

$$\left\{\begin{aligned} \mathbf{q} &= \mathbf{T}\mathbf{u} && \text{with} \quad \mathbf{T} \triangleq \left(\mathbf{R}^T\mathbf{W}\mathbf{R}\right)^{-1}\mathbf{R}^T\mathbf{W} && (4.119\text{a})\\ \tilde{\mathbf{u}} &= \mathbf{F}\mathbf{u} && \text{with} \quad \mathbf{F} \triangleq \mathbf{R}\mathbf{T} && (4.119\text{b})\\ \boldsymbol{\mu} &= (\mathbf{I} - \mathbf{F})\,\mathbf{u} && && (4.119\text{c}) \end{aligned}\right.$$

Matrix $\mathbf{T}$ is the essential operator of the virtual point transformation. It performs a linear combination of the n measured DOFs so that m virtual point coordinates are retained. A minimum condition for the existence of $\mathbf{T}$ is to have an IDM matrix of rank m, which can theoretically be realized by $n = m$ displacement DOFs. This implies in practice that the physical DOFs $\mathbf{u}$ are located and orientated properly to describe all m modes independently.[9] As a consequence, $\boldsymbol{\mu} = \mathbf{0}$ and $\tilde{\mathbf{u}} = \mathbf{u}$, meaning that no filtering is applied.

Any $n > m$ implies that modal filtering is performed on the displacement space, which means that the residual $\boldsymbol{\mu}$ may become nonzero. The filtered displacement $\mathbf{u} \rightarrow \tilde{\mathbf{u}}$ can be obtained directly using the $n \times n$ filter matrix $\mathbf{F}$ defined by (4.119b), which differs from identity if $n > m$. The virtual point transformation procedure becomes a least square fit that minimizes the weighted error in the residuals:

$$\mathbf{q} = \underset{\mathbf{q} \in \mathbb{R}^m}{\arg\min} \left(\boldsymbol{\mu}^T \mathbf{W} \boldsymbol{\mu} \right). \tag{4.120}$$

If $\mathbf{W}$ is chosen to be identity, the standard pseudo-inverse is found, which leads to a minimization of the quadratic residuals of the displacements. In general, if $\mathbf{W}$ is a diagonal matrix, it assigns weighting to the individual displacements in the error minimization. This provides control over the importance of a certain measured sensor DOF for the transformation or allows to temporarily exclude a particular DOF by setting its weighting to zero. Alternatively, if $\mathbf{W}$ is chosen to represent a (dynamic) stiffness matrix, one is nullifying some local residual energy. Finally, it shall be noted that $\mathbf{W}$ can be defined per frequency line, allowing to make different choices for the low-, medium-, and high-frequency range, for instance.

Most important is that the virtual point motion $\mathbf{q}$ can be obtained from an overdetermined set of measured translations $\mathbf{u}$ using an IDM matrix $\mathbf{R}$. The transformation employs a spatial reduction that does not require inversion of measurement data. Several properties can be assigned to the quality of the transformation, which is further elaborated in Sect. 4.6.5.

Interface Force Reduction

The reduction of interface forces follows a similar procedure, although in a reversed manner. Unlike displacements, forces are not uniquely defined by virtual point forces and moments $\mathbf{m}$ (from here on simply called virtual loads). In fact, the other way around is true.

► To exemplify this, consider a rigid beam hinged around a point of rotation, i.e., the virtual point. If the beam rotates, every point on the beam has a displacement $\mathbf{u}$ which can be uniquely determined from the rotation of the virtual point $\mathbf{q}$. In other words, $\mathbf{q} \rightarrow \mathbf{u}$ is a unique transformation, whereas $\mathbf{q}$ can be determined/averaged from $\mathbf{u}$, provided that a sufficient amount of displacements is measured (two in case of a simple rotation).

Let us now consider a set of forces acting on the same beam. If a resultant moment $\mathbf{m}$ is known in the virtual point, there is no unique transformation $\mathbf{m} \rightarrow \mathbf{f}$, as multiple combinations of forces exist that generate the same moment.

[9] To determine a 6-DOF virtual point with just $n = 6$ translations, one should consider a triad of points and register, for instance, the directions XYZ, XY, and Z, respectively.

However, one can always calculate a resultant moment $\mathbf{m}$ in the virtual point due to the applied forces $\mathbf{f}$, if their locations and directions are known. ◄

Hence, the reduction of forces is performed inversely. A similar IDM matrix can be constructed, relating the measured forces to virtual point loads. This is again denoted by $\mathbf{R}$; distinction with the displacement IDMs will be made later on. Note, however, that the matrix needs to be transposed in order to write the relation between m virtual loads and n (interface) forces[10]:

$$\mathbf{m} = \mathbf{R}^T\mathbf{f}. \tag{4.121}$$

As m is typically smaller than n, (4.121) is underdetermined and the inversion shall take the form of a weighted right inverse of $\mathbf{R}^T$. This inverse seeks for the minimum-quadratic set of forces that realizes a given vector $\mathbf{m}$. In order to prioritize the elements in the force vector, a symmetric weighting matrix $\mathbf{W}$ is again introduced:

$$\tilde{\mathbf{f}} = \mathbf{WR}\left(\mathbf{R}^T\mathbf{WR}\right)^{-1}\mathbf{m}. \tag{4.122}$$

The resulting set of forces $\tilde{\mathbf{f}}$ is automatically admissible, i.e., in the space of $\mathbf{R}$. Analogue to the displacements, let us introduce the following substitutions:

$$\begin{cases} \tilde{\mathbf{f}} = \mathbf{T}^T\mathbf{m} & \text{with} \quad \mathbf{T}^T \triangleq \mathbf{WR}\left(\mathbf{R}^T\mathbf{WR}\right)^{-1} & (4.123\text{a}) \\ \tilde{\mathbf{f}} = \mathbf{F}\mathbf{f} & \text{with} \quad \mathbf{F} \triangleq \mathbf{T}^T\mathbf{R}^T = \mathbf{RT} & (4.123\text{b}) \\ \boldsymbol{\nu} = (\mathbf{I} - \mathbf{F})\,\mathbf{f}. & & (4.123\text{c}) \end{cases}$$

Equations (4.123b) and (4.123c) allow to calculate the IDM-filtered forces and residuals (here denoted by $\boldsymbol{\nu}$) in case that the forces are the input. For the virtual point transformation of admittance, these are normally not used.

Virtual Point Assembly

The interface reduction steps as derived above allow us to assemble only the part of the dynamics of substructures A and B that can be described by the subspaces of their respective IDMs. The remaining displacements and forces are left free. From here on, the transformation matrices for the displacements and forces are clearly denoted by $\mathbf{T_u}$ and $\mathbf{T_f}$.

Let us now continue the assembly of (4.115) by coupling only the part of the displacements and forces that is admissible with respect to the IDMs. The following definitions are used:

$$\mathbf{u} = \begin{bmatrix} \mathbf{u}^{\mathrm{A}}{}_2 \\ \mathbf{u}^{\mathrm{B}}{}_2 \end{bmatrix}; \quad \mathbf{g} = \begin{bmatrix} \mathbf{g}^{\mathrm{A}}{}_2 \\ \mathbf{g}^{\mathrm{B}}{}_2 \end{bmatrix}; \quad \mathbf{Y} = \begin{bmatrix} \mathbf{Y}^{\mathrm{A}}{}_{22} & \mathbf{0} \\ \mathbf{0} & \mathbf{Y}^{\mathrm{B}}_{22} \end{bmatrix}$$

[10]The calculation of $\mathbf{m}$ in (4.121) shows resemblance with the modal participation factor calculated from a vector of applied forces.

$$\mathbf{q} = \begin{bmatrix} \mathbf{q}^{A}{}_{2} \\ \mathbf{q}^{B}{}_{2} \end{bmatrix}; \quad \mathbf{m} = \begin{bmatrix} \mathbf{m}^{A}{}_{2} \\ \mathbf{m}^{B}{}_{2} \end{bmatrix}; \quad \mathbf{T_u} = \begin{bmatrix} \mathbf{T_u}^{A} & \mathbf{0} \\ \mathbf{0} & \mathbf{T_u}^{B} \end{bmatrix}; \quad \mathbf{T_f} = \begin{bmatrix} \mathbf{T_f}^{A} & \mathbf{0} \\ \mathbf{0} & \mathbf{T_f}^{B}. \end{bmatrix}$$

The virtual point transformations of (4.119a) and (4.123a) are now used to write compatibility and equilibrium. Whereas the measured displacements $\mathbf{u}_2^{A}$, $\mathbf{u}_2^{B}$ are incompatible by nature, the virtual points are in fact compatible so it is permitted to require $\mathbf{q}_2^{A} = \mathbf{q}_2^{B}$. This means that the IDMs used to express the kinematics of substructure A and B are defined the same way, for instance: both describe three global-frame translations and three rotations with respect to a unique point in space, i.e., the virtual point. The virtual point DOFs are collocated and can thus be used to write compatibility and equilibrium.

The same holds for the virtual interface loads, allowing to write the following dual coupling equations:

$$\begin{cases} \mathbf{u} = \mathbf{Y}(\mathbf{f}+\mathbf{g}) = \mathbf{Y}\left(\mathbf{f} - \mathbf{T_f}^T\mathbf{B}^T\boldsymbol{\lambda}\right) \\ \mathbf{Bq} = \mathbf{BT_u u} = \mathbf{0} \end{cases} \quad \text{with} \quad \mathbf{B} \triangleq \left[-\mathbf{I}\ \mathbf{I}.\right] \tag{4.123}$$

Lagrange multipliers are substituted for the virtual point interface forces of A and B, enforcing equilibrium on the (rigid) part of the forces in the subspace of the IDMs. Likewise, compatibility is enforced by the second equation onto the admissible displacements. The assembly procedure continues in a familiar LM-FBS fashion, namely, by solving for $\boldsymbol{\lambda}$:

$$\begin{aligned} \mathbf{BT_u Y}\left(\mathbf{f} - \mathbf{T_f}^T\mathbf{B}^T\boldsymbol{\lambda}\right) &= \mathbf{0} \\ \mathbf{BT_u Yf} &= \mathbf{BT_u Y T_f}^T\mathbf{B}^T\boldsymbol{\lambda} \\ \boldsymbol{\lambda} &= \left(\mathbf{BT_u Y T_f}^T\mathbf{B}^T\right)^{-1}\mathbf{BT_u Yf} \end{aligned} \tag{4.124}$$

Now that the virtual point coupling forces are known, the full virtual point coupled solution is found by substitution back into (4.123):

$$\mathbf{u} = \mathbf{Yf} - \mathbf{Y T_f}^T\mathbf{B}^T\left(\mathbf{BT_u Y T_f}^T\mathbf{B}^T\right)^{-1}\mathbf{BT_u Yf}. \tag{4.125}$$

Let us now examine (4.125). At first sight, one recognizes the same ingredients as for standard LM-FBS coupling. The response consists of an uncoupled response, i.e., the first term on the right-hand side, and a coupled response to the applied forces **f**. The coupling, however, takes a "detour" through a total of four virtual point transformation matrices. To better comprehend this procedure, let us expand (4.125):

$$\mathbf{u} = \underbrace{\mathbf{Yf}}_{\substack{\text{uncoupled}\\\text{response}}} - \underbrace{\mathbf{Y T_f}^T\mathbf{B}^T}_{\substack{\text{response to}\\\text{virt. point }\boldsymbol{\lambda}}}\underbrace{\left(\mathbf{B}\overbrace{\mathbf{T_u Y T_f}^T}^{\substack{\text{virt. point}\\\text{admittance}}}\mathbf{B}^T\right)^{-1}}_{\substack{\text{virt. point}\\\text{int. admittance}}}\underbrace{\mathbf{BT_u Yf}}_{\substack{\text{incomp.}\\\text{in }\mathbf{q}}}.$$

Reading from right to left, we first recognize the incompatibility caused by the uncoupled responses. These are transformed to the virtual point space by the operator $\mathbf{T_u}$. Between the brackets, we observe the combined interface admittance. This is put in the virtual point space by pre- and post-multiplication of the measured FRFs by $\mathbf{T_u}$ and $\mathbf{T_f}^T$. Hence, the term in the center constitutes the *virtual point admittance* of the interface, which will be elaborated on later. The Boolean matrices on both sides of this term function as expected, namely, to select and sum up the admittance at the interface DOFs in the virtual point space. Lastly, the term to the left of the inverted admittance governs the response to the virtual point interface forces that are in front of it. These forces are transformed back to the measured DOF space by the operator $\mathbf{T_f}^T$.

This section has demonstrated how interface FRFs, incompatible by the nature of experimental measurement, can be coupled by use of appropriate virtual point transformations. Having clarified all the terms in (4.125), let us now define two approaches to proceed.

Virtual Point Coupling

The coupling detour explained above can be simplified by grouping the respective pairs of $\mathbf{B}$ and $\mathbf{T}$ matrices. This yields two new, yet familiar operators that carry out the actions of virtual point transformation and DOF selection at once:

$$\begin{cases} \mathbf{B}_c \triangleq \mathbf{B}\mathbf{T_u} & \text{compatibility at virtual point} \quad (4.127a) \\ \mathbf{B}_e \triangleq \mathbf{B}\mathbf{T_f} & \text{equilibrium at virtual point} \quad (4.127b) \end{cases}$$

Note that these matrices can be fed straight into the LM-FBS algorithm. The substructures' admittance matrices can be left untouched and, as a consequence, the resulting coupling FRFs are found in the original dual space of the measured DOFs.

Virtual Point FRFs

In the introduction of this chapter, the need for a nodal model was highlighted: a dynamic model, composed from collocated FRFs, that is compatible for coupling with FE or other experimental models. Indeed, for the purpose of dynamic substructuring or transfer path analysis, it may be preferred to transform the entire measurement into a virtual point admittance matrix. Such matrix can be stored in a database and easily retrieved for modeling purposes, in the capacity of a dynamic super-element.

The virtual point admittance was already identified above, namely, as the combined subsystems' admittances pre- and post-multiplied with their respective transformation matrices. This may as well be computed on a substructure level, for instance, to obtain the nodal admittance FRFs of substructure s:

$$\mathbf{Y}^{(s)}_{\mathbf{qm}} = \mathbf{T}^{(s)}_{\mathbf{u}} \mathbf{Y}^{(s)} \mathbf{T}^{(s)}_{\mathbf{f}}{}^{T}. \tag{4.128}$$

The subscript $\star_{\mathbf{qm}}$ is used here to denote the virtual point DOF space of the new admittance matrix. As will be demonstrated later, the IDM matrices $\mathbf{R}$ have a block diagonal structure and so do their resulting **T** matrices. This means that, given an arbitrary structure s, the subset of FRFs used for transformation of coupling point 1 is not being reused for transformation of coupling point 2. In addition, the block diagonal structure makes it easy to keep measured points of interest or additional excitation points untouched, by mixing in identity elements in the substructure's transformation matrices. Let us demonstrate this for the DOFs of the substructures A and B:

$$\begin{bmatrix} \mathbf{u}_1^{\mathrm{A}} \\ \mathbf{q}_2^{\mathrm{A}} \end{bmatrix} = \begin{bmatrix} \mathbf{I} & \mathbf{0} \\ \mathbf{0} & \mathbf{T}_{\mathbf{u}_2}^{\mathrm{A}} \end{bmatrix} \begin{bmatrix} \mathbf{u}_1^{\mathrm{A}} \\ \mathbf{u}_2^{\mathrm{A}} \end{bmatrix} \quad \Longrightarrow \quad \mathbf{q}^{\mathrm{A}} = \mathbf{T}_{\mathbf{u}}^{\mathrm{A}} \mathbf{u}^{\mathrm{A}}$$

$$\begin{bmatrix} \mathbf{q}_2^{\mathrm{B}} \\ \mathbf{u}_3^{\mathrm{B}} \end{bmatrix} = \begin{bmatrix} \mathbf{T}_{\mathbf{u}_2}^{\mathrm{B}} & \mathbf{0} \\ \mathbf{0} & \mathbf{I} \end{bmatrix} \begin{bmatrix} \mathbf{u}_2^{\mathrm{B}} \\ \mathbf{u}_3^{\mathrm{B}} \end{bmatrix} \quad \Longrightarrow \quad \mathbf{q}^{\mathrm{B}} = \mathbf{T}_{\mathbf{u}}^{\mathrm{B}} \mathbf{u}^{\mathrm{B}}.$$

The coordinate sets $\mathbf{q}^{\mathrm{A}}$ and $\mathbf{q}^{\mathrm{B}}$ now contain a mix of original internal DOFs and virtual point interface DOFs. With respect to the interface conditions, one can now define Boolean matrices in the traditional fashion, as all interface DOFs are already collocated due to their matching IDMs. The principles as demonstrated here can be employed in a customized fashion to serve the needs of the substructuring or TPA job in prospect.

4.6.3 Interface Displacement Modes

In the preceding section, it was shown that the virtual point transformation relies on a spatial reduction of coordinates. The interface connectivity is reduced to virtual points using a predetermined set of interface displacement modes (IDMs) **R**. This section elaborates on the construction of the 6-DOF IDMs for a virtual point. The example of Fig. 4.26 is used to illustrate the theory. Virtual point v is surrounded by $n_k = 3$ triaxial acceleration sensors, registering a total of nine translational displacements[11] in the local (x, y, z) frame of the sensors. A hammer impact is indicated by a thick red arrow.

Displacement IDM

The sensor IDM matrix **R** (the subscript is dropped in this section) translates the local-frame displacements to six virtual point displacements and rotations plus a residual. Let us write out this equation for a single triaxial sensor k. The orientation of the sensor is determined by three measurement directions, specified as orthonormal unit vectors:

$$[\ \mathbf{e}_x^k \ \ \mathbf{e}_y^k \ \ \mathbf{e}_z^k\ ] = \mathbf{E}^k$$

[11] In fact the sensors measure accelerations, but for simplicity of notation displacements are considered here.

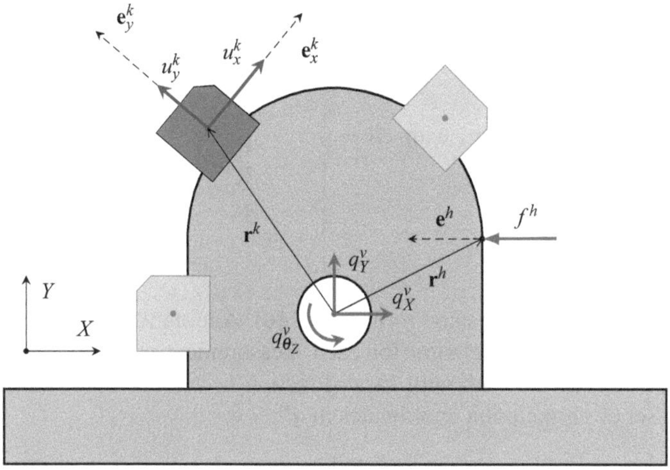

Fig. 4.26 The IDMs associated with the virtual point (green) can be constructed from the positions and directions of the measured displacements (blue) and impacts (red)

The distance from the sensor to the virtual point is given by vector $\mathbf{r}^k$. The respective local displacements along these directions are denoted by $\mathbf{u}^k$. The six DOFs of virtual point v comprised of the set $\mathbf{q}^v$. The following kinematic relation can be established between the virtual point $\mathbf{q}^v$ and the sensor displacements $\mathbf{u}^k$:

$$\begin{bmatrix} u_x^k \\ u_y^k \\ u_z^k \end{bmatrix} = \begin{bmatrix} e_{x,X}^k & e_{x,Y}^k & e_{x,Z}^k \\ e_{y,X}^k & e_{y,Y}^k & e_{y,Z}^k \\ e_{z,X}^k & e_{z,Y}^k & e_{z,Z}^k \end{bmatrix} \begin{bmatrix} 1 & 0 & 0 & 0 & r_Z^k & -r_Y^k \\ 0 & 1 & 0 & -r_Z^k & 0 & r_X^k \\ 0 & 0 & 1 & r_Y^k & -r_X^k & 0 \end{bmatrix} \begin{bmatrix} q_X^v \\ q_Y^v \\ q_Z^v \\ q_{\theta_X}^v \\ q_{\theta_Y}^v \\ q_{\theta_Z}^v \end{bmatrix} + \begin{bmatrix} \mu_x^k \\ \mu_y^k \\ \mu_z^k \end{bmatrix}. \tag{4.129}$$

This relation states how much displacement is measured on the channels of sensor k for a given displacement and rotation of the corresponding virtual point v. Any residual displacement ends up in $\boldsymbol{\mu}$.

Introducing $\bar{\mathbf{R}}^{kv}$ as the 3×6 global-frame IDM matrix associated with sensor k and virtual point v, we can write this relation for each sensor:

$$\begin{aligned} \mathbf{u}^k &= \mathbf{E}^{k^T} \bar{\mathbf{R}}^{kv} \mathbf{q}^v + \boldsymbol{\mu}^k \\ \mathbf{u}^k &= \mathbf{R}^{kv} \mathbf{q}^v + \boldsymbol{\mu}^k \quad \text{with} \quad \mathbf{R}^{kv} \triangleq \mathbf{E}^{k^T} \bar{\mathbf{R}}^{kv}. \end{aligned} \tag{4.130}$$

$\mathbf{R}^{kv}$ is the local-frame IDM matrix. For a typical configuration of three sensors per virtual point, the $\mathbf{R}^{kv}$ matrices can be stacked to form the following system of equations:

$$\mathbf{u} = \mathbf{R}\mathbf{q} + \boldsymbol{\mu} \quad \text{with} \quad \mathbf{R} = \begin{bmatrix} \mathbf{R}^{1,1} & & & \\ \mathbf{R}^{2,1} & & & \\ \mathbf{R}^{3,1} & & & \\ & \mathbf{R}^{4,2} & & \\ & \mathbf{R}^{5,2} & & \\ & \mathbf{R}^{6,2} & & \\ & & \ddots & \\ & & & \mathbf{R}^{n_k,n_v} \end{bmatrix}. \tag{4.131}$$

Equation (4.131) has the same form as (4.116). Alternatively, one may prefer to describe the virtual point relation for each measurement channel individually. A single displacement DOF u_i^k with local direction vector $\mathbf{e}_i^k$ can be related to the associated set of virtual point coordinates in $\mathbf{q}^v$:

$$u_i^k = \begin{bmatrix} \mathbf{e}_i^{k\,T} & (\mathbf{r}_i^k \times \mathbf{e}_i^k)^T \end{bmatrix} \mathbf{q}^v + \mu_i^k = \mathbf{R}_i^{kv}\mathbf{q}^v + \mu_i^k. \tag{4.132}$$

The single-DOF row vector between the block brackets is of size 1×6. The full matrix $\mathbf{R}$ can be constructed row-wise per sensor channel i and column-wise for the different virtual points.

Regardless of the construction method, the resulting IDM matrix $\mathbf{R}$ is block diagonal, or at least contain IDMs that are uncoupled with respect to the various virtual points and sensor groups. Although not treated here, $\mathbf{R}$ may be augmented with flexible IDMs if desired.[12] In any case, care should be taken that $\mathbf{R}$ is full rank, which implies that sufficient measurement points are considered to compute an inverse or pseudo-inverse $\mathbf{T}$ using (4.119a).

Force IDM

The transformation of the forces can be performed in an analogue way. The following relation can be written for the set of virtual forces $\mathbf{m}^v$ as a result of a single impact f^h in the direction of $\mathbf{e}^h$ and at a distance $\mathbf{r}^h$ from the virtual point v:

$$\begin{bmatrix} m_X^v \\ m_Y^v \\ m_Z^v \\ m_{\theta_X}^v \\ m_{\theta_Y}^v \\ m_{\theta_Z}^v \end{bmatrix} = \begin{bmatrix} 1 & 0 & 0 \\ 0 & 1 & 0 \\ 0 & 0 & 1 \\ 0 & -r_Z^h & r_Y^h \\ r_Z^h & 0 & -r_X^h \\ -r_Y^h & r_X^h & 0 \end{bmatrix} \begin{bmatrix} e_X^h \\ e_Y^h \\ e_Z^h \end{bmatrix} f^h \tag{4.133}$$

Note that the virtual point forces are indeed the direct result of applied forces; hence, we do not need to add a residual.

[12]For a definition and application of flexible IDMs, see Pasma et al. (2018).

Now, to remain consistent with (4.132), the transpose of $\mathbf{R}^{hv}$ is defined:

$$\mathbf{m}^{v} = \begin{bmatrix} \mathbf{e}^h \\ \mathbf{r}^h \times \mathbf{e}^h \end{bmatrix} f^{\,h} = \mathbf{R}^{hv^T} f^{\,h}. \tag{4.134}$$

Every impact adds a single column to the transposed IDM matrix $\mathbf{R}^T$. Assuming nine impacts per virtual point, the system for the complete set of DOFs reads

$$\mathbf{m} = \mathbf{R}^T \mathbf{f} \quad \text{with} \quad \mathbf{R} = \begin{bmatrix} \mathbf{R}^{1,1} & & \\ \vdots & & \\ \mathbf{R}^{9,1} & & \\ & \mathbf{R}^{10,2} & \\ & \vdots & \\ & \mathbf{R}^{18,2} & \\ & & \ddots \\ & & & \mathbf{R}^{n_h,n_v} \end{bmatrix}. \tag{4.135}$$

Note that the IDM matrix of the forces takes the same form as the IDM matrix of the displacements from (4.131). In fact, if one decides to excite only on the sensor faces (which is generally not advised, see Sect. 4.6.5), one has obtained the exact same IDM matrix, apart from some sign changes.

With more than six independent excitations per virtual point, (4.135) becomes underdetermined, which means that there is no unique combination of the excitation forces $\mathbf{f}$ from which a certain $\mathbf{m}$ can be constructed. The solution is found using the pseudo-inverse of (4.123a). Instead of minimizing a residual (which was the case for the displacements), this operation finds a set of excitation forces that minimizes the $\mathbf{W}^{-1}$-norm of $\mathbf{f}$:

$$\mathbf{f} = \text{argmin} \| \, \mathbf{f}^T \mathbf{W}^{-1} \mathbf{f} \, \|.$$

This again has advantageous properties for interface weakening and error suppression, as discussed in Sect. 4.6.2.

4.6.4 Measurement Quality Indicators

An important feature of the virtual point transformation is the ability to quantify the consistency of the measurement. Note that the classical coherence function can be used to assess properties of single FRFs, but does not tell if a set of n FRFs is dynamically plausible. The quality functions introduced here are a by-product of the transformation and reveal important practical aspects of the whole measurement.

Sensor Consistency

The sensor consistency function is an indicator of the consistency of the measured response channels with respect to a certain load case. It compares the measured responses with the responses after projection onto the IDMs. It was originally proposed in de Klerk et al. (2008) as an indicator of the "rigidness" of the interface, but has additional, generally useful properties.

For the evaluation of the consistency of the sensors around virtual point 1, let us consider their responses to a load case $\mathbf{f}_2$, denoted by $\mathbf{u}_{1,2}$. This load vector is composed from one or more excitations at coupling point 2, reasonably distant from virtual point 1. This way, the responses in $\mathbf{u}_1$ due to $\mathbf{f}_2$ are probably more global (showing some signal in all directions) than when considered for one ore more excitations near the sensors around point 1. The filtered responses are found by premultiplying the responses with the IDM projection matrix $\mathbf{F_u}$ for the respective coordinates:

$$\begin{cases} \mathbf{u}_{1,2} \triangleq \mathbf{u}_{1,\mathbf{f}_2} = \mathbf{Y}_{12}\mathbf{f}_2 & \text{(4.136a)} \\ \tilde{\mathbf{u}}_{1,2} \triangleq \tilde{\mathbf{u}}_{1,\mathbf{f}_2} = \mathbf{F}_{11}\mathbf{Y}_{12}\mathbf{f}_2. & \text{(4.136b)} \end{cases}$$

As mentioned before, $\mathbf{F_u}$ is a rank-m filter operator. It differs from identity if the transformation is overdetermined, namely, for $n > m$. The filtered responses are in that case limited by the m modes of the IDM; $\mathbf{F_u}$ implements the reduction and expansion step, (4.119a)–(4.119b) in a single operator.

The responses $\tilde{\mathbf{u}}_{1,2}$ shall now be compared with $\mathbf{u}_{1,2}$ to evaluate to what extent the responses are affected by IDM filtering. Two criteria are discussed next.

Overall Sensor Consistency

To get a quick feeling of the consistency of all sensor channels, the norm of both vectors can be compared. This yields a frequency-dependent function bounded by zero (no consistency) and one (full consistency):

$$\rho_{\mathbf{u}_{1,2}}(\omega) = \frac{\|\tilde{\mathbf{u}}_{1,2}(\omega)\|}{\|\mathbf{u}_{1,2}(\omega)\|}. \tag{4.137}$$

A function value $\rho(\omega) = 1$ means that the amplitude of all filtered sensor channels is left unchanged, suggesting that all sensor responses can be fully described by the m IDMs (6 for a rigid interface) at the specified frequency. Mathematically, it implies that the singular value decomposition $\mathbf{Y}_{12} = \mathbf{U}_{11}\mathbf{\Sigma}_{12}\mathbf{V}_{22}^T$ only comprises m significant singular values and that the associated left singular modes $\mathbf{U}_{11}$ are all in the subspace of $\mathbf{R_u}$. In practice, it indicates that all sensors are correctly positioned, aligned, and calibrated, as otherwise at least one channel would "swing" out of line and reduce the overall value of ρ.

If the value drops down above a certain frequency, it is often an indication of the presence of a flexible interface mode. This can be verified by looking at mode $m + 1$

in $\mathbf{U}_{11}$. If the associated singular value is significant, one could consider to augment the IDM basis by a flexible mode, e.g., based on mode $m+1$ in $\mathbf{U}_{11}$.

Specific Sensor Consistency

For evaluation of the consistency per measurement channel, we introduce a variation of the spectral coherence function. Essentially, this is an expansion of the spectral coherence function Bendat and Piersol (1980) when evaluated for a total of two spectra[13]:

$$\mathrm{coh}(x, y) \triangleq \frac{(x+y)(x^*+y^*)}{2(xx^*+yy^*)} \qquad x, y \in \mathbb{C} \qquad \Longrightarrow \qquad \begin{cases} 1 & \text{if } x = y \\ \frac{1}{2} & \text{if } x \perp y \\ 0 & \text{if } x = -y. \end{cases} \tag{4.138}$$

This equation compares two values with respect to their amplitude and phase. It is again bounded between zero and one and can thus be used to objectively compare two complex response spectra with each other. For evaluation of the specific consistency of measurement channel u_i, we write

$$\rho_{u_{i,2}}(\omega) = \mathrm{coh}\left(\tilde{u}_{i,2}(\omega), u_{i,2}(\omega)\right) \qquad u_i \in \mathbf{u}_1. \tag{4.139}$$

If the overall sensor consistency is poor over the full frequency range, it may very well be a problem of just one incorrect sensor direction or calibration value.[14] (4.139) allows to evaluate the consistency of each measurement channel with respect to the full transformation. This way, the problematic sensor can be identified and the entries in $\mathbf{R_u}$ corrected if needed.

Impact Consistency

Similar to the sensor consistency, it is possible to define a consistency function for the applied forces. This value is particularly useful to assess the accuracy of the excitation positions and directions—crucial aspects of impact hammer measurement.

We now consider the effect of the force filter matrix $\mathbf{F_f}$ on the responses. In order to provide a methodology analogue to the sensor consistency, a linear combination of responses is considered for the evaluation, denoted by scalar y_2. Vector $\mathbf{w}_2$ is introduced to compute a (weighted) sum over the response channels of virtual point 2:

$$y_2 = \mathbf{w}_2^T \mathbf{u}_2 \qquad \mathbf{w}_2 \in \mathbb{R}^{n_2}.$$

Note that the quantity y_2 has very little dynamical significance; it only allows to assess the effect of the force filtering with respect to some responses in $\mathbf{u}_2$, selected

[13]Note that other similarity functions may be used here that penalize differences in amplitude and phase of the two spectra.

[14]For instance, if one sensor is rotated by 90° along the z-axis, a rigid interface motion in the x, y-plane will appear as a partly flexible motion. The effect will be visible as a low consistency over the full frequency range.

and combined using $\mathbf{w}_2$. Let us again define an original and force-filtered set of responses, namely, for all forces $f_j \in \mathbf{f}_1$ corresponding to virtual point 1:

$$\begin{cases} \mathbf{y}_{2,1} \triangleq [\ \dots\ y_{2,f_j}\ \dots\ ] = \mathbf{w}_2^T \mathbf{Y}_{21} & (4.140a) \\ \tilde{\mathbf{y}}_{2,1} \triangleq [\ \dots\ \tilde{y}_{2,f_j}\ \dots\ ] = \mathbf{w}_2^T \mathbf{Y}_{21}.\mathbf{F}_{11} & (4.140b) \end{cases}$$

Here, $\mathbf{y}_{2,1}$ is a row vector comprising the summed responses as a result of all singleton impacts associated with the columns of $\mathbf{Y}_{21}$. Row vector $\tilde{\mathbf{y}}_{2,1}$ comprises a similar set of responses, yet for the forces filtered by the force IDM matrix. This can be reasoned as follows: for every single impact $f_j \in \mathbf{f}_1$, $\mathbf{F}_{11}$ looks for an equivalent combination of impacts $\mathbf{f}_1$ that produces the same load on the virtual point. This allows to evaluate if there is any difference in response between the direct application of a single impact and the compound excitation suggested by the IDM matrix.

Overall Impact Consistency

Comparing the norm of the original and filtered set of responses, a measure is obtained for the overall consistency of the impacts:

$$\rho_{\mathbf{f}_{1,2}}(\omega) = \frac{\|\tilde{\mathbf{y}}_{2,1}^T(\omega)\|}{\|\mathbf{y}_{2,1}^T(\omega)\|}. \tag{4.141}$$

This frequency-dependent function quantifies how well the full set of impact forces $\mathbf{f}_1$ can be represented by three translational forces and three moments (in case of a 6-DOF IDM matrix). If some excitations points have been positioned too far away from the virtual point, they may load the interface in a flexible manner, which is reflected by a low value. More often actually, the value drops because of position and direction errors in the impacts.

Specific Impact Consistency

The specific impact consistency is used to spot individual impacts that are "non-consistent" with respect to the transformation. Using the definitions of (4.138) and (4.140a–4.140b), the following function can be defined:

$$\rho_{f_{j,2}}(\omega) = \operatorname{coh}\left(\tilde{y}_{2,j}(\omega), y_{2,j}(\omega)\right) \qquad f_j \in \mathbf{f}_1. \tag{4.142}$$

A low consistency value can have multiple causes, for instance,

- The impact position was off;
- The impact was not in the expected direction;
- The impact was not properly executed (e.g., double pulse, too little energy);
- The impact resulted in a signal overload at one or more channels in $\mathbf{u}_2$.

Like the specific sensor consistency, the specific impact consistency may assist in finding troublesome impacts and fixing entries in the force IDM matrix. This can greatly improve the transformation results.

Virtual Point Reciprocity and Passivity

Evaluating sensor and impact consistency is useful to troubleshoot errors in the measurement setup and validation of the assumptions with respect to interface rigidness. However, a high sensor/impact consistency does not guarantee a truthful virtual point FRF matrix $\mathbf{Y_{qm}}$. Two additional properties are discussed for quality assessment of the two-sided transformed data.

FRF Reciprocity

Reciprocity of the original measurement data is in most cases not very meaningful as the location of sensors and impact positions do not nicely coincide. In contrast, the virtual point motion and loads are actually collocated, meaning that the virtual point FRF matrix should be reciprocal. Reciprocity can thus be used to assess the transformation quality.

Let i and j denote two different DOFs from the set of virtual point DOFs. Then a frequency-dependent reciprocity function between zero and one is defined by

$$\chi(\omega)_{ij} = \operatorname{coh}\left(Y_{ij}(\omega), Y_{ji}(\omega)\right) \qquad Y_{ij}, Y_{ji} \in \mathbf{Y_{qm}}. \tag{4.143}$$

This function quantifies the similarity in both amplitude and phase between the two reciprocal virtual point FRFs. Note that for the diagonal of the matrix, $i = j$ and $\chi_{ii} = 1$ by definition. The reciprocity function is therefore only useful for evaluation of the off-diagonal terms.

Driving Point Passivity

One of the lesser known properties of the FRF matrix is driving point passivity. Driving point FRFs, i.e., the FRFs on the diagonal of the matrix, need to be minimum-phase functions. This means that an applied force at DOF i shall always result in a displacement *in the same direction* at that point.[15] If the function is not minimum phase, it implies that energy is added after the impulse, which contradicts system passivity Sjövall and Abrahamsson (2007), Liljerehn and Abrahamsson (2014). As a consequence, the phase of driving point FRFs shall be bounded as follows:

$$\angle \mathbf{Y}_{ii} \begin{cases} \in [-180, 0] & \text{for receptance FRFs} \\ \in [-90, 90] & \text{for mobility FRFs} \\ \in [\;0, 180] & \text{for accelerance FRFs} \end{cases} \tag{4.144}$$

This property allows to assess the diagonal of the virtual point FRF matrix, which are the only entries that must demonstrate minimum-phase responses. Conveniently, those are also the only entries for which the reciprocity criterion does not apply. Hence, the two criteria are complementary, in some practical way.

---

[15] Sign changes only occur at a minimum distance of half the wavelength of the underlying modes. This can be understood from modal synthesis.

4.6.5 Instrumentation in Practice

This section discusses some more practical issues related to the virtual point experiment. Proper positioning of the measurement and excitation points is essential to obtain a high-quality model. Care should be taken to ensure that all six DOFs per virtual point can be described independently. This has implications for both sensor placement and impact positions. The following preparation and post-processing steps are presented in the order as they normally appear in time.

Sensor Placement

The use of triaxial accelerometers has become standard practice in experimental testing. A single accelerometer hosts three sensing devices in orthogonal directions, which can be assumed to be collocated. Although one pair of those sensors measures a total of six channels, it is normally not sufficient to describe all six DOFs of the virtual point. One linear dependence will appear in the virtual point DOFs: the triaxial sensors are unable to describe the rotation over the axis spanned between both sensors, as illustrated in Fig. 4.27b. This is regardless of the position of the sensors relative to the virtual point. Introduction of a third triaxial sensor, such that the three sensors span a surface (Fig. 4.27c), enables the three sensors with a total of nine DOFs to describe all six DOFs of the virtual point properly.

The additional benefit of the third triaxial sensor is the overdetermination of the interface problem. Applying the virtual point transformation as discussed in Sect. 4.6.2, the effects of uncorrelated measurement noise, as well as bias errors due to misalignment of the sensors are reduced, which is generally considered as good practice. The use of at least three triaxial sensors per virtual point can therefore be held as rule of thumb, if the aim is to describe the virtual point by six independent coordinates.

Finally, considering the rigid interface assumption, it seems logical to place the sensors as close as possible to the virtual point. The smaller the distances are, the lesser the effects of flexible interface motion compared to the rigid interface motion.

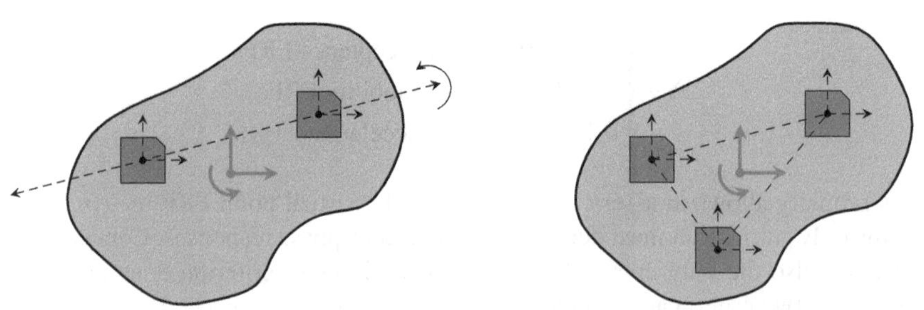

(a) Two sensors spanning a line: one dependency exists between the rotational axes.

(b) Three sensors spanning a surface: all rotations are fully determined.

Fig. 4.27 Sensor placement for measurement of accelerations in an arbitrary structure

On the other hand, for smaller distances, the virtual point transformation gets more sensitive to absolute errors on the position. In general, one should approach this aspect with some engineering judgment or a priori knowledge about the system and measurement equipment.

Excitation Positions

Unlike the six accelerations measured by two triaxial sensors, six well-positioned hammer impacts are potentially sufficient to fully determine the six generalized loads of the virtual point. Still, for the same reason, it is preferable to use more excitation points and overdetermine the force transformation. Similar to the sensor placement, three impacts in each direction (x, y, z) can be used as minimum, creating nine columns in the FRF matrix for each coupling point. However, as one is not restricted by the available measurement equipment (as is the case for the simultaneous sensor positions), it is advised to excite at much more points. Care should be taken to include excitation directions that do not point straight to the virtual point, in order to generate "moment" along the rotational axes.

In some studies, the faces of the acceleration sensors were suggested as possible impact locations de Klerk (2009), van der Seijs et al. (2013). As the impact locations and directions are equal to the locations and orientations of the measured responses of the sensor, the same IDM matrix can potentially be used for both transformations. However, practice shows that FRFs obtained at the sensors' faces exhibit poor coherence, especially for the cross-directional FRFs of one sensor. Also, the sensor is easily driven in overload. As the virtual point transformation does not require physical driving point measurement (driving point FRFs are correctly rendered by the two transformations), this type of excitation is discouraged.

With respect to the distance of impact points from the virtual point, a similar reasoning applies as for the sensors. However, remind that the uncertainty of hammer impacts is highly subject to the skill of the experimentalist.[16] Position errors are always made, typically in the order of a few millimeters. To minimize their effect for the transformation, it is advised to excite at some more distance from the virtual point. This is possible as long as the local rigidness assumption is still justified, which is normally a matter of the targeted frequency bandwidth.

4.7 Real Applications

This section shows an example of experimental modeling of two substructures, followed by coupling by means of frequency-based substructuring. The example is taken from the IMAC paper van der Seijs et al. (2017).

[16]This only applies to manual impact hammer testing. Automated hammer or shaker measurements probably reduce these uncertainties, but are far more elaborate in practice. Considering the substantial amount of measurement points for our applications, impact hammer measurement is the preferred method for using the virtual point transformation.

4.7.1 Experimental Modeling of a Substructure Using the Virtual Point Transformation

In this section, nodal FRF models will be obtained from impact hammer measurements on the two substructures as depicted in Fig. 4.28. Let us introduce them one by one and briefly touch upon some design considerations:

- Substructure A is welded together from three pieces of solid aluminum (30×30 mm). It forms an evenly sided triangle and loosely resembles the character "A", but was made asymmetric to avoid double resonance modes. It comprises a number of 10 mm diameter holes, at the corner points and along the length of the members, evenly spaced at distances of 75 mm. It hosts a vibration source (not further discussed here) and can therefore represent the active source system in a TPA or source characterization problem. The combined weight is circa 2.5 kg.
- Substructure B is constructed from two plates of stainless steel with a solid piece of steel welded in between. The plates are produced using precise laser cutting. Five holes are placed spanning a total distance of 300 mm, again with 75 mm spacing in between. A honeycomb-like pattern of cuts was introduced to reduce weight, as well as to provide a pattern to align sensors for an observability investigation. As such, substructure B represents a receiving side into which the source vibrations of substructure A may propagate. The total weight is circa 10 kg.

FRF Measurement

All substructure FRFs have been obtained by impact hammer testing. Figure 4.28 depicts how hammer impact points (red arrows) and triaxial accelerometers (gray cubes) have been positioned and oriented on substructures A and B. Besides some internal points, the main interests for both substructures are the three coupling points.

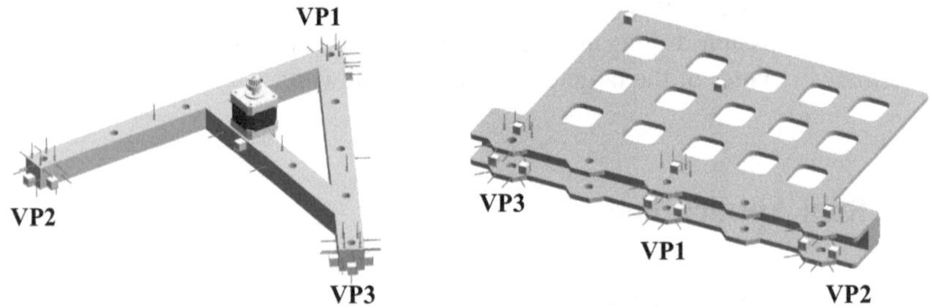

(a) Substructure A. The three coupling points are each instrumented by 3 tri-axial accelerometers and 16 impact points.

(b) Substructure B. The three coupling points are each instrumented by 3 tri-axial accelerometers and 16 impact points; 2 additional sensors register target responses in the structure.

Fig. 4.28 Acceleration sensors (indicated by gray cubes) and impact locations (red arrows) visualized on the substructures

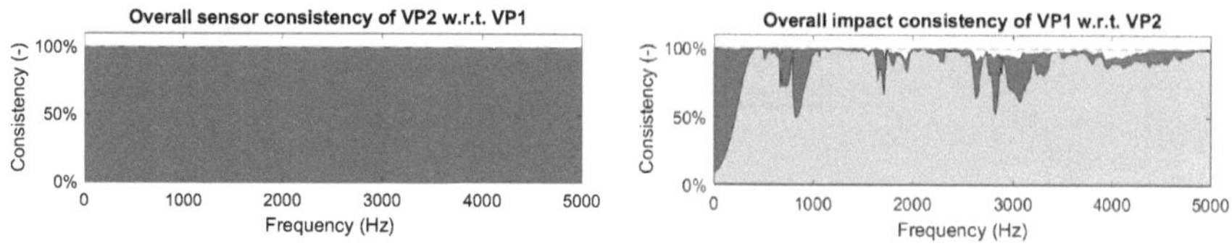

(a) Overall sensor consistency of 9 sensor channels around VP2 with respect to excitations around VP1.

(b) Overall impact consistency of 16 (light blue) and 13 (blue) impacts out of 16 around VP1 with respect to responses around VP2.

Fig. 4.29 Sensor and impact consistency checks for substructure A

Each coupling point has been instrumented by three triaxial accelerometers of type PCB 356B21. To determine forces and moments, 16 impact hammer positions are chosen per coupling point. Altogether, this results in sufficient overdetermination of the virtual point transformations.

FRF Consistency

In order to evaluate the above assumption on rigidity and obtain insight into the contribution of single force impacts or displacements to the VP dynamics, several consistency checks can be done. Let us illustrate the various consistency checks for substructure A. Figure 4.29a shows the overall sensor consistency of VP2 for excitations around VP1. This operation takes the accelerances of all nine sensor channels (FRF matrix rows) around VP2 ($\mathbf{u}$) for a combination of hammer impacts (FRF matrix columns) around VP1, transforms these to the virtual point $\mathbf{q}$, and expands the accelerances back to their original sensor channels ($\tilde{\mathbf{u}}$). The score of 100% over the full bandwidth of 5000 Hz indicates that all sensor channels are perfectly consistent, i.e., $\tilde{\mathbf{u}} = \mathbf{u}$. This is obvious as the region between the three sensors is very stiff; values below 100% would probably indicate incorrect placement of a sensor.

Figure 4.29b shows the overall impact consistency for VP1 with respect to responses around VP2. The light-blue area was computed for all 16 impact points, which is clearly not optimal. Looking into the specific impact consistency for each 16 impacts, three impacts had significant lower score than average. By discarding these 3 from the set of 16, the full 6-DOF set of virtual point forces/moments can still be determined. The dark-blue area was computed for the optimized set, clearly showing an improved overall impact consistency.

FRF Reciprocity

The VP transformation allows to validate reciprocity of the obtained virtual point FRFs, as computed by Eq. (4.128). Note that this is possible as the VP displacements (i.e., translational and rotational accelerations) are perfectly "vectorially associated" with the corresponding VP loads (i.e., forces and moments). In other words, the virtual point FRFs behave as if they were computed for nodes of an FE model.

Figure 4.30 shows two typical virtual point FRFs: response VP2Y over force VP3Y of substructure A (left) and response VP2Z over force VP3Z of substructure

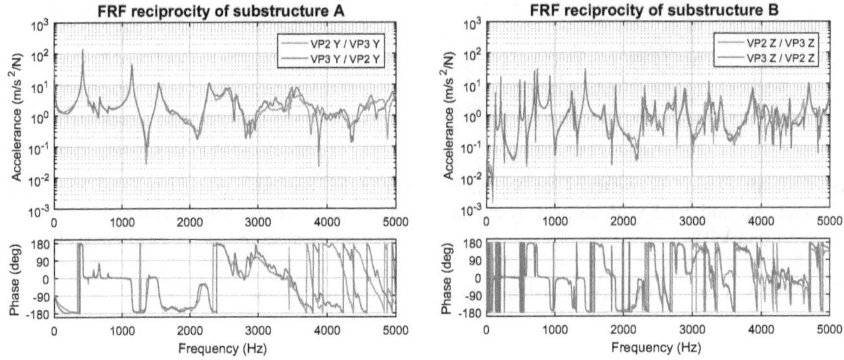

(a) Substructure A: Y-direction of VP2 to VP3 and its reciprocal.

(b) Substructure B: Z-direction of VP2 to VP3 and its reciprocal.

Fig. 4.30 Reciprocity of the virtual point FRFs of the experimental models of substructures A and B

B (right). The FRFs reciprocal FRFs are displayed in red. It can be observed that reciprocity is indeed satisfied, especially up to 2 kHz.

4.7.2 Experimental Dynamic Substructuring of Two Substructures

Now that VP-transformed FRFs are available for substructures A and B, both structures are coupled using the LM-FBS algorithm. To do so, the substructure FRF matrices of A and B are put in block diagonal form and an appropriate Boolean matrix $\mathbf{B}$ is written as follows (not discussed here, see Sect. 2.3 for guidelines):

$$\tilde{\mathbf{Y}} = \mathbf{Y} - \mathbf{Y}\mathbf{B}^T\left(\mathbf{B}\mathbf{Y}\mathbf{B}^T\right)^{-1}\mathbf{B}\mathbf{Y} \qquad \mathbf{Y} \triangleq \begin{bmatrix} \mathbf{Y}^{\mathrm{A}} & \mathbf{0} \\ \mathbf{0} & \mathbf{Y}^{\mathrm{B}} \end{bmatrix}. \tag{4.145}$$

The two-point coupling configuration is considered, which means that coupling is performed by requiring strict coordinate compatibility and force equilibrium for the FRFs of virtual points 2 and 3. We now focus on the frequency range of 0–1600 Hz.

Some results of the substructured FRFs of AB are depicted in Fig. 4.31. First in Fig. 4.31a, a driving point FRF on the coupling interface is shown, namely, for VP2 in Z-direction. The phase is shown as well to assess the passivity[17] of the FRF. Figure 4.31b shows a transfer FRF from an internal force impact point on structure A to an acceleration response internally on structure B. Both points are not part of a coupling VP; hence, the transfer FRF is realized by coupling over the interface. The

[17]For an accelerance driving point FRF, the phase should be bounded by 0 and +180°.

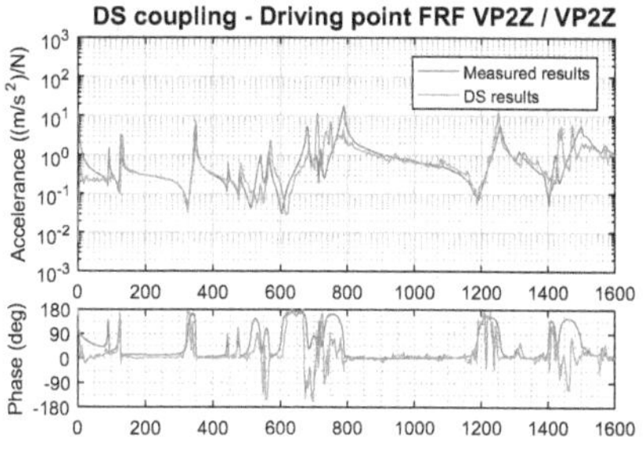

(a) Driving-point FRF for VP3 in Z-direction.

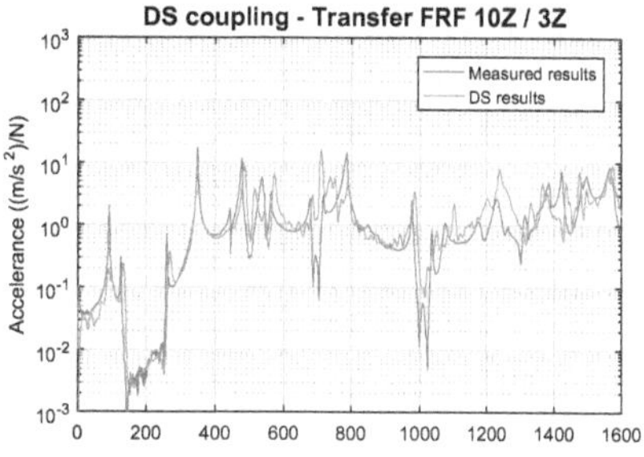

(b) Transfer FRF for an internal impact point on A to an internal acceleration response on B, both in Z-direction.

Fig. 4.31 Application of dynamic substructuring: assembled FRFs of AB (blue) in two-point coupling configuration, compared against the validation measurement (red)

substructured FRFs (blue) are compared the measured FRFs of the coupled structure AB.

The first substructuring results, although not perfect yet, are by all means encouraging. It can be seen how resonance frequencies are created at roughly the right frequencies. The phase around antiresonances is not fully stable; however, the overall amplitude of the FRFs matches quite well. Note that no filtering or processing has been applied to the measured FRF data, except for transformation to virtual points.

4.8 Finite Element Model Updating for Substructuring

In this era of simulation, computational solid mechanics is playing an increasingly important role in the design and performance of engineering systems. Automobiles, aircraft, bridges, and high-rise buildings are examples of engineered systems that have become more and more reliant on computational models and simulation results to predict their performance, reliability, and safety. For these models to be truly credible, they need to be validated against experimental data obtained by physical testing. Because of many, far from trivial, modeling issues in computational structural dynamics, the model validation is most often made in conjunction with a *model calibration* (also known as *model updating* or *white-box system identification*) to get a better fit to test data. In this calibration, the model is parameterized for parameters that are to some degree not certain and these are allowed to adjust for better fit between simulation and test data to possibly better fulfill setup validation criteria.

A finite element model for a mechanical system is most often developed from first principles. Such are Hooke's law for material behavior and Newton's laws for equilibrium. In the linear regime, it is usually found that components manufactured from a solid piece of material are almost trivially modeled to a very high accuracy with the finite element method. However, the modeling becomes much harder when parts (substructures) are put together into an assembly by riveting, bolting, gluing, or welding. It is often at the modeling of joints and component interfaces that the modeler's skill and experience come to a severe test. A vast part of the contribution to system damping is often attributed to the physical processes occurring in the interfaces between parts. This poses one challenging problem in the dynamic substructuring since it is at such interfaces the system is normally divided to make the submodels more manageable. It normally takes a very good insight into physics and precise information about the details of the system to get the damping models right at first time. The system damping is therefore most often obtained by testing. After a model calibration, this damping can usually be mapped to the system model without precise knowledge about the physics of the damping processes.

This section describes to some detail a frequency response-based calibration process that has been found to work on large-scale finite element models and an industrial example will be given. It starts with defining the concept of a model structure which regards the parameterization of the calibration problem and some concepts related to that. It continues with the imprecision of the estimation of the calibration parameters that is the result of unavoidable noise in testing. Thereafter, it briefly treats the data processing that leads to the data split that is used for validation with calibration and cross-validation for assessment of model prediction accuracy. It finishes with the specifics of calibration and cross-validation, and in particular treats some concepts and tricks that have been found useful in the calibration of large-scale FE models with many calibration parameters. However, it does not cover the important topics of best preparing for an accurate and informative validation test, how to best conduct the test or to best process the test data for possible inaccuracies and outliers.

Finite element model structure. A finite element model is often specified by an input file with a given structure with data for the analysis. In that file, the system's physical properties are assigned numerical values. The constitutive models for the used materials are given material property data. The structure's geometry is given by nodal point locations of the established discretization grid and the selected element types. The modeler might have chosen lower dimension elements to represent all or parts of the structure. Such may by simple scalar elements such as discrete springs or lumped masses, it may be one-dimensional rod or beam elements, and it may be two-dimensional plate or shell elements. Such simplistic elements also need to be described in the input file as plate thicknesses, beam cross-sectional properties, lumped mass weight, etc. All such numerical data are required to describe the structural model. The numerical data given can be seen as one specific realization from a general parametric model structure given by the structure of the input file. Let us call that model structure $\mathfrak{M}(\mathbf{p})$ with parameter vector $\mathbf{p}$. Two models $\mathfrak{M}_1$ and $\mathfrak{M}_2$ are meant to be equal, *i.e.*, $\mathfrak{M}_1(\mathbf{p}_1) = \mathfrak{M}_2(\mathbf{p}_2)$, if they share the same input-to-output relation but not necessarily the same parameters. If the parameters of the model parameter vector $\mathbf{p}$ were let free, they may be adjusted to the parameter setting that gives a better fit to the input-to-output relation obtained as experimental data. If the ideal situation would occur in which the experimental data were free from noise and bias errors, an exact fit of a model of a proper model structure would be possible. If that situation would occur (in practice it never will), it is said that a *consistent model structure* or a *true model structure* is at hand and the associated parameter setting of the associated perfect model is denoted $\mathbf{p}^{\star}$. This parameter setting is sometimes referred to as the *oracle parameter* setting.

An optimization scheme can be used to adjust the model parameters to best fit the test data, that is, the process of model calibration. The calibrated model is thus one specific realization from the given model structure. If it is found, then find, eventually after unsuccessful calibration or validation attempts, that it was impossible to calibrate or validate the model to an acceptable accuracy one needs to choose another model structure or give up and accept defeat. A validation attempt that resulted in a negative outcome can often be scrutinized to give some hints to what model structure changes that are required. A change to another constitutive model for the material constituents, such as changing from an isotropic material description to an orthotropic, is an example of changing the model structure. Another possibility might be to do a more precise modeling of parts that join structural components together. There are endless other possibilities. However, all such model structure changes have to be made with good scientific/engineering judgement to be truly successful. Besides data of the structural model, the input file of the finite element simulation code also includes a load model and a boundary condition model. The process of validating the load model is a topic of its own and is not further treated here. It is assumed that the loading in the associated calibration and validation experiments are measured with such high precision resulting in load uncertainties that are too small to matter.

However, the boundary condition model may require some considerations. The boundary conditions for the structure under study might be different in the test laboratory than in the real-world application for which the model should be made valid. In commonly used finite element models, the structural boundaries are most often considered to be perfectly free or perfectly rigid. These are conditions that are never met in practice or in the test laboratory. Any nonideal boundary condition may cause bias effects for the model calibration if not properly accounted for. An assessment of the mobility of the structural boundary that is connected to the surrounding via hardware may be required. The remaining surfaces of the structure that tend toward open-air normally play an insignificant role to the observed structural behavior, but might be of importance for very lightweight (meaning small weight in relation to the enveloped volume) structures. For submerged structures, the wet surface boundary conditions can have a great impact on the structural response and should be modeled properly.

4.8.1 Finite Element Model Parameter Statistics and Cross-Validation

In the model calibration, the test data that are available are used in the estimation of model parameters. These test data are always associated with some randomness. If an experiment is repeated under seemingly identical conditions in multiple realizations, more-or-less different results will be obtained that will affect the parameter estimates. If the tests are set up carefully and performed with proper hardware and skilled test engineers in a low-noise environment, the estimation differences will be small but still present. Identical calibration procedures applied to these different realizations will render different calibration outcomes. The calibration parameters estimated from test data will then seemingly have random character because of variations in test data even if they represent some property that is constant at test. Some basic aspects of the associated statistical data obtained from the parameter estimation in the calibration process are described in the following.

Test data for parameter estimation and validation. The topic of parameter estimation deals with the problem of extraction information from observations that have randomness. The observations are then realizations of stochastic processes where noise is really the random part that affects the parameter estimation. Suppose that the observations are represented by processed test data as random data in $\mathbf{z} \in R^N$. These observations may be processed time-domain data $\mathbf{r} \in R^{n_r}$, that is, the response output recorded in the vibration testing. In vibration testing, often with a high channel-count data acquisition system, high sampling rates and long duration testing, the collected discrete-time data vector sizes may be in the order of billions. These data are most often processed with frequency analysis and averaging into transfer function estimates or further via system identification to eigensystem data as system eigenfrequencies and eigenvectors. This process is then a mapping of $\mathbf{r}$ into $\mathbf{z}$. Although the statistical properties of the collected output data $\mathbf{r}$ can often be easily assessed, the statistical properties of the processed data are often less known. However, let us assume that the joint probability density function (pdf) of the elements of $\mathbf{z}$ is

$$\mathrm{pdf}(z_1, z_2, \ldots, z_N|\mathbf{p}), \tag{4.146}$$

i.e., the probability function depends on the parameter setting $\mathbf{p}$ that is believed to represent real-world physics. But that parameter setting is hidden and implicitly embedded in test data. The probability of the test outcome to be within the domain $\mathbf{Z}$ ($\mathbf{z} \in \mathbf{Z}$) is then

$$P(\mathbf{z} \in \mathbf{Z}) = \int_{\mathbf{z} \in \mathbf{Z}} \mathrm{pdf}(z_1, z_2, \ldots, z_N|\mathbf{p}) dz_1 dz_2 \ldots dz_N. \tag{4.147}$$

In (4.147), $\mathbf{p}$ is the n_p-dimensional parameter vector that describes properties of the process that we observe. These parameters are unknown, and the purpose of the observation is to estimate the vector $\mathbf{p}$ using observation data $\mathbf{z}$. This is accomplished by an estimator $\hat{\mathbf{p}}(\mathbf{z})$ which is a mapping function from R^N to R^{n_p}. If the observed value of $\mathbf{z}$ is $\mathbf{z}^*$, then consequently the resulting estimate of the parameters is $\hat{\mathbf{p}}^* = \hat{\mathbf{p}}(\mathbf{z}^*)$.

Various estimator functions $\hat{\mathbf{p}}(\mathbf{z})$ are possible. A particular estimator that maximizes the probability of the observed data is the *maximum likelihood estimator* [4.5].

It is based on the joint probability density function (4.146) for the random observations $\mathbf{z}$. The probability that the observed realization indeed should take value $\mathbf{z}^*$ is thus proportional to $\text{jpdf}(z_1^*, z_2^*, \ldots, z_N^* | \mathbf{p})$. This is a deterministic function of $\mathbf{p}$ once the numerical values $\mathbf{z}^*$ are inserted. This function is called the *likelihood function*. It reflects the likelihood that the observed realization should indeed take place. A reasonable estimator of $\mathbf{p}$ could then be to select it so that the observed realization becomes as likely as possible. That is to seek

$$\hat{\mathbf{p}}_{\text{ML}}(\mathbf{z}) = \arg \max_{\mathbf{p}}(\text{pdf}(\mathbf{z}^* | \mathbf{p})), \tag{4.148}$$

where the maximization is performed for the processed test data $\mathbf{z}^*$, which is one set of data of the infinitely many realizations possible. The function $\mathbf{p}_{\text{ML}}$ is known as the maximum likelihood estimator (MLE) for $\mathbf{p}$.

The Cramer–Rao Inequality. The credibility of an estimator can be assessed by its mean-square error matrix

$$\mathbf{C} = E([\bar{\mathbf{p}} - \mathbf{p}_0][\bar{\mathbf{p}} - \mathbf{p}_0]^T). \tag{4.149}$$

Here, $\mathbf{p}_0$ denotes the true (but hidden) value of $\mathbf{p}$, and (4.149) is evaluated under the assumption that the joint probability density of $\mathbf{z}$ is $\text{pdf}(\mathbf{z}|\mathbf{p}_0)$.

An estimator that make $\mathbf{C}$ small seems to be a natural choice. It is then interesting to note that there is a lower limit to the values of $\mathbf{C}$ that can be obtained with various unbiased estimators. This is given by the *Cramer–Rao inequality* that states the following.

Theorem 5.1 (Cramer–Rao inequality) *Let* $\hat{\mathbf{p}}(\mathbf{z})$ *be an estimator of* $\mathbf{p}$ *such that* $E[\hat{\mathbf{p}}(\mathbf{z})] = \mathbf{p}_0$, *where E evaluates the mean, assuming that the joint probability of* $\mathbf{z}$ *is pdf*$(\mathbf{z}|\mathbf{p}_0)$, *and suppose that* $\mathbf{z}$ *may take values in a subset of* R^N, *whose boundary does not depend on* $\mathbf{p}$. *Then*

$$E([\bar{\mathbf{p}} - \mathbf{p}_0][\bar{\mathbf{p}} - \mathbf{p}_0]^T) \geq \mathfrak{F}^{-1}, \tag{4.150}$$

where

$$\begin{aligned}\mathfrak{F} = E([d(\log pdf(\mathbf{z}|\mathbf{p}))/d\mathbf{p}][d(\log pdf(\mathbf{z}|\mathbf{p}))/d\mathbf{p}]^T)|_{\mathbf{p}=\mathbf{p}_0} = \\ -E([d^2(\log pdf(\mathbf{z}^*|\mathbf{p}))/d\mathbf{p}^2])|_{\mathbf{p}=\mathbf{p}_0}.\end{aligned} \tag{4.151}$$

Proof See reference [4.8]. □

Since $\mathbf{p}$ is an n_p-dimensional vector, $d(\log \text{pdf}(\mathbf{z}|\mathbf{p}))/d\mathbf{p}$ is a n_p-dimensional column vector and the Hessian in (4.151) is an n_p-by-n_p matrix. This matrix $\mathfrak{F}$ is known as the *Fisher information matrix*. Notice that the evaluation of $\mathfrak{F}$ normally requires knowledge of $\mathbf{p}_0$, so the exact value of $\mathfrak{F}$ may not be available to us.

The Asymptotic Properties of the Maximum Likelihood Estimator. It is usually very difficult to exactly calculate the statistical properties of the parameter estimates of any given estimator. However, for the MLE estimator there is a classical theorem, see reference [4.9], that is valid for data that are statistically independent and in a number that tends to infinity. The theorem states the following.

Theorem 5.2 *Suppose that the random data* $\mathbf{p} \in R^N$ *are statistically independent so that*

$$jnf(z_1, z_2, \ldots, z_N|\mathbf{p}) = \prod_{n=1}^{N} jnf(z_n|\mathbf{p}). \tag{4.152}$$

Suppose also that, in particular, the distribution of $\mathbf{z}$ *for a given setting of the parameters* $\mathbf{p} = \mathbf{p}_0$ *is given by* $jnf(\mathbf{z}|\mathbf{p}_0)$. *Then the random parameter variable* $\mathbf{p}_{ML}(\mathbf{z})$ *tends to* $\mathbf{p}_0$ *with probability 1 as N tends to infinity, and the random variable* $\sqrt{N}(\mathbf{p}_{ML} - \mathbf{p}_0)$ *converges in distribution to the normal distribution with zero mean and covariance matrix given as the Cramér–Rao lower bound, i.e., the covariance matrix is the inverse of the Fisher information matrix.*

Proof See Ljung (1998). □

Thus, when the number of data N tends to infinity, the MLE $\mathbf{p}_{ML}$ are distributed $N(\mathbf{p}_0, \mathfrak{F}^{-1})$. According to the Cramer–Rao theorem, this is the best an estimator can do and therefore it is said that the MLE is an *efficient estimator*.

The Case of Normally Distributed Data. Consider the special case when the data z_n^X, as obtained from evaluation of test data (eXperiment), can be predicted without bias with a parameterized model in its calibrated setting $\mathbf{p}^*$. Let the predicted data from the model be $\breve{z}_n$. The relation

$$z_n^X = \breve{z}_n(\mathbf{p}^*) + \varepsilon_n \tag{4.153}$$

introduces ε_n being the residual that cannot be explained by the model. Assume that the residuals are statistically independent variables distributed $\varepsilon_n \in N(0, \sigma_n^2)$ with known standard deviation σ_n. In that case, the Fisher information matrix, and thus the parameter covariance lower bound, can be shown to be

$$\mathfrak{F}(\mathbf{p}) = \sum_{n=1}^{N} \sigma_n^{-2} [\frac{\partial \breve{z}_n(\mathbf{p})}{\partial \mathbf{p}}][\frac{\partial \breve{z}_n(\mathbf{p})}{\partial \mathbf{p}}]^T. \tag{4.154}$$

This result simplifies matters much. As can be seen, test data are not explicitly part of the equation, but only implicitly through the data variance σ_n^2. The identifiability of the parameters of a given model can thus be evaluated provided assumptions on residual variance. Different model structures can thus be compared against each other to find out which gives the best parameter identifiability. For a derivation of Eq. (4.154), see Ljung (1998).

Model Distinguishability and model selection. In structural dynamics modeling of complex systems, the possiblity to vary the model structure seems endless. It is then natural to ask whether the test to be performed on a physical item will make it possible to decide which model structure is the best. This question is that of the *distinguishability* of model structures, which receives a partial answer in the same idealized framework as model identifiability. One thus assumes that the physical item *is* a model with true model structure $\widehat{\mathfrak{M}}$ represented by a model has the model structure $\mathfrak{M}$. The parameter vector associated with $\mathfrak{M}$ is $\mathbf{p}$ and that associated with $\widehat{\mathfrak{M}}$ is $\widehat{\mathbf{p}}$. Since the physical item and its model are not necessarily consistent with the same model structure, it may become impossible to tune the parameters $\mathbf{p}$ of the model so as to obtain the same input–output behavior as that of the real world. It is this impossibility that may permit the falsification of model structure $\mathfrak{M}_1$ in favor of another model structure $\mathfrak{M}_2$. More precisely, $\mathfrak{M}$ will be *structurally distinguishable* (s.d.) from $\widehat{\mathfrak{M}}$ if, for almost any feasible values of $\widehat{\mathbf{p}}$, there are no feasible values of $\mathbf{p}$ such that $\mathfrak{M}(\mathbf{p}) = \widehat{\mathfrak{M}}(\widehat{\mathbf{p}})$. Note the asymmetry of the previous definition. The fact that $\mathfrak{M}$ is s.d. from $\widehat{\mathfrak{M}}$ does not imply that the converse is true. One class of model structures may include the other as a subclass. Whenever $\mathfrak{M}$ is s.d. from $\widehat{\mathfrak{M}}$ and $\widehat{\mathfrak{M}}$ is s.d. from $\mathfrak{M}$, they are said to be *mutually structurally distinguishable*. In *model selection*, the objective is to find that model structure $\mathfrak{M}^*$ that is the least distinguishable from the true model structure $\widehat{\mathfrak{M}}$. This is made by maximizing the plausibility that the model structures are indistinguishable, often by use of Bayesian statistics, see, e.g., Beck and Yuen (2004).

Data split for calibration and validation. The full set of processed test data $\mathbf{z}^x$ is normally split into one subset $\mathbf{z}^x_\delta$ for calibration and another subset $\mathbf{z}^x_\gamma$ for validation with $\mathbf{z}^x = \mathbf{z}^x_\delta \cup \mathbf{z}^x_\gamma$. The analysis data $\breve{\mathbf{z}}$ are set up accordingly into $\breve{\mathbf{z}}_\delta$ and $\breve{\mathbf{z}}_\gamma$ and the deviation vectors $\boldsymbol{\delta} = \mathbf{z}^x_\delta - \breve{\mathbf{z}}_\delta$ and $\boldsymbol{\gamma} = \mathbf{z}^x_\gamma - \breve{\mathbf{z}}_\gamma$ can be used to form metrics for calibration and validation respectively. In cross-validation, described next, multiple data split realizations are used to obtain model prediction and model parameter statistics.

Model Cross-Validation. The cross-validation (XV) technique is normally used for estimating the prediction capacity of a model. It is a statistical tool that has been used for a long time, but since it requires intense computing it has not been used much for large-scale problems until recent years. It normally requires much more computation than a single calibration since it makes use of repeated calibrations on subsets of available data. XV targets the expected squared prediction deviation of the n:th data, i.e.,

$$PD_n = E[(z^X_n - \breve{z}_n(\mathbf{p}))^2]. \tag{4.155}$$

The expectation refers to the mean value obtained after repeated sampling of same experimental data entity z^X_n. Closely related to the PD is the residual squared deviation (RSD) of the prediction defined as

$$RSD = (\mathbf{z}^X - \breve{\mathbf{z}}(\mathbf{p}))^T(\mathbf{z}^X - \breve{\mathbf{z}}(\mathbf{p}))/N_z, \tag{4.156}$$

where N_z is the number of data, i.e., the length of the data vectors $\mathbf{z}^X$ and $\check{\mathbf{z}}$. The RSD may be evaluated separately for validation data and calibration data as

$$RSD_\gamma = \boldsymbol{\gamma}^T(\mathbf{p})\boldsymbol{\gamma}(\mathbf{p})/N_\gamma \tag{4.157}$$

and

$$RSD_\delta = \boldsymbol{\delta}^T(\mathbf{p})\boldsymbol{\delta}(\mathbf{p})/N_\delta, \tag{4.158}$$

where usually RSD_γ is larger than RSD_δ since the model has been calibrated with an effort to minimize $\boldsymbol{\delta}^T\boldsymbol{\delta}$. To truly evaluate the PD_i would require experimental data that are fully independent on calibration data and thus require repeated tests. Usually, however, test repetitions are not always available for reasons of cost or time. To circumvent that and give a reasonable substitute of the true prediction deviation, XV uses part of the available data to calibrate the model, and a different part to evaluate the RSD of it. This splitting of data is repeated multiple times with different data in the validation and calibration data sets each time to give useful RSD statistics. The following describes one possible strategy to do the data splitting.

K-Fold Cross-Validation. In K-fold XV, the available data set is split into K parts (folds) of roughly the same size. Let $k = 1, 2, \ldots, K$ be the index of the k:th such fold with the N_z data values divided into $N_{\gamma k}$ validation data and $N_{\delta k} = N_z - N_{\gamma k}$ calibration data. The union of the two data sets thus constitutes the entire data set. The validation sets are mutually unique and the union of them spans the entire data set. The validation deviation of the kth set is $RSD_{\gamma k}$ and the corresponding calibration deviation is $RSD_{\delta k}$.

For the XV, the calibration is made K times using the deviation metric $Q_k(\mathbf{p}) = \boldsymbol{\delta}_k^T\boldsymbol{\delta}_k$ to obtain a parameter estimate associated with the kth split as

$$\mathbf{p}_k = \underset{\mathbf{p}}{\operatorname{argmin}}\ Q_k(\mathbf{p}) \tag{4.159}$$

resulting in a validation RSD for the k:th fold to be

$$RSD_{\gamma k} = \boldsymbol{\gamma}_k^T(\mathbf{p}_k)\boldsymbol{\gamma}_k(\mathbf{p}_k)/N_{\gamma k} \tag{4.160}$$

with the XV estimate of the statistical mean prediction deviation as

$$RSD_{\mathrm{XV}} = \frac{1}{K}\sum_{k=1}^{K} RSD_{\gamma k}. \tag{4.161}$$

A popular K-fold XV scheme is to do ten data splits, thus after 10 calibrations leading to the associated tenfold XV statistics. Another special case of K-fold XV, when the number of folds K is taken to the extreme, is the *leave-one-out cross-validation*. In leave-one-out XV, $K = N_z$ splits are made which leaves one single data for validation while the remaining $N_{\delta k} = N_z - 1$ data are used for calibration in

N calibration runs. While the leave-one-out XV have certain advantages, a distinct disadvantage in a FEM setting is the large number of costly calibrations that have to be made which makes the leave-one-out cross-validation practically infeasible. Other popular XV splitting techniques are based on bootstrap data splitting or Monte Carlo data splitting.

Cross-Validation Estimates of Parameter Statistics. The XV procedures above are seen to involve multiple runs for finding the calibration parameter setting that minimizes the calibration deviation metric for various splittings of available data. It is natural to use the multiple parameter estimates for statistical evaluation. Evaluation of mean and covariance of the parameter estimates $\mathbf{p}_k$ can be done with ease. This can then be made without explicit knowledge of the noise properties of experimental data. Compared with the parameter covariance estimate by the Cramer–Rao bound, which requires explicit knowledge of the noise properties, this is a distinct advantage.

4.8.2 Finite Element Model Calibration

Most calibration problems are solved by gradient-based minimization techniques. A calibration scheme that uses a gradient-based minimizer needs to work with a smooth deviation metric for high likelihood of success. That is to obtain convergence in the search for the parameter's optimum settings from start settings of the parameters. A well-calibrated model should give high accuracy in simulation of test output quantities, and ideally predictions with high credibility of other output quantities not tested. In a frequency-domain context, this often translates to that model which accurately captures the strong response a structural resonance and possibly also the small response at its antiresonances. A metric that does not discriminate against deviations at frequencies where the structural response is small is the quadratic functional $Q(\mathbf{p})$ in

$$Q = \boldsymbol{\delta}^H \boldsymbol{\delta} \qquad \boldsymbol{\delta} = \log(\text{vect}(\mathbf{H}^A(\mathbf{p}))./\text{vect}(\mathbf{H}^X)). \tag{4.162}$$

Here $\mathbf{H}^A$ and $\mathbf{H}^X$ are the frequency response functions established by FE analysis and provided by experiments, respectively, see Eq. (3.85). The function vect(.) is the vectorizing operation that makes all frequency response function elements of the transfer function, at all discrete frequencies used for evaluation, into a column vector and the/ operator denotes the element-by-element division.

Since finite element model calibration tends to be very computationally demanding, calibration criteria that lead to computational efficiency are strong of the essence. If operations can be spared and therefore reduce calculation times, it can mean that the calibration issue can move from being an interesting theoretical concept to being practically useful. However, all computations need to be optimized to provide as much useful information as possible with as little effort as possible.

Such optimization targets the sampling strategy for the discrete frequencies that are selected for frequency response function evaluation. The half-bandwidth $\Delta\omega_i$ of a damped structural resonance at frequency ω_i and modal damping ζ_i is given by $\Delta\omega_i = \zeta_i\omega_i$. One observes that the half-bandwidth increases linearly with increasing resonance frequency. It then seems to be a good frequency sampling strategy to utilize frequency steps that increase linearly with frequency and thus not oversample the high-frequency modes or under-sample the low-frequency modes. Such sampling keeps the number of samples over one half-bandwidth constant over the range. That is to take steps such that the logarithm of the frequency steps over the frequency range is constant which makes the frequency steps in direct proportion to frequency. That sampling strategy seems reasonable, provided that relative damping of all modes in the range are equal, which rarely happens for experimentally found eigenmodes. However, the damping can be equalized by a procedure that is treated below. The influence of the density of discrete frequency steps and damping level can be seen in Fig. 4.32. It can be noted that the smoothness of the calibration metric is affected by damping and varying frequency steps.

Damping equalization. A central issue for FRF-based model calibration is that of model damping. Since, in general, damping has been found to be very difficult to

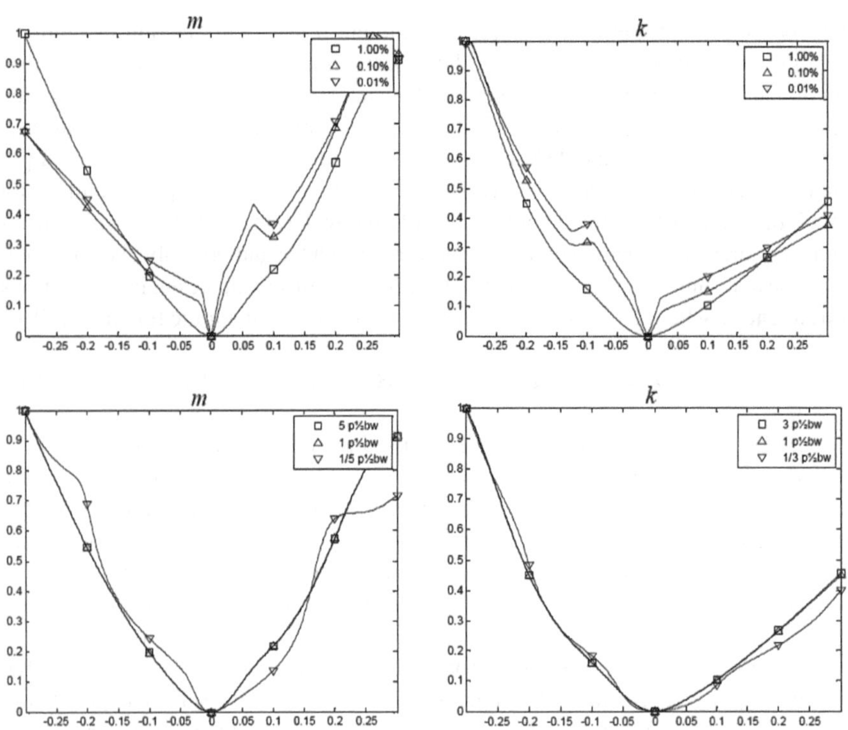

Fig. 4.32 Normalized deviation metric versus parameter variation from nominal of a stiffness parameter k and a mass parameter m for (upper) three system damping levels at 0.01, 0.1, and 1%, and (lower) various frequency sampling rates in number of samples per half-bandwidth (p1/2bw)

model using first principles, it is most often assigned a simple representation for modeling convenience. Such are the Rayleigh damping and the modal damping models. These simple representations of all physical dissipation mechanisms that contribute to the system damping often render a model with prediction accuracy that is sufficient for its intended purpose. In case of modal damping modeling, the model's damping is set using the outcome of experimental modal analysis of a modal test of the structure under investigation, or using engineering judgement in the mapping of modal damping data from other similar systems. The modal damping found in experiments is normally used for FE simulation without further attempts to understand their physical background. Physically based parameterization of damping phenomena such as friction, radiation, and dissipation is therefore uncommon. The nature of the damping mechanisms is normally such that the modal damping varies from mode to mode. That makes a mapping of experimentally obtained modal damping into modal damping of FE modes cumbersome. The difficulty arises since the mapping of modal damping relies on mode shape pairing, meaning that the same amount of modal damping should be assigned to modes that are similar with respect to their deformation pattern.

Mode pairing of experimentally found modes and FEM modes are usually made through correlation analysis using MAC correlation quantification. Such MAC-based pairing is normally not straightforward, especially for systems with high modal density and with sparsely distributed experimental sensor layout. Eigenmode pairing for the purpose of damping mapping would be unnecessary if the modal damping was the same for all modes. To overcome the problem of mode pairing, a method of damping equalization has therefore been suggested, see Abrahamsson and Kammer (2015). If all modes have the same amount of damping, there is no need for mode matching. The damping equalization is achieved by imposing the same modal damping on all experimentally found system modes by perturbation of a mathematical model of the experimental data found from system identification using raw frequency response function data.

Using efficient system identification methods, such as the state-space subspace method N4SID, these experimental data can be used to obtain a mathematical state-space model. The experimentally found system transfer function $\mathbf{H}^X_{\text{raw}}$ can then be represented by the identified system

$$\dot{\mathbf{x}} = \mathbf{A}\mathbf{x} + \mathbf{B}\mathbf{s}, \quad \mathbf{r} = \mathbf{C}\mathbf{x} + \mathbf{D}\mathbf{s}, \tag{4.163}$$

such that

$$\mathbf{H}^X = \mathbf{C}(\mathrm{i}\omega\mathbf{I} - \mathbf{A})^{-1}\mathbf{B} + \mathbf{D}. \tag{4.164}$$

The experimental state-space system can be brought to diagonal form by a similarity transformation as seen in Eq. 3.139. We thus have that

$$\dot{\mathbf{z}} = \Sigma\mathbf{z} + \Lambda^T\mathbf{B}\mathbf{s}, \quad \mathbf{r} = \mathbf{C}\mathbf{T}\mathbf{z} + \mathbf{D}\mathbf{s} \tag{4.165}$$

with $\Sigma = \text{diag}(\sigma_i)$ in which σ_i are the complex-valued system poles as given by the experimental data. The relative modal damping ζ_i, obtained from these poles, are

$$\zeta_i = -\text{sgn}(\mathfrak{Im}(\sigma_i))\mathfrak{Re}(\sigma_i)/\mathfrak{Im}(\sigma_i). \tag{4.166}$$

In the process of damping equalization, the real parts of the poles are perturbed such that the damping is made equal for all modes. The modal dampings are then set to a single fixed value ζ_0, i.e.,

$$\zeta_i = \zeta_0. \tag{4.167}$$

The effect of such damping equalization is that the oscillatory imaginary part of the poles is preserved and the real damping part is modified such that the perturbed system poles are now

$$\tilde{\sigma}_i = -\zeta_0|\mathfrak{Re}(\sigma_i)| + \mathrm{i}\mathfrak{Im}(\sigma_i), \tag{4.168}$$

and the modified state-space realization is

$$\dot{\mathbf{z}} = \tilde{\mathbf{A}}\mathbf{z} + \Lambda^T\mathbf{B}\mathbf{s}, \quad \mathbf{r} = \mathbf{C}\mathbf{T}\mathbf{z} + \mathbf{D}\mathbf{s} \tag{4.169}$$

with

$$\tilde{\mathbf{A}} = \text{diag}(\tilde{\sigma}_i). \tag{4.170}$$

This in turn gives a modified transfer function for the experimental model, such that the transfer function used for calibration with damping equalization is

$$\mathbf{H}^X = \mathbf{C}\mathbf{T}(\mathrm{i}\omega\mathbf{I} - \tilde{\mathbf{A}})^{-1}\mathbf{T}^{-1}\mathbf{B} + \mathbf{D}. \tag{4.171}$$

At this stage, it should be obvious that the application of the system identification procedure on the raw test data $\mathbf{H}^X_{\text{raw}}$ has led to a mathematical model which can be evaluated for any frequency ω. In particular, it means that equal log-frequency increments can be used for transfer function evaluation. In addition to that, also fictitious modifications of the system under test can be made. A particular use of such modification is that the system damping level can be adjusted, leaving stiffness and inertia properties intact, such that all system modal damping are set equal. The model calibration of the FE model can then be made toward this fictitious experimental model for calibration of parameters that relate to mass and stiffness only.

For the FE-based system representation, the modal damping allows for a simple representation. For a system with given mass and stiffness matrices $\mathbf{M}$ and $\mathbf{K}$, the viscous damping matrix $\mathbf{V}$ is (see Craig 2000)

$$\mathbf{V} = \mathbf{M}\mathbf{X}\text{diag}(m_i)^{-1}\text{diag}(2\zeta_0 m_i\omega_i)^{-1}\text{diag}(m_i)^{-1}\mathbf{X}^T\mathbf{M} \tag{4.172}$$

with eigenfrequencies ω_i, modal masses m_i, and the modal matrix $\mathbf{X}$ given by the undamped system's eigenvalue problem

$$\mathbf{KX} = \mathbf{MX}\text{diag}(\omega_i) \quad \text{diag}(m_i) = \mathbf{X}^T\mathbf{MX}. \tag{4.173}$$

In a calibration procedure, the mass- and stiffness-related parameters $\mathbf{p}$ of the FE model $\{\mathbf{K}(\mathbf{p}), \mathbf{M}(\mathbf{p})\}$ can be searched for that render the transfer function $\mathbf{H}^A$ given by Eqs. 3.85 and 3.110 and that let the criterion function of Eq. 4.162 to be minimal. The discrete frequencies used to evaluate Eq. 4.171 do not have to match the discrete frequencies used in testing.

In model calibration processed test, data are available and a parameterized FE model can be used to calculate the same entities. The calibration task is then to minimize the metric Q, i.e., by searching for the parameter values that give a "best fit" to experimental data. This is normally a nonlinear minimization procedure in the sense that the metric is nonlinear in the parameters. The following chapter treats some well-known procedures to do such minimization. It leads up to the celebrated Levenberg–Marquardt algorithm much used in model calibration.

Minimizing a Quadratic Functional. In all calibration situations, it is possible to formulate a quadratic deviation functional from the deviation metric, termed the residual at $\mathbf{p}$ as

$$Q(\mathbf{p}) = \frac{1}{2}\sum_{i=1}^{m} \delta_i^2(\mathbf{p}) = \frac{1}{2}\boldsymbol{\delta}^T\boldsymbol{\delta}. \tag{4.174}$$

Calibration problems of the type $\mathbf{p}^* = \text{argmin}(Q(\mathbf{p}))$ with the quadratic structure of Q are *nonlinear parameter estimation* problems. If the calibrated model should be thrustworthy, $Q(\mathbf{p}^*)$ should be small, and the number of deviation vector elements M should be greater than the number of variables n_p. If the latter was not true, then an arbitrary model could give a close fit to the data.

Although the function in (4.174) can be minimized by a general unconstrained minimization method, in most situations, the properties of it make it worthwhile to use methods designed specifically for nonlinear least squares minimization problems. In particular, the quadratic form leads to a Hessian matrix of Q that has a special structure. Let the $M \times n_p$ Jacobian matrix of $\boldsymbol{\delta}(\mathbf{p})$ be denoted $\mathbf{J}(\mathbf{p})$, and let the matrix $\mathbf{H}_i(\mathbf{p})$ denote the Hessian matrix with respect to the deviation vector element $\delta_i(\mathbf{p})$. Then, the Hessian of Q is

$$\mathbf{H}(\mathbf{p}) = \mathbf{J}(\mathbf{p})^T\mathbf{J}(\mathbf{p}) + \sum_{i=1}^{M} \delta_i(\mathbf{p})\mathbf{H}_i(\mathbf{p}) \equiv \mathbf{J}(\mathbf{p})^T\mathbf{J}(\mathbf{p}) + \mathbf{S}. \tag{4.175}$$

From (4.175), we observe that the Hessian of the least squares objective function consists of a special combination of first- and second-order information. Least squares methods are typically based on the premise that eventually, as the iterative search for the minimum go on, the first-order term $\mathbf{J}(\mathbf{p})^T\mathbf{J}(\mathbf{p})$ of (4.175) will dominate over the second-order term $\mathbf{S}$ as the deviations δ_i shrink. This assumption is not justified when the residual at the solution $Q(\mathbf{p})$ is large, i.e., in the order of the largest eigenvalue of $\mathbf{J}(\mathbf{p})^T\mathbf{J}(\mathbf{p})$ or greater. In such a case, which might be due to a poor model structure or noisy test data, one might as well use a general unconstrained minimization method. However, for many calibration problems, the residual is indeed small enough to justify the use a special method that uses the smallness of $\mathbf{S(p)}$ to the advantage.

The Newton Method. Let $\mathbf{p}_k$ be the current estimate of the solution at the kth algorithmic iteration step. Any quantity subscripted by k will indicate that it is evaluated at this step. A Taylor series expansion of to second order about the current point $\mathbf{p}_k$ with a perturbation $\mathbf{q}$ gives

$$Q(\mathbf{p}_k + \mathbf{q}) \approx Q_k + \mathbf{J}_k^T\mathbf{q} + \frac{1}{2}\mathbf{q}^T\mathbf{H}_k\mathbf{q}. \tag{4.176}$$

The search is for the optimizing step $\mathbf{q}^*$ that minimizes Q, i.e., $\mathbf{p}^* = \mathbf{p}_k + \mathbf{q}^*$. The minimum will be achieved if $\mathbf{q}^*$ is a minimizer of the quadratic function

$$\Phi(\mathbf{q}) = \mathbf{J}_k^T\mathbf{q} + \frac{1}{2}\mathbf{q}^T\mathbf{H}_k\mathbf{q}, \tag{4.177}$$

where $\mathbf{q}_k^*$ satisfies the linear system

$$d\Phi(\mathbf{q})/d\mathbf{q} = \mathbf{J}_k + \mathbf{H}_k\mathbf{q}_k^* = 0 \tag{4.178}$$

A minimization algorithm in which the stationary point is defined by (4.178) is called a *Newton method*, and the solution $\mathbf{q}_k^*$ of (4.178) is called the Newton direction.

If $\mathbf{H}_k$ is positive definite, only one iteration is required to reach the minimum of the function (4.176) from any starting point $\mathbf{p}_k$, i.e., $\mathbf{p}^* = \mathbf{p}_k + \lambda_k\mathbf{q}_k$ with the step length parameter $\lambda_k = 1$. Therefore, a good convergence rate can be expected from the Newton method when the quadratic model (4.176) is accurate and higher order Taylor series terms are indeed small. For a general nonlinear function Q, the Newton method converges quadratically to $\mathbf{p}^*$ if $\mathbf{p}_k$ is sufficiently close to $\mathbf{p}^*$ and the Hessian matrix is positive definite at $\mathbf{p}^*$. Also the step lengths λ_k then converge to unity. However, for many problems in model calibration, the deviation metric Q is highly nonlinear in the parameters and other higher order methods suit better.

The Gauss–Newton Method. Gauss developed the Newton method further. From (4.174), the Newton equation (4.177) become

$$(\mathbf{J}_k^T\mathbf{J}_k + \mathbf{S}_k)\mathbf{q}_k = -\mathbf{J}_k^T\boldsymbol{\delta}_k. \tag{4.179}$$

Let $\mathbf{q}_k$ denote the solution of (4.178), also called the Newton direction $\mathbf{q}_\mathrm{N}$. The kth step minimization of the function is then made by a one-dimensional search for the minimum along $\mathbf{p} = \mathbf{p}_{k-1} + \lambda\mathbf{p}_\mathrm{N}$ for the search factor λ. The minimum at $\mathbf{p}_{k-1} + \lambda^*\mathbf{p}_\mathrm{N}$ is then given by $\lambda^* = \text{argmin } Q(\mathbf{p}(\lambda))$. If Q_k tends to zero as $\mathbf{p}_k$ approaches the minimizing solution, the matrix $\mathbf{S}_k$ also tends to zero. Thus, the Newton direction can be approximated by the solution of the equation

$$\mathbf{J}_k^T\mathbf{J}_k\mathbf{q}_k = -\mathbf{J}_k^T\boldsymbol{\delta}_k. \tag{4.180}$$

Note that Eq. (4.180) involves only the first derivatives of $\boldsymbol{\delta}$. The solution of (4.180) is the solution of the least squares problem

$$\mathbf{q}_k^* = \arg\min_{\mathbf{q}_k} ||\mathbf{J}_k\mathbf{q}_k + \boldsymbol{\delta}_k||_2^2 \tag{4.181}$$

and is unique if $\mathbf{J}_k$ has full rank. The vector that solves (4.181) is called the Gauss–Newton direction and will be denoted $\mathbf{q}_{\text{GN}}$. A method in which this vector is used as a search direction is known as a *Gauss–Newton method*.

If $\mathbf{J}_k$ is of full rank, the Gauss–Newton direction approaches the Newton direction as $\mathbf{S}_k$ tends to zero in the sense that if $||\mathbf{S}_k|| < \varepsilon$ for a sufficiently small positive scalar ε, then

$$||\mathbf{q}_{\text{N}} - \mathbf{q}_{\text{GN}}||/||\mathbf{q}_{\text{N}}|| = O(\varepsilon). \tag{4.182}$$

Consequently, if $||\boldsymbol{\delta}(\mathbf{p}^*)||$ is zero and the columns of $\mathbf{J}(\mathbf{p}^*)$ are linearly independent, the Gauss–Newton method can ultimately achieve a quadratic rate of convergence, despite the fact that only first derivatives are used to compute $\mathbf{q}_{\text{GN}}$.

In implementations of the Gauss–Newton method, great care is taken to estimate the rank of $\mathbf{J}_k$. It is seen in Eq. (4.180) that a rank-deficient $\mathbf{J}_k$ cause a singular equation system, and a close-to rank-deficient matrix $\mathbf{J}_k$ will make the equation system ill-conditioned rendering errors in determining the search direction. Ill-conditioning is a common feature of nonlinear least squares parameter identification problems if parameter identifiability has not been ascertained. It often manifests itself by that the deviation metric is practically independent of variation of one or more model parameters or along a variation of a combination of parameters. Algorithm robustification is normally made by involving the singular value decomposition in the solution of (4.180) in which the determination of the rank of $\mathbf{J}_k$ plays an important role in estimation good singular value rejection. The rank estimation is determined by approximation methods. When Q is actually close to an ill-conditioned quadratic function, the best strategy is normally to allow the maximum possibly estimation of the rank. However, when $\mathbf{J}_k$ is nearly rank-deficient, a generous estimate of the rank tends to cause very large elements in the solution of (4.180) for the search direction. This causes large parameter variation for even small steps along the search direction along which the quadratic deviation function vary very little. This is an unwanted feature in the parameter estimation process in which the solution should preferably stay at the nominal configuration for insensitive parameters. This has motivated the introduction of parameter regularization as is done in the *Levenberg–Marquardt method* that is described next.

The Levenberg–Marquardt Method. A good alternative to the Gauss–Newton method for ill-conditioned Hessian is the Levenberg–Marquardt method. It is also an alternative to the commonly used regularization method of augmenting the criterion function by penalizing deviation from the nominal parameter setting $\mathbf{p}_0$ into

$$Q_{\text{reg}} = Q + \kappa(\mathbf{p} - \mathbf{p}_0)^T(\mathbf{p} - \mathbf{p}_0), \tag{4.183}$$

where κ is a regularization parameter. In the Levenberg–Marquardt method, the downside of modifying the criterion function by augmentation is avoided, but the upside of regularization of the search direction is still utilized. The Levenberg–Marquardt search direction is defined as the solution of the equation

$$(\mathbf{J}_k^T \mathbf{J}_k + \kappa_k \mathbf{I})\mathbf{q}_k = -\mathbf{J}_k^T \boldsymbol{\delta}_k, \tag{4.184}$$

where the regularizing parameter κ_k is a nonnegative real scalar. It can be noted that it differs from the calculation of the Gauss–Newton search direction of (4.180) in that the approximation to the Hessian is augmented with the Hessian of the regularizing term in (4.183).

A unit step is almost always taken along $\mathbf{q}_k$, i.e., $\mathbf{p}_{k+1}$ is given by $\mathbf{p}_{k+1} = \mathbf{p}_k + \mathbf{q}_k$. It can be shown that, for some scalar κ_0 related to κ_k, the vector $\mathbf{q}_k$ is the solution of the constrained subproblem

$$\mathbf{q}_k^* = \arg \min_{||\mathbf{q}_k||_2 \leq \kappa_0} ||\mathbf{J}_k \mathbf{q}_k + \boldsymbol{\delta}_k||_2^2. \tag{4.185}$$

By that a unit step in the Levenberg–Marquardt direction is taken at each iteration step, it makes it a so-called *trust-region method*. As such a good value of κ_k must be chosen in order to ascertain descent. If κ_k is zero, $\mathbf{q}_k$ is the Gauss–Newton direction and as $\kappa_k \to \infty$, $||\mathbf{q}_k|| \to 0$ then $\mathbf{q}_k^*$ becomes identical to the search direction of the well-known *steepest-descent method*. This implies that $Q(\mathbf{p}_k + \mathbf{q}_k) < Q_k$ for a sufficiently large regularization parameter κ_k. As an alternative, the regularization parameter κ_k may be fixed to $\kappa_k = \kappa_0$ and the iterate minimum be found by one-dimensional line search from $\mathbf{p}_k$ along the direction $\mathbf{q}_{\text{LM}} \equiv \mathbf{q}_k^*$.

The usefulness of the Levenberg–Marquardt algorithm for the calibration of computational structural model is because that these models are often overparameterized. This, or other reasons, normally makes some parameters very little identifiable from test data. That manifests itself by large parameter covariance estimates and is related to the Fisher information. By the use of the Levenberg–Marquardt method, the marginally identifiable parameters do not change much from iteration to iteration and the calibrated solution is close to the initial parameter setting for such parameters, which is, by many, considered to be a sympathetic property of the method. This is in contrast to the results obtained by the Gauss–Newton method under the same circumstances.

All methods here considered are seen to use function value and gradient information only. None uses the Hessian and are thus *Hessian-free*, which is a huge benefit over Hessian-based methods, since the numerical methods of obtaining the Hessian come with a very high cost. An example of using this calibration procedure is given as an industrial size application example in Sect. 5.8.

4.9 Substructuring with State-Space Models

As shown in Sect. 3.4, a state-space model $\dot{\mathbf{x}} = \mathbf{A}\mathbf{x} + \mathbf{b}\mathbf{s},\ \mathbf{r} = \mathbf{C}\mathbf{x} + \mathbf{D}\mathbf{s}$ submitted to a transformation of variables $\mathbf{x} = \mathbf{T}\mathbf{z}$ preserves its input-to-output relation. One such transformation, which is suitable for coupling of components on state-space form, is described next. With that, a transformation can be made to efficiently reduce the redundant degrees of freedom that are common to the coupled components. The elimination can be done using compatibility at the component intersection and is described after an outlook to identification procedures that lead to a state-space model deduced from test data.

4.9.1 *State-Space System Identification*

State-space system identification (SSSI) methods are usually split in two categories out of which one is the *stochastic SSSI* category (a.k.a. output-only system identification) relating to a sub-calls of state-space models of the full state-space model of Eq. (3.82). For this subclass, one has

$$\begin{aligned} \mathbf{x} &= \mathbf{A}\mathbf{x} + \mathbf{E}\mathbf{w} \\ \mathbf{r} &= \mathbf{C}\mathbf{x} + \mathbf{v} \end{aligned} \tag{4.186}$$

for which no known stimuli $\mathbf{s}$ exist. In stochastic SSSI approximating assumptions need to be done for the unknown stochastic system loading $\mathbf{w}$ and the stochastic sensor noise $\mathbf{v}$ as a way forward in the system identification. The stochastic SSSI is not further treated here as calibration needs more precise models with lesser approximations. The other category is the *deterministic SSSI* relating to the state-space equation

$$\begin{aligned} \mathbf{x} &= \mathbf{A}\mathbf{x} + \mathbf{B}\mathbf{s} \\ \mathbf{r} &= \mathbf{C}\mathbf{x} + \mathbf{D}\mathbf{s} \end{aligned} \tag{4.187}$$

for which both the system stimuli $\mathbf{s}$ and response $\mathbf{r}$ are considered being fully known test data. For the deterministic SSSI category, the system transfer function $\mathbf{H}^x(\omega)$ can be estimated from experimental data. Central to the two categories is the estimation of the system's observability matrix

$$\mathfrak{O}^x = \begin{bmatrix} \mathbf{C} \\ \mathbf{CA} \\ \mathbf{CA}^2 \\ \vdots \\ \mathbf{CA}^{n_x-1} \end{bmatrix} \equiv \begin{bmatrix} \mathbf{C} \\ \underline{\mathfrak{O}} \end{bmatrix} \equiv \begin{bmatrix} \overline{\mathfrak{O}} \\ \mathbf{CA}^{n_x-1} \end{bmatrix}. \tag{4.188}$$

Since the true model order n_s of the tested system is hidden (for a continuous system, the order is infinite), a system identification user needs to specify the model order n_x that he believes best capture the model as seen through test data. Discrete-time data for the SSSI are from the n_t steps of the stimuli sequences ($\mathbf{s}_k, k = 1, 2, \ldots, n_t$) and response sequences ($\mathbf{r}_k, k = 1, 2, \ldots, n_t$). The data projection method used for the estimation of $\mathfrak{O}^x$ is at the core of the method but is beyond the scope of this text with good reading provided by van Overschee and de Moor (1996). For a given $\mathfrak{O}^x$, an estimation of the system's output matrix $\mathbf{C}$ is given by the upper n_r rows of $\mathfrak{O}^x$. The shift property $\underline{\mathfrak{O}} = \overline{\mathfrak{O}}\mathbf{A}$ is exploited to give an estimation of the system's dynamic matrix $\mathbf{A}$ as

$$\mathbf{A} = \overline{\mathfrak{O}}^{\dagger}\underline{\mathfrak{O}}, \tag{4.189}$$

where $()^{\dagger}$ denotes the Moore–Penrose pseudo-inverse. For a given couple $\{\mathbf{A}, \mathbf{C}\}$, one notes that the system transfer function (3.85) is linear in $\{\mathbf{B}, \mathbf{D}\}$ and estimates $\{\mathbf{B}^{*}, \mathbf{D}^{*}\}$ for these can be obtained by the least squares solution from data given at discrete frequencies $\omega_k, k = 1, 2, \ldots, K$ as

$$\{\mathbf{B}^{*}, \mathbf{D}^{*}\} = \underset{\{\mathbf{B},\mathbf{D}\}|\{\mathbf{A},\mathbf{C}\}}{\arg\min} \left\| \begin{bmatrix} \mathbf{C}(i\omega_1\mathbf{I} - \mathbf{A})^{-1}\mathbf{B} + \mathbf{D} - \mathbf{H}^x(\omega_1) \\ \mathbf{C}(i\omega_2\mathbf{I} - \mathbf{A})^{-1}\mathbf{B} + \mathbf{D} - \mathbf{H}^x(\omega_2) \\ \vdots \\ \mathbf{C}(i\omega_K\mathbf{I} - \mathbf{A})^{-1}\mathbf{B} + \mathbf{D} - \mathbf{H}^x(\omega_K) \end{bmatrix} \right\|_2^2. \tag{4.190}$$

Since, for a given set $\{\mathbf{A}, \mathbf{B}\}$, the system transfer function (3.85) is linear in $\{\mathbf{C}, \mathbf{D}\}$ a better estimation of these $\{\mathbf{C}^{*}, \mathbf{D}^{*}\}$ can be obtained by solving the complementary least squares problem

$$\{\mathbf{C}^{*}, \mathbf{D}^{*}\} = \underset{\{\mathbf{C},\mathbf{D}\}|\{\mathbf{A},\mathbf{B}\}}{\arg\min} \left\| \begin{bmatrix} \mathbf{C}(i\omega_1\mathbf{I} - \mathbf{A})^{-1}\mathbf{B} + \mathbf{D} - \mathbf{H}^x(\omega_1) \\ \mathbf{C}(i\omega_2\mathbf{I} - \mathbf{A})^{-1}\mathbf{B} + \mathbf{D} - \mathbf{H}^x(\omega_2) \\ \vdots \\ \mathbf{C}(i\omega_K\mathbf{I} - \mathbf{A})^{-1}\mathbf{B} + \mathbf{D} - \mathbf{H}^x(\omega_K) \end{bmatrix} \right\|_2^2. \tag{4.191}$$

The procedure using Eqs. (4.190) and (4.191) in an iterative fashion usually renders a quick convergence to a good model $\{\mathbf{A}, \mathbf{B}, \mathbf{C}, \mathbf{D}\} = \{\mathbf{A}, \mathbf{B}^{*}, \mathbf{C}^{*}, \mathbf{D}^{*}\}$.

4.9.2 Physically Consistent State-Space Models

A second-order structural dynamics finite element model as that given in Eq. (3.103) is physically sound. Its bases are well-known first principles and well-established discretization techniques. The corresponding state-space companion can therefore also be considered as physically sound.

On the other hand, state-space models identified from test data can suffer from errors made in testing and errors and approximations introduced by signal processing of test data as the result of pure mistakes, nonideal sensors, and others. Such errors may lead to state-space model behavior that is not motivated by the physics of the tested real-world system. Typical such errors are related to stimuli measurements. Vibration testing is usually made by either impact stimuli by use of hammer-like impactors or by electromagnetic or hydraulic shakers. In hammer testing, the system is hit repeatedly at the same location in the same angle of attack toward the system surface and averages are made to minimize noise effects on estimated frequency response functions. Ideally, these impacts are exact repetitions which they never are in practice. For shaker-induced vibrations, the stimuli force obtained by the motion of the vibrating piston of the shaker, through a tension/compression rod called a stinger, is measured by a force cell. The stinger should ideally transmit only the force component in the direction of the stinger and the force cell should be ideal with sensor output-only linearly proportional to that force transmitted by the stinger. These ideal conditions do not occur in practice, and unmeasured shear forces and bending/twisting couples are always transmitted from the shaker through the stinger sensor into the tested system and spuriously affect the measured system response. These are just a few examples of the many potential error sources that may lead to identified state-space models that are not truly valid for their intended purpose. Due to the unavoidable imperfections introduced by the testing procedure the fundamental laws of physics can be violated by the identified model. If that is the case, that model is thus *physically inconsistent*. If proper actions are not taken state-space models describing the relation between force/couple input and motion output (such as displacement, velocity, or acceleration) may violate various physical laws such as follows.

Stability law. A linear system is stable if the response due to any excitation is bounded. A system is said to be asymptotically stable if its free motion converges to a fixed equilibrium state. All system poles of an asymptotically stable system have negative real parts. An unstable system, on the other hand, has one or more system poles that are positive real which leads to exponential response growth from arbitrarily small stimulus. A system is said to be marginally stable if it is neither asymptotically stable nor unstable. An undamped structural dynamics system $\mathbf{M}\ddot{\mathbf{u}} + (\mathbf{K} + \lambda\mathbf{K}_g)\mathbf{u} = \mathbf{F}$ with $\lambda < \lambda_{cr}$ is stable but not asymptotically stable since it will vibrate about an equilibrium point for eternity after impact. On the other hand, a damped system $\mathbf{M}\ddot{\mathbf{u}} + \mathbf{V}\dot{\mathbf{u}} + \mathbf{K}\mathbf{u} = \mathbf{F}$ with a symmetric positive definite viscous damping matrix $\mathbf{V}$ is asymptotically stable. An identified state-space model, stemming from test data of a stable system, that has system poles with spurious positive part thus needs trimming. A well-established technique is to mirror any unstable model poles in the imaginary axis and thus bring those poles being negative real instead of positive real. This trimming will usually just lead to a small effect on the system's frequency-domain transfer functions.

Passivity law. System passivity relates to the flow of energy over a system boundary. The energy conservation principle stipulates that

$$E_s(t) + E_{in}(t) - E_{out}(t) = E_s(t_0) \ \forall t > t_0 \tag{4.192}$$

with E_s being the system's internal energy and E_{in} and E_{out} are the energies provided by outside energy sources or absorbed by outside energy sinks. These energies are all positive quantities. The system is isolated by a system boundary that clearly encapsules the inside (the system) from its outside. For a system to be *long time passive in the mechanical sense* (LTP) the outside energy influx E_{in} is purely mechanical and the system energy at $t = t_0$ is bounded. Assume that there exists a stationary periodic solution (period T), to a stationary periodic system stimuli such that

$$\begin{aligned} E_s(t+nT) + E_m(t+nT) - E_{out}(t+nT) = \\ E_s(t+nT+T) + E_m(t+nT) + \Delta E_m(T) - E_{out}(t+nT) - \Delta E_{out}(T) \end{aligned} \tag{4.193}$$

with $E_m = E_{in}$ being the energy provided by the mechanical input. Over one full cycle T, the internal energy has thus been changed the amount of $\Delta E_m(T) - \Delta E_{out}(T)$. For the LTP system, the energy is bounded $0 \leq E_s(0) < \infty$ and thus it is required for long time stationarity that this amount of change needs to be positive and thus

$$\Delta E_m(T) > \Delta E_o(T) \geq 0. \tag{4.194}$$

If not positive, it would require nonpositive energy $E_s(t+nT+T) < 0$ after long time $t + nT + T$, which is unphysical. A periodic solution under the condition that $\Delta E_m(T) = 0$ can only persist until the system's initial energy $E_s(t_0)$ (such as elastic energy, kinetic energy, potential energy, chemical energy, electromagnetic energy and heat) is fully drained. For a system to be LTP, it is thus necessary for the net energy provided by the mechanical stimuli over a fyll cycle to be positive.

The instantaneous power $P(t)$ supplied by a force stimuli $\mathbf{F}(t)$ with energy-conjugate velocity response $\mathbf{v}(t)$ is $P(t) = \mathbf{F}^T\mathbf{v}(t)$. For a stationary harmonic (and thus periodic) force $\mathbf{F}(t) = \hat{\mathbf{F}}(t)e^{i\omega t}$ and stationary harmonic response $\mathbf{v}(t) = \hat{\mathbf{v}}(t)e^{i\omega t}$, the mechanical energy provided to the system over a full cycle $T = 2\pi/\omega$ is

$$\Delta E_m(T) = \int_{\tau=0}^{T} P(\tau)d\tau = \mathfrak{Re}\{\hat{\mathbf{F}}^H\hat{\mathbf{v}}\} = \mathfrak{Re}\{\hat{\mathbf{F}}^H\mathbf{H}\hat{\mathbf{F}}\}, \tag{4.195}$$

where $\mathbf{H}$ is the system transfer function for force-to-velocity. The real part of $\hat{\mathbf{F}}^H\mathbf{H}\hat{\mathbf{F}}$ evaluates to

$$\Delta E_m(T) = \mathfrak{Re}\{\hat{\mathbf{F}}\}^T\mathfrak{Re}\{\mathbf{H}\}\mathfrak{Re}\{\hat{\mathbf{F}}\} + \mathfrak{Im}\{\hat{\mathbf{F}}\}^T\mathfrak{Re}\{\mathbf{H}\}\mathfrak{Im}\{\hat{\mathbf{F}},\} \tag{4.196}$$

which is seen to be positive for all positive definite $\mathfrak{Re}\{\mathbf{H}\}$. For an LTP system, the real part of the transfer function thus needs to be positive definite. This puts a constraint on any system model that should truly mimic such a system. Since the system is passive only if the real part of the mobility frequency response function

$\mathbf{C}(i\omega\mathbf{I}+\mathbf{A})^{-1}\mathbf{B}$ is positive definite, the passivity constraint is related to the positive real (PR) lemma, see Anderson (1967). The lemma states that for a system model to be PR for all frequencies ω there needs to be a symmetric positive definite matrix $\mathbf{P}$ for some $\mathbf{L}$ such that there is a solution to the Lyapunov equation

$$\mathbf{PA}+\mathbf{A}^T\mathbf{P}=-\mathbf{L}^T\mathbf{L}<0 \tag{4.197}$$

under the condition that

$$\mathbf{B}^T\mathbf{P}=\mathbf{C} \tag{4.198}$$

holds. It has been experienced to be far from trivial to find a solution to (4.196) and (4.197) for the general MIMO system, and no successful attempts have been made to the authors' knowledge. Therefore, attempts have been made to prove PR under less general SISO conditions instead. The general problem is hard to solve since it requires the search for a solution $\{\mathbf{P},\mathbf{L}\}$ if such at all exists (the PR case) or to prove that no such solution exists (the non-PR case).

Reciprocity law. Many mechanical systems obey the principle of reciprocity which manifests itself in symmetric coefficient matrices in the second-order differential equations. One exception is the spinning mechanical system that generates unsymmetric Coriolis forces and another exception is the aeroelastic system for which the aeroelastic forces cause unsymmetry. A common practice to avoid that a state-space model identified from test data violates the law of reciprocity is to process the data such that reciprocity results. Say that MIMO transfer function data $\mathbf{H}(\omega)$ are taken from a system that is judged to be reciprocal with energy-conjugate stimuli and responses. The matrix $\mathbf{H}(\omega)$ is thus square. A simple procedure to modify test data to be symmetric, and form reciprocal transfer functions $\mathbf{H}_r(\omega)$, is then to make averaging of non-diagonal elements of the transfer function matrix as

$$\mathbf{H}_r(\omega)=(\mathbf{H}(\omega)+\mathbf{H}^T(\omega))/2. \tag{4.199}$$

Other methods rely on weighted averages in which test data quality indicators, such as test data coherence of the individual elements in $\mathbf{H}$, are used.

Displacement–velocity consistency. Newton's second law gives a direct relation between force and acceleration. A state-space model for acceleration response can therefore include a direct throughput term in the output equation, i.e., the output equation for acceleration may read $\mathbf{r}_a=\mathbf{C}_a\mathbf{x}+\mathbf{D}_a\mathbf{s}$ with $\mathbf{D}_a\neq\mathbf{0}$. However, there is no direct relation from force to displacement or velocity; these are indirectly related to force and needs to be integrated from the dynamic equation $\dot{\mathbf{x}}=\mathbf{A}\mathbf{x}+\mathbf{B}\mathbf{s}$. The output equation for displacement output thus must read $\mathbf{r}_d=\mathbf{C}_d\mathbf{x}$. Since the velocity output is related to displacement output as $\mathbf{r}_v=\dot{\mathbf{r}}_d=\mathbf{C}_d(\mathbf{A}\mathbf{x}+\mathbf{B}\mathbf{s})=\mathbf{C}_d\mathbf{A}\mathbf{x}+\mathbf{C}_d\mathbf{B}\mathbf{s}$, it must also hold that $\mathbf{C}_d\mathbf{B}=\mathbf{0}$. With $\mathbf{C}_d\mathbf{B}=\mathbf{0}$, one also notes that the acceleration output may be written $\mathbf{r}_a=\dot{\mathbf{r}}_v=\mathbf{C}_d\mathbf{A}(\mathbf{A}\mathbf{x}+\mathbf{B}\mathbf{s})=\mathbf{C}_d\mathbf{A}^2\mathbf{x}+\mathbf{C}_d\mathbf{A}\mathbf{B}\mathbf{s}$ which leads to the conclusion that $\mathbf{C}_d\mathbf{A}\mathbf{B}$ is not necessarily zero.

For accelerance data $\mathbf{H}^x$, the procedure to improve the estimates of the pair $\{\mathbf{B}, \mathbf{C}\}$ described in Sect. 4.9.1 can be modified to take account for such physically motivated constraint by iteratively improve the estimates of first $\mathbf{B}$ and then $\mathbf{C}_d$ in a augmented least squares sense in the sequence

$$\mathbf{B}^* = \underset{\mathbf{B}|\{\mathbf{A},\mathbf{C}_d\}}{\arg\min} || \begin{bmatrix} \mathbf{C}_d\mathbf{A}^2(i\omega_1\mathbf{I}-\mathbf{A})^{-1}\mathbf{B}+\mathbf{C}_d\mathbf{A}\mathbf{B}-\mathbf{H}^x(\omega_1) \\ \mathbf{C}_d\mathbf{A}^2(i\omega_2\mathbf{I}-\mathbf{A})^{-1}\mathbf{B}+\mathbf{C}_d\mathbf{A}\mathbf{B}-\mathbf{H}^x(\omega_2) \\ \vdots \\ \mathbf{C}_d\mathbf{A}^2(i\omega_k\mathbf{I}-\mathbf{A})^{-1}\mathbf{B}+\mathbf{C}_d\mathbf{A}\mathbf{B}-\mathbf{H}^x(\omega_k) \\ \mathbf{C}_d\mathbf{B} \end{bmatrix} ||_2^2 \tag{4.200}$$

and

$$\mathbf{C}_d^* = \underset{\mathbf{C}_d|\{\mathbf{A},\mathbf{B}\}}{\arg\min} || \begin{bmatrix} \mathbf{C}_d\mathbf{A}^2(i\omega_1\mathbf{I}-\mathbf{A})^{-1}\mathbf{B}+\mathbf{C}_d\mathbf{A}\mathbf{B}-\mathbf{H}^x(\omega_1) \\ \mathbf{C}_d\mathbf{A}^2(i\omega_2\mathbf{I}-\mathbf{A})^{-1}\mathbf{B}+\mathbf{C}_d\mathbf{A}\mathbf{B}-\mathbf{H}^x(\omega_2) \\ \vdots \\ \mathbf{C}_d\mathbf{A}^2(i\omega_k\mathbf{I}-\mathbf{A})^{-1}\mathbf{B}+\mathbf{C}_d\mathbf{A}\mathbf{B}-\mathbf{H}^x(\omega_k) \\ \mathbf{C}_d\mathbf{B} \end{bmatrix} ||_2^2. \tag{4.201}$$

Mobility and accelerance constraints. If the mechanical system under study is not free to undergo rigid body motion its mobility and accelerance transfer functions in statics ($\omega = 0$) need to be zero. With the velocity output equation $\mathbf{r}_v = \mathbf{C}_v\mathbf{x}$ and the acceleration output equation $\mathbf{r}_a = \dot{\mathbf{r}}_v = \mathbf{C}_v\dot{\mathbf{x}} = \mathbf{C}_v\mathbf{A}\mathbf{x} + \mathbf{C}_v\mathbf{B}\mathbf{s}$, the mobility matrix at statics thus needs to be $\mathbf{C}_v\mathbf{A}^{-1}\mathbf{B} = \mathbf{0}$ and the accelerance matrix at statics needs to be $2\mathbf{C}_v\mathbf{B} = \mathbf{0}$

Augmented least squares solutions that enforce such constraints for accelerance data $\mathbf{H}^x$ can be obtained by the iterative solutions of

$$\mathbf{B}^* = \underset{\mathbf{B}|\{\mathbf{A},\mathbf{C}_d\}}{\arg\min} || \begin{bmatrix} \mathbf{C}_v\mathbf{A}(i\omega_1\mathbf{I}-\mathbf{A})^{-1}\mathbf{B}+\mathbf{C}_v\mathbf{B}-\mathbf{H}^x(\omega_1) \\ \mathbf{C}_v\mathbf{A}(i\omega_2\mathbf{I}-\mathbf{A})^{-1}\mathbf{B}+\mathbf{C}_v\mathbf{B}-\mathbf{H}^x(\omega_2) \\ \vdots \\ \mathbf{C}_v\mathbf{A}(i\omega_k\mathbf{I}-\mathbf{A})^{-1}\mathbf{B}+\mathbf{D}+\mathbf{C}_v\mathbf{B}-\mathbf{H}^x(\omega_k) \\ \mathbf{C}_v\mathbf{A}^{-1}\mathbf{B} \\ \mathbf{C}_v\mathbf{B} \end{bmatrix} ||_2^2 \tag{4.202}$$

and

$$\mathbf{C}_d^* = \underset{\mathbf{C}_d|\{\mathbf{A},\mathbf{B}\}}{\arg\min} || \begin{bmatrix} \mathbf{C}_d\mathbf{A}(i\omega_1\mathbf{I}-\mathbf{A})^{-1}\mathbf{B}+\mathbf{C}_v\mathbf{B}-\mathbf{H}^x(\omega_1) \\ \mathbf{C}_d\mathbf{A}(i\omega_2\mathbf{I}-\mathbf{A})^{-1}\mathbf{B}+\mathbf{C}_v\mathbf{B}-\mathbf{H}^x(\omega_2) \\ \vdots \\ \mathbf{C}_d\mathbf{A}(i\omega_k\mathbf{I}-\mathbf{A})^{-1}\mathbf{B}+\mathbf{C}_v\mathbf{B}-\mathbf{H}^x(\omega_k) \\ \mathbf{C}_v\mathbf{A}^{-1}\mathbf{B} \\ \mathbf{C}_v\mathbf{B} \end{bmatrix} ||_2^2. \tag{4.203}$$

4.9.3 State-Space Realization on Substructuring Form

Consider two state-space components (I and II) that are subjected to coupling. The motion of the interface between the components is represented by the displacement vector $\mathbf{r} \in \Re^{n_c}$. To the displacement, the vectorially associated (energy-conjugate) stimuli vectors $\mathbf{s}^i \in \Re^{n_c}$ that act on the two components are $\mathbf{s}^I$ and $\mathbf{s}^{II}$. For displacement output, a proper state-space representation of the two models $i = I, II$ is

$$\begin{aligned}\dot{\mathbf{x}}^i &= \mathbf{A}^i\mathbf{x}^i + \mathbf{B}^i\mathbf{s}^i \\ \mathbf{r} &= \mathbf{C}^i\mathbf{x}^i.\end{aligned} \tag{4.204}$$

Displacement compatibility at the common interface leads to that the n_c displacement outputs $\mathbf{r}$ are same for both components. For an energy-conjugate system, a state-space model structure $\{\mathbf{A}^i, \mathbf{B}^i, \mathbf{C}^i\}$ without a direct throughput term is proper since such a system lacks direct throughput in consistency with the following lemma.

Lemma 1 *An energy-conjugate state-space system* $\{\mathbf{A}, \mathbf{B}, \mathbf{C}, \mathbf{D}\}$ *for which the instantaneous power P provided by the stimuli $\mathbf{s}$ amounts to $P = \mathbf{s}^T\dot{\mathbf{r}}$, the direct throughput matrix needs to be* $\mathbf{D} = \mathbf{0}$.

Proof With the response $\mathbf{r} = \mathbf{C}\mathbf{x} + \mathbf{D}\mathbf{s}$, the response rate is $\dot{\mathbf{r}} = \mathbf{C}\dot{\mathbf{x}} + \mathbf{D}\dot{\mathbf{s}} = \mathbf{C}\mathbf{A}\mathbf{x} + \mathbf{C}\mathbf{B}\mathbf{s} + \mathbf{D}\dot{\mathbf{s}}$ and thus the instantaneous power is $P = \mathbf{s}^T(\mathbf{C}\mathbf{A}\mathbf{x} + \mathbf{C}\mathbf{B}\mathbf{s}) + \mathbf{s}^T\mathbf{D}\dot{\mathbf{s}}$. To be independent on the stimuli rate $\dot{\mathbf{s}}$, it requires that $\mathbf{s}^T\mathbf{D} = \mathbf{0}$ for which the only possible solution for arbitrary stimulus $\mathbf{s}$ is that $\mathbf{D} = \mathbf{0}$. □

Let further the two systems be such that they could alternatively be represented by second-order differential equations. Then, the relation between the matrices of the triple $\{\mathbf{A}, \mathbf{B}, \mathbf{C}\}$ is given by the next lemma.

Lemma 2 *For a state-space realization* $\dot{\mathbf{x}} = \mathbf{A}\mathbf{x} + \mathbf{B}\mathbf{s}$, $\mathbf{r} = \mathbf{C}\mathbf{x}$ *of a second-order system* $\mathbf{M}\ddot{\mathbf{u}} + \mathbf{V}\dot{\mathbf{u}} + \mathbf{K}\mathbf{u} = \mathbf{P}_s\mathbf{s}$ *with response* $\mathbf{r} = \mathbf{P}_d\mathbf{u}$ *being a linear combination of* $\mathbf{u}$ *only and with nonsingular mass matrix* $\mathbf{M}$, (a) *the relation between* $\mathbf{B}$ *and* $\mathbf{C}$ *is such that* $\mathbf{C}\mathbf{B} = \mathbf{0}$, *and* (b) *the relation between* $\mathbf{A}$, $\mathbf{B}$ *and* $\mathbf{C}$ *is such that* $\mathbf{C}\mathbf{A}\mathbf{B} \neq \mathbf{0}$.

Proof For a second-order system, one possible state-space realization pair is, see Eq. (3.110),

$$\mathbf{B} = \begin{bmatrix}\mathbf{0} \\ \mathbf{M}^{-1}\mathbf{P}_s\end{bmatrix}, \quad \mathbf{C} = \mathbf{P}_d\begin{bmatrix}\mathbf{I} & \mathbf{0}\end{bmatrix},$$

from which it follows that, also after an arbitrary similarity transformation $\mathbf{x} = \mathbf{T}\mathbf{z}$, the relation of the pair $\{\mathbf{T}^{-1}\mathbf{B}, \mathbf{C}\mathbf{T}\}$ is such that

$$(\mathbf{C}\mathbf{T})(\mathbf{T}^{-1}\mathbf{B}) = \mathbf{P}_d\begin{bmatrix}\mathbf{I} & \mathbf{0}\end{bmatrix}\begin{bmatrix}\mathbf{0} \\ \mathbf{M}^{-1}\mathbf{P}_s\end{bmatrix} = \mathbf{0},$$

which proofs part (a). With the structure of $\mathbf{A}$ given by Eq. (3.110), it can easily be verified that $\mathbf{CAB} = \mathbf{P}_d\mathbf{M}^{-1}\mathbf{P}_s$. Since $\mathbf{M}$ is full rank thus $\mathbf{CAB} \neq \mathbf{0}$, unless under the trivial condition that $\mathbf{P}_d = \mathbf{0}$ and/or $\mathbf{P}_s = \mathbf{0}$, which concludes the proof of (b). □

4.9.4 State-Space Model Coupling

Let the interface loads of the components be decomposed of two parts, the external force $\alpha_i\mathbf{s}$ and the interaction force between the components $\beta_i\bar{\mathbf{s}}$. The interface of the two components is then loaded by the total external stimuli $\mathbf{s} = \alpha_I\mathbf{s} + \alpha_{II}\mathbf{s}$ (thus $\alpha_I + \alpha_{II} = 1$). For the interaction forces $\beta_I\bar{\mathbf{s}}$ and $\beta_{II}\bar{\mathbf{s}}$, it holds that $\beta_I = -\beta_{II}$ according to Newton's law on action and interaction. Further, assume that the models can be brought to coupling form with a model structure like

$$\dot{\mathbf{x}}^i = \begin{bmatrix} \mathbf{0} & \mathbf{I} & \mathbf{0} \\ \mathbf{A}^i_{vd} & \mathbf{A}^i_{vv} & \mathbf{A}^i_{vb} \\ \mathbf{A}^i_{bd} & \mathbf{A}^i_{bv} & \mathbf{A}^i_{bb} \end{bmatrix} \mathbf{x}^i + \begin{bmatrix} \mathbf{0} \\ \mathbf{B}^i_{vv} \\ \mathbf{0} \end{bmatrix} (\alpha_i\mathbf{s} + \beta_i\bar{\mathbf{s}}) \qquad \mathbf{r} = \begin{bmatrix} \mathbf{I} & \mathbf{0} & \mathbf{0} \end{bmatrix} \mathbf{x}^i \tag{4.205}$$

with state vector

$$\mathbf{x}^i = \begin{pmatrix} \mathbf{r} \\ \dot{\mathbf{r}} \\ \mathbf{x}^i_b \end{pmatrix} \tag{4.206}$$

and with nonsingular $\mathbf{B}^i_{vv}$ with inverse $\mathbf{B}^{-i}_{vv}$. The two systems are thus

$$\begin{pmatrix} \dot{\mathbf{r}} \\ \mathbf{B}^{-I}_{vv}\ddot{\mathbf{r}} \\ \dot{\mathbf{x}}^I_b \end{pmatrix} = \begin{bmatrix} \mathbf{0} & \mathbf{I} & \mathbf{0} \\ \mathbf{B}^{-I}_{vv}\mathbf{A}^I_{vd} & \mathbf{B}^{-I}_{vv}\mathbf{A}^I_{vv} & \mathbf{B}^{-I}_{vv}\mathbf{A}^I_{vb} \\ \mathbf{A}^I_{bd} & \mathbf{A}^I_{bv} & \mathbf{A}^I_{bb} \end{bmatrix} \mathbf{x}^I + \begin{bmatrix} \mathbf{0} \\ \alpha_I\mathbf{s} + \beta_I\bar{\mathbf{s}} \\ \mathbf{0} \end{bmatrix} \tag{4.207}$$

$$\mathbf{r} = \begin{bmatrix} \mathbf{I} & \mathbf{0} & \mathbf{0} \end{bmatrix} \mathbf{x}^I$$

and

$$\begin{pmatrix} \dot{\mathbf{r}} \\ \mathbf{B}^{-II}_{vv}\ddot{\mathbf{r}} \\ \dot{\mathbf{x}}^{II}_b \end{pmatrix} = \begin{bmatrix} \mathbf{0} & \mathbf{I} & \mathbf{0} \\ \mathbf{B}^{-II}_{vv}\mathbf{A}^{II}_{vd} & \mathbf{B}^{-II}_{vv}\mathbf{A}^{II}_{vv} & \mathbf{B}^{-II}_{vv}\mathbf{A}^{II}_{vb} \\ \mathbf{A}^{II}_{bd} & \mathbf{A}^{II}_{bv} & \mathbf{A}^{II}_{bb} \end{bmatrix} \mathbf{x}^{II} + \begin{bmatrix} \mathbf{0} \\ \alpha_{II}\mathbf{s} + \beta_{II}\bar{\mathbf{s}} \\ \mathbf{0} \end{bmatrix} \tag{4.208}$$
$$\mathbf{r} = \begin{bmatrix} \mathbf{I} & \mathbf{0} & \mathbf{0} \end{bmatrix} \mathbf{x}^{II}.$$

Adding the two systems together results in the coupled system equations with the expanded state vector $\bar{\mathbf{x}}^T = [\mathbf{r}\ \dot{\mathbf{r}}\ \mathbf{x}^I_b\ \mathbf{x}^{II}_b]$ as

$$
\begin{pmatrix} \dot{\mathbf{r}} \\ (\mathbf{B}_{vv}^{-I} + \mathbf{B}_{vv}^{-II})\ddot{\mathbf{r}} \\ \mathbf{x}_b^I \\ \mathbf{x}_b^{II} \end{pmatrix} = \tag{4.209}
$$

$$
\begin{bmatrix} \mathbf{0} & \mathbf{I} & \mathbf{0} & \mathbf{0} \\ \mathbf{B}_{vv}^{-I}\mathbf{A}_{vd}^{I} + \mathbf{B}_{vv}^{-II}\mathbf{A}_{vd}^{II} & \mathbf{B}_{vv}^{-I}\mathbf{A}_{vv}^{I} + \mathbf{B}_{vv}^{-II}\mathbf{A}_{vv}^{II} & \mathbf{B}_{vv}^{-I}\mathbf{A}_{vb}^{I} & \mathbf{B}_{vv}^{-II}\mathbf{A}_{vb}^{II} \\ \mathbf{A}_{bd}^{I} & \mathbf{A}_{bv}^{I} & \mathbf{A}_{bb}^{I} & \mathbf{0} \\ \mathbf{A}_{bd}^{II} & \mathbf{A}_{bv}^{II} & \mathbf{0} & \mathbf{A}_{bb}^{II} \end{bmatrix} \bar{\mathbf{x}}
$$

$$
+ \begin{bmatrix} \mathbf{0} \\ (\alpha_I + \alpha_{II})\mathbf{s} + (\beta_{II} + \beta_{II})\bar{\mathbf{s}} \\ \mathbf{0} \\ \mathbf{0} \end{bmatrix}
$$

$$
\mathbf{r} = \begin{bmatrix} \mathbf{I} \; \mathbf{0} \; \mathbf{0} \; \mathbf{0} \end{bmatrix} \bar{\mathbf{x}}.
$$

Since $\alpha_I + \alpha_{II} = 1$ and $\beta_I + \beta_{II} = 0$, this results in the coupled state-space model

$$
\dot{\bar{\mathbf{x}}} = \begin{bmatrix} \mathbf{0} & \mathbf{I} & \mathbf{0} & \mathbf{0} \\ \bar{\mathbf{A}}_{vd} & \bar{\mathbf{A}}_{vv} & \bar{\mathbf{A}}_{vb}^{I} & \bar{\mathbf{A}}_{vb}^{II} \\ \mathbf{A}_{bd}^{I} & \mathbf{A}_{bv}^{I} & \mathbf{A}_{bb}^{I} & \mathbf{0} \\ \mathbf{A}_{bd}^{II} & \mathbf{A}_{bv}^{II} & \mathbf{0} & \mathbf{A}_{bb}^{II} \end{bmatrix} \bar{\mathbf{x}} + \begin{bmatrix} \mathbf{0} \\ \bar{\mathbf{B}}_{vv} \\ \mathbf{0} \\ \mathbf{0} \end{bmatrix} \mathbf{s} \qquad \mathbf{r} = \begin{bmatrix} \mathbf{I} \; \mathbf{0} \; \mathbf{0} \; \mathbf{0} \end{bmatrix} \bar{\mathbf{x}}, \tag{4.210}
$$

for which it is assumed that $\mathbf{B}_{vv}^{I}$, $\mathbf{B}_{vv}^{II}$ and $\mathbf{B}_{vv}^{-I} + \mathbf{B}_{vv}^{-II} \equiv \Gamma^{-1}$ are nonsingular and its following matrix partitions can therefore be formed

$$
\bar{\mathbf{A}}_{vd} = \Gamma[\mathbf{B}_{vv}^{-I}\mathbf{A}_{vd}^{I} + \mathbf{B}_{vv}^{-II}\mathbf{A}_{vd}^{II}] \tag{4.211}
$$

$$
\bar{\mathbf{A}}_{vv} = \Gamma[\mathbf{B}_{vv}^{-I}\mathbf{A}_{vv}^{I} + \mathbf{B}_{vv}^{-II}\mathbf{A}_{vv}^{II}] \tag{4.212}
$$

$$
\bar{\mathbf{A}}_{vb}^{I} = \Gamma\mathbf{B}_{vv}^{-I}\mathbf{A}_{vb}^{I} \tag{4.213}
$$

$$
\bar{\mathbf{A}}_{vb}^{II} = \Gamma\mathbf{B}_{vv}^{-II}\mathbf{A}_{vb}^{II} \tag{4.214}
$$

$$
\bar{\mathbf{B}}_{vv} = \Gamma. \tag{4.215}
$$

Then, a state-space model structure on coupling form is realizable, under certain conditions on **A** and **C**, for the energy-conjugate state-space system follows from the proof of the following lemma.

Lemma 3 (Coupling form lemma)
A state-space system $\dot{\mathbf{x}} = \mathbf{A}\mathbf{x} + \mathbf{B}\mathbf{s}$, $\mathbf{r} = \mathbf{C}\mathbf{x}$ $(\mathbf{A} \in R^{n \times n}, \mathbf{B} \in R^{n \times n_c}, \mathbf{C} \in R^{n_c \times n})$ *stemming from a second-order system* $\mathbf{M}\ddot{\mathbf{u}} + \mathbf{V}\dot{\mathbf{u}} + \mathbf{K}\mathbf{u} = \mathbf{P}_s\mathbf{s}$ *can be transformed*

to a system with state-space triple $\{\bar{\mathbf{A}}, \bar{\mathbf{B}}, \bar{\mathbf{C}}\}$ *on the coupling form according to Eq. (4.205) provided that* $[\mathbf{C}; \mathbf{CA}]$ *is of full row rank* $2n_c$ *and that* $n_c \leq n$.

Proof Let $\mathbf{T}$ be a transformation for $\{\mathbf{A}, \mathbf{B}, \mathbf{C}\}$ with inverse $\mathbf{Z}$ be such that

$$\mathbf{T} = \begin{bmatrix} \mathbf{C} \\ \mathbf{CA} \\ \mathbf{N} \end{bmatrix} \text{ and } \mathbf{Z} \equiv \begin{bmatrix} \mathbf{Z}_1 \; \mathbf{Z}_2 \; \mathbf{Z}_3 \end{bmatrix}. \tag{4.216}$$

Let further $\mathbf{N} \in R^{n-2n_c \times n}$ be the projection of the nullspace $\mathbf{N}_B \in R^{n-n_c \times n}$ of $\mathbf{B}$ on the nullspace $\mathbf{N}_C \in R^{n-2n_c \times n}$ of $[\mathbf{C}; \mathbf{CA}]$ as

$$\mathbf{N} = \mathbf{N}_C \mathbf{N}_B^T (\mathbf{N}_B \mathbf{N}_B^T)^{-1} \mathbf{N}_B, \tag{4.217}$$

with the nullspaces $\mathbf{N}_C$ and $\mathbf{N}_B$ defined by

$$[\mathbf{C}; \mathbf{CA}]\mathbf{N}_C = \mathbf{0} \text{ and } \mathbf{N}_B \mathbf{B} = \mathbf{0}. \tag{4.218}$$

The nullspace $\mathbf{N}_C$ is of full row rank $n - 2n_c$ and thus the combined nullspace $\mathbf{N}$ is also of full row rank $n - 2n_c$ because $\mathbf{B}$ is not in the nullspace of $[\mathbf{C}; \mathbf{CA}]$ since $\mathbf{CAB} \neq \mathbf{0}$. Since both $[\mathbf{C}; \mathbf{CA}]$ and $\mathbf{N}$ are full row rank and $\mathbf{N}$ is a nullspace of $[\mathbf{C}; \mathbf{CA}]$, it implies that $\mathbf{T}$ is full rank and thus nonsingular. Using that $\mathbf{TZ} = \mathbf{I}$, one has

$$\begin{bmatrix} \mathbf{CZ}_1 & \mathbf{CZ}_2 & \mathbf{CZ}_3 \\ \mathbf{CAZ}_1 & \mathbf{CAZ}_2 & \mathbf{CAZ}_3 \\ \mathbf{NZ}_1 & \mathbf{NZ}_2 & \mathbf{NZ}_3 \end{bmatrix} = \mathbf{I},$$

which, in particular, imply that $\mathbf{CAZ}_1 = \mathbf{0}$, $\mathbf{CAZ}_2 = \mathbf{I}$, and $\mathbf{CAZ}_3 = \mathbf{0}$. The first block row of the system matrix on coupling form $\bar{\mathbf{A}} = \mathbf{TAT}^{-1} = \mathbf{TAZ}$ is thus

$$\mathbf{CA} \begin{bmatrix} \mathbf{Z}_1 \; \mathbf{Z}_2 \; \mathbf{Z}_3 \end{bmatrix} = \begin{bmatrix} \mathbf{CAZ}_1 \; \mathbf{CAZ}_2 \; \mathbf{CAZ}_3 \end{bmatrix} = \begin{bmatrix} \mathbf{0} \; \mathbf{I} \; \mathbf{0} \end{bmatrix},$$

which concludes the proof of the coupling form structure of $\bar{\mathbf{A}}$. The structure of $\bar{\mathbf{B}}$ is given by

$$\bar{\mathbf{B}} = \mathbf{TB} = \begin{bmatrix} \mathbf{C} \\ \mathbf{CA} \\ \mathbf{N} \end{bmatrix} \mathbf{B} = \begin{bmatrix} \mathbf{CB} \\ \mathbf{CAB} \\ \mathbf{N}_C \mathbf{N}_B^T (\mathbf{N}_B \mathbf{N}_B^T)^{-1} \mathbf{N}_B \mathbf{B} \end{bmatrix} = \begin{bmatrix} \mathbf{0} \\ \mathbf{CAB} \\ \mathbf{0} \end{bmatrix} \equiv \begin{bmatrix} \mathbf{0} \\ \bar{\mathbf{B}}_{vv} \\ \mathbf{0} \end{bmatrix},$$

since $\mathbf{CB} = \mathbf{0}$ and $\mathbf{N}_B \mathbf{B} = \mathbf{0}$ which thus proves the coupling form structure of $\bar{\mathbf{B}}$. The structure of $\bar{\mathbf{C}}$ on the other hand is given from the definition of $\mathbf{Z}$ by

$$\bar{\mathbf{C}} = \mathbf{CT}^{-1} = \mathbf{CZ} = \begin{bmatrix} \mathbf{CZ}_1 & \mathbf{CZ}_2 & \mathbf{CZ}_3 \end{bmatrix} = \begin{bmatrix} \mathbf{I} & \mathbf{0} & \mathbf{0} \end{bmatrix},$$

which ends the proof. □

References

Abrahamsson, T., & Kammer, D. C. (2015). Finite element model calibration using frequency responses with damping equalization. *Mechanical Systems and Signal Processing, 62–63*, 218–234.

Allen, M. S., & Mayes, R. L. (2007). Comparison of FRF and modal methods for combining experimental and analytical substructures. In *25th international modal analysis conference (IMAC XXV)*.

Allen, M. S., & Mayes, R. L. (2007). Comparison of frf and modal methods for combining experimental and analytical substructures.

Allen, M. S., Kammer, D. C., & Mayes, R. L. (2010). Uncertainty in experimental/analytical substructuring predictions: A review with illustrative examples. In *Proceedings of the international conference on noise and vibration engineering (ISMA)* (pp. 1833–1850). Belgium: Leuven.

Allen, M. S., Mayes, R. L., & Bergman, E. J. (2010). Experimental modal substructuring to couple and uncouple substructures with flexible fixtures and multi-point connections. *Journal of Sound and Vibration, 329*(23), 4891–4906.

Anderson, B. D. O. (1967). A system theory criterion for positive real matrices. *SIAM Journal on Control and Optimization, 5*(2), 171–182.

Avitabile, P. (2003). Twenty years of structural dynamic modification – a review. *Sound & Vibration Magazine***January**, 14–25.

Beck, J. L., & Yuen, K. V. (2004). Model selection using response measurements: Bayesian probabilistic approach. *Journal of Engineering Mechanics, 130*, 192–203. https://doi.org/10.1061/(ASCE)0733-9399.

Bello, M., Sestieri, A., D'Ambrogio, W., & La Gala, F. (2003). Development of a rotation transducer based on bimorph PZTs. *Mechanical Systems & Signal Processing, 17*(5), 1069–1081. https://doi.org/10.1006/mssp.2003.1607.

Bendat, J. S., & Piersol, A. G. (1980). *Engineering Applications of Correlation and Spectral Analysis*. New York: Wiley.

Bregant, L., & Sandersons, M. (2000). Rotational degrees of freedom: an historical overview on techniques and methods. In *Proceedings of the international conference on noise and vibration engineering (ISMA)* (pp. 973–980). Belgium: Leuven.

Carne, T. G., & Dohrmann, C. R. (2006). Improving experimental frequency response function matrices for admittance modeling.

Cloutier, D., & Avitabile, P. (2011). Dynamic uncoupling of a system model for component identification. In *IMAC-XXIX: International Modal Analysis Conference, Jacksonville, FL*. Bethel, CT: Society for Experimental Mechanics.

Craig, R. R. (2000). Coupling of substructures for dynamic analyses: An overview. In *Proceedings of AIAA/ASME/ASCE/AHS/ASC structures, structural dynamics, and materials conference and exhibit* (pp. 1573–1584).

Crowley, J. R., Klosterman, A. L., Rocklin, G. T., & Vold, H. (1984). Direct structural modification using frequency response functions. *Proceedings of IMAC II*.

D'Ambrogio, W., & Fregalent, A. (2009). Decoupling procedures in the general framework of frequency based substructuring. In *IMAC-XXVII: International modal analysis conference, Orlando, FL*. Bethel, CT: Society for Experimental Mechanics.

D'Ambrogio, W., & Fregalent, A. (2014). Are rotational dofs essential in substructure decoupling? In *IMAC-XXXII: International modal analysis conference, Orlando, FL*. Bethel, CT: Society for Experimental Mechanics.

D'Ambrogio, W., & Fregolent, A. (2015). Ignoring rotational dofs in decoupling structures connected through flexotorsional joints. In *IMAC-XXXIII: International modal analysis conference, Orlando, FL*. Bethel, CT: Society for Experimental Mechanics.

de Klerk, D. (2009). *Dynamic response characterization of complex systems through operational identification and dynamic substructuring*. Ph.D. thesis, Delft University of Technology, The Netherlands. http://resolver.tudelft.nl/uuid:2e4dd63e-68d5-41a4-8201-1e784f2f752c.

de Klerk, D., & Voormeeren, S. (2008). Uncertainty propagation in experimental dynamic substructuring. In *Proceedings of the XXVI international modal analysis conference (IMAC), Orlando, FL*. Bethel, CT: Society for Experimental Mechanics.

de Klerk, D., Rixen, D. J., de Jong, J. (2006). The frequency based substructuring (FBS) method reformulated according to the dual domain decomposition method. In *International modal analysis conference, IMAC-XXV*. St Louis, MO: SEM.

de Klerk, D., Rixen, D., Voormeeren, S., & Pasteuning, F.: Solving the RDoF problem in experimental dynamic substructuring. In: *Proceedings of the XXVI international modal analysis conference (IMAC), Orlando, FL*. Bethel, CT: Society for Experimental Mechanics.

Duarte, M. L. M., & Ewins, D. J. (2000). Rotational degrees of freedom for structural coupling analysis via finite-difference technique with residual compensation. *Mechanical Systems & Signal Processing*, *14*(2), 205–227. https://doi.org/10.1006/mssp.1999.1241.

Elliott, A. S., Moorhouse, A. T., & Pavić, G. (2012). Moment excitation and the measurement of moment mobilities. *Journal of Sound & Vibration*, *331*(11), 2499–2519. https://doi.org/10.1016/j.jsv.2012.01.022.

Epp, D. S., Ozdoganlar, O. B., Chaplya, P. M., Hansche, B. D., & Carne, T. G. (2004). A base excitation test facility for dynamic testing of microsystems. In *IMAC-XXII: International modal analysis conference, Orlando, FL*. Bethel, CT: Society for Experimental Mechanics.

Ewins, D. J., & Inman, D. J. (2001). *Structural dynamics @ 2000: current status and future directions*. Baldock, Hertfordshire, England: Research Studies Press.

Géradin, M., & Rixen, D. (2015). *Mechanical vibrations* (3rd ed.). Theory and application to structural dynamics. Chichester: Wiley.

Géradin, M., & Rixen, D. J. (2017). Impulse-based substructuring in a floating frame to simulate high frequency dynamics in flexible multibody dynamics. *Multibody System Dynamics* (pp. 1–31).

Gialamas, T., Tsahalis, D., Otte, D., Van der Auwaraer, H., & Manolas, D. (2001). Substructuring technique: improvement by means of singular value decomposition (svd). *Applied Acoustics*, *62*(10), 1211–1219.

Gordis, J. H. (1995). Integral equation formulation for transient structural synthesis. *AIAA Journal*, *33*(2), 320–324.

Groult, B. (2008). *Extension d'une mèthode de modification structurale pour la conception de dispositifs dissipatifs intègrant des matèriaux viscoèlastiques*. Ph.D. thesis, Ecole Centrale Paris, Laboratoire de MÈcanique des Sols, Structures et MatÈriaux.

Haeussler, M., Klaassen, S., & Rixen, D. (2018). Comparison of substructuring techniques for experimental identification of rubber isolators dynamic properties. In P. S. et al. (ed.) *International conference on noise and vibration engineering, ISMA*. Belgium: KUL, Leuven.

Harvie, J., & Avitabile, P. (2013). Effects of precise FRF measurements for frequency based substructuring. In *Topics in experimental dynamic substructuring, Proceedings of the 31st IMAC, a conference on structural dynamics* (vol. 2, Chap. 22, pp. 277–285). New York: Springer. https://doi.org/10.1007/978-1-4614-6540-9_22.

Helderweirt, S., Van der Auweraer, H., Mas, P., Bregant, L., & Casagrande, D. (2001). Application of accelerometer-based rotational degree of freedom measurements for engine subframe modelling. In *Proceedings of the XIX international modal analysis conference (IMAC), St. Louis, MO* (vol. 2, pp. 1298–1304). Bethel, CT: Society for Experimental Mechanics.

Höller, C., & Gibbs, B. M. (2015). Indirect determination of the mobility of structure-borne sound sources. *Journal of Sound and Vibration*, *344*, 38–58.

Imregun, M., Robb, D., & Ewins, D. (1987). Structural modification and coupling dynamic analysis using measured FRF data. In *IMAC-V: international modal analysis conference, London, England* (pp. 1136–1141). Bethel, CT: Society for Experimental Mechanics.

Ind, P., & Ewins, D. (2003). Impedance based decoupling and its application to indirect modal testing and component measurement: A numerical investigation. In *IMAC-XXI: International modal analysis conference, Kissimmee, FL*. Bethel, CT: Society for Experimental Mechanics

Jetmundsen, B., Bielawa, R. L., & Flannelly, W. G. (1988). Generalized frequency domain substructure synthesis. *Journal of the American Helicopter Society*, *33*(1), 55–64.

Kalaycıoğlu, T., & Özgüven, H. N. (2018). Experimental verification of a recently developed frf decoupling method for nonlinear systems. In *IMAC-XXXVI: International modal analysis conference, Orlando, FL.* Bethel, CT: Society for Experimental Mechanics.

Keersmaekers, L., Mertens, L., Penne, R., Guillaume, P., & Steenackers, G. (2015). Decoupling of mechanical systems based on in-situ frequency response functions: The link-preserving, decoupling method. *Mechanical Systems and Signal Processing.*

Klerk, D. D., Rixen, D. J., & Voormeeren, S. N. (2008). General framework for dynamic substructuring: History, review and classification of techniques. *AIAA Journal, 46*(5), 1169–1181.

Klosterman, A. L. (1971). *A combined experimental and analytical procedure for improving automotive system dynamics*. Ph.D. thesis, University of Cincinnati, Department of Mechanical Engineering.

Liljerehn, A., & Abrahamsson, T. (2014). Dynamic sub-structuring with passive state-space components. In *Proceedings of the international conference on noise and vibration engineering (ISMA)* (pp. 3879–3890). Belgium: Leuven.

Lim, T.C., & Steyer, G. C. (1992). An improved numerical procedure for the coupling of dynamic components using frequency response functions. In: *IMAC-IX: international modal analysis conference, Florence, Italy* (pp. 355–355). SEM Society for Experimental Mechanics INC.

Liu, W., & Ewins, D. (1999). The importance assessment of RDOF in FRF coupling analysis. In *Proceedings of the XVII international modal analysis conference (IMAC), Kissimmee, FL* (pp. 1481–1487). Bethel, CT: Society for Experimental Mechanics.

Liu, J., Li, M., Qin, L., & Liu, J. (2013). Principle research on a single mass piezoelectric six-degrees-of-freedom accelerometer. *Sensors, 13*(8), 10844–10855. https://doi.org/10.3390/s130810844.

Ljung, L. (1998). System identification: theory for the user. Pearson Education.

Mahmoudi, A. E., Meyer, C. H., & Rixen, D. J. (2019). Comparison of different approaches to include rubber bushings into frequency-based substructuring coupling process. Experimental Techniques **submitted**.

Martinez, D.R., Carne, T.G., Gregory, D.L., Miller, A.K.: Combined experimental/analytical modeling using component mode synthesis (1984). Compilation and indexing terms, Copyright 2007 Elsevier Inc. All rights reserved 84060103006 COMPUTED MODAL FREQUENCY ACCURACY MODE SHAPES SIMPLE BEAM STRUCTURE FREE-FREE MODES EXPERIMENTAL MODAL DATA BASE FINITE ELEMENT MODEL

Martinez, D. R., Miller, A. K. & Carne, T. G. (1985). Combined experimental and analytical modeling of shell/payload structures.

Mayes, R. L., & Arviso, M. (2010). Design studies for the transmission simulator method of experimental dynamic substructuring. In *International seminar on modal analysis, ISMA*. Belgium: KUL, Leuven.

Mayes, R. L., & Stasiunas, E. C. (2007). Lightly damped experimental substructures for combining with analytical substructures.

Mayes, R. L., Allen, M. S., & Kammer, D. C. (2012). Eliminating indefinite mass matrices with the transmission simulator method of substructuring.

Mayes, R., & Ross, M. (2012). Advancements in hybrid dynamic models combining experimental and finite element substructures. *Mechanical Systems and Signal Processing, 31*, 56–66.

Meggitt, J., & Moorhouse, A. (2018). In-situ dynamic sub-structure decoupling of resiliently coupled assemblies. Mechanical Systems and Signal Processing. In press (2018).

Meggitt, J. W. R., Elliott, A. S., Moorhouse, A. T., & Lai, H. K. (2015). In situ determination of dynamic stiffness for resilient elements. In *Proceedings of the institution of mechanical engineers, Part C: Journal of mechanical engineering science* (p. 0954406215618986).

Meggitt, J., Moorhousea, A. T., & Elliott, A. S. (2018). On the problem of describing the coupling interface between sub-structures: An experimental test for 'completeness'. In *IMAC-XXXVI: International modal analysis conference, Orlando, FL.* Bethel, CT: Society for Experimental Mechanics.

Montalvao, D., Ribeiro, A., Maia, N., & Silva, J. (2004). Estimation of the rotational terms of the dynamic response matrix. *Shock and Vibration*, *11*(3–4), 333–350. https://doi.org/10.1155/2004/780837.

Moorhouse, A., Elliot, A. S., & Heo, Y. (2013). Intrinsic characterisation of structure-borne sound sources and isolators from in-situ measurements. In *Proceedings of Meetings on Acoustics ICA2013* (vol. 19, p. 065053). ASA.

Moorhouse, A. T. (2003). Compensation for discarded singular values in vibro-acoustic inverse methods. *Journal of Sound & Vibration*, *267*(2), 245–252. https://doi.org/10.1016/S0022-460X(02)01432-3.

Nicgorski, D. (2008). *Investigation on experimental issues related to frequency response function measurements for frequency based substructuring*. Master's thesis, University of Massachusetts Lowell.

Nicgorski, D., & Avitabile, P. (2010). Experimental issues related to frequency response function measurements for frequency-based substructuring. *Mechanical Systems and Signal Processing*, *24*(5), 1324–1337.

Nicgorski, D., & Avitabile, P. (2010). Conditioning of FRF measurements for use with frequency based substructuring. *Mechanical Systems & Signal Processing*, *24*(2), 340–351. https://doi.org/10.1016/j.ymssp.2009.07.013.

Nicgorski, D., & Avitabile, P. (2010). Experimental issues related to frequency response function measurements for frequency-based substructuring. *Mechanical Systems & Signal Processing*, *24*(5), 1324–1337. https://doi.org/10.1016/j.ymssp.2009.09.006.

O'Callahan, J. C., Avitabile, P.,& Riemer, R. (1989) System equivalent reduction expansion process (SEREP). In *Proceedings of the VII international modal analysis conference (IMAC), Boston, MA* (vol. 1, pp. 29–37). Bethel, CT: Society for Experimental Mechanics.

Otte, D., Leuridan, J., Grangier, H., & Aquilina, R. (1991). Prediction of the dynamics of structural assemblies using measured FRF-data: Some improved data enhancement techniques. In *IMAC-IX: international modal analysis conference, Florence, Italy* (pp. 909–918). Bethel, CT: Society for Experimental Mechanics.

Pasma, E. A., van der Seijs, M. V., Klaassen, S. W. B., & van der Kooij, M. W. (2018). Frequency Based Substructuring with the Virtual Point Transformation, Flexible Interface Modes and the Transmission Simulator.

Pasteuning, F. (2007). *Experimental dynamic substructuring and its application in automotive research*. Master's thesis, Delft University of Technology, The Netherlands.

Petersson, B. (1987). On the use of giant magnetostrictive devices for moment excitation. *Journal of Sound & Vibration*, *116*(1), 191–194. https://doi.org/10.1016/S0022-460X(87)81332-9.

Rixen, D. J. (2008). How measurement inaccuracies induce spurious peaks in frequency based substructuring. In *IMAC-XXVII: International modal analysis conference, Orlando, FL*. Bethel, CT: Society for Experimental Mechanics.

Rixen, D. (2008). How measurement inaccuracies induce spurious peaks in frequency based substructuring. In *Proceedings of the XXVI international modal analysis conference (IMAC), Orlando, FL*. Bethel, CT: Society for Experimental Mechanics.

Rixen, D. J. (2010). Substructuring technique based on measured and computed impulse response functions of components. In P.S. et al. (Ed.) *ISMA*. K.U. Leuven.

Rixen, D. J. (2010). Substructuring using impulse response functions for impact analysis. In *IMAC-XXVIII: International modal analysis conference, Jacksonville, FL*. Bethel, CT: Society for Experimental Mechanics.

Rixen, D., Godeby, T., Pagnacco, E. (2006). Dual assembly of substructures and the fbs method: Application to the dynamic testing of a guitar. In *International conference on noise and vibration engineering, ISMA*. Belgium: KU Leuven.

Rixen, D. J., & van der Valk, P. L. (2013). An impulse based substructuring approach for impact analysis and load case simulations. *Journal of Sound and Vibration*, *332*, 7174–7190.

Sanderson, M. A., & Fredö, C. R. (1995). Direct measurement of moment mobility: Part I: A theoretical study. *Journal of Sound & Vibration*, *179*(4), 669–684. https://doi.org/10.1006/jsvi.1995.0043.

Sjövall, P., & Abrahamsson, T. (2007). Substructure system identification from coupled system test data. *Mechanical Systems and Signal Processing*, *22*, 1–260.

Sjövall, P., & Abrahamsson, T. (2007). Component system identification and state-space model synthesis. *Mechanical Systems & Signal Processing*, *21*(7), 2697–2714. https://doi.org/10.1016/j.ymssp.2007.03.002.

Sjövall, P., McKelvey, T., & Abrahamsson, T. (2006). Constrained state-space system identification with application to structural dynamics. *Automatica*, *42*, 1539–1546.

Tol, Ş., & Özgüven, H. N. (2015). Dynamic characterization of bolted joints using frf decoupling and optimization. *Mechanical Systems and Signal Processing*, *54*, 124–138.

Urgueira, A. P. V. (1989). *Dynamic analysis of coupled structures using experimental data*. Ph.D. thesis, Imperial College, London.

van Overschee, P., & de Moor, B. (1996). *Subspace Identification for Linear Systems*. Kluwer Academic Publishers.

van der Seijs, M., de Klerk, D., Rixen, D., & Rahimi, S. (2013). Validation of current state frequency based substructuring technology for the characterisation of steering gear – vehicle interaction. In *IMAC-XXXI: International modal analysis conference, Garden Grove, California USA* (pp. 11–14). Bethel, CT: Society for Experimental Mechanics.

van der Seijs, M. V., de Klerk, D., Rixen, D. J., & Rahimi, S. (2013). Validation of current state frequency based substructuring technology for the characterisation of steering gear – vehicle interaction. In *Topics in experimental dynamic substructuring, Proceedings of the 31st IMAC, a conference on structural dynamics* (Vol. 2, Chap. 20, pp. 253–266). New York: Springer. https://doi.org/10.1007/978-1-4614-6540-9_20.

van der Seijs, M. V., Pasma, E. A., van den Bosch, D. D., & Wernsen, M. W. M. (2017). A benchmark structure of validation of experimental substructuring, transfer path analysis and source characterisation techniques. In *Dynamics of coupled structures, Proceedings of the 35th IMAC, a conference and exhibition on structural dynamics* (Vol. 1, Chap. 6, pp. 73–82). New York: Springer.

van der Seijs, M. V., van den Bosch, D. D., Rixen, D. J., & de Klerk, D. (2013). An improved methodology for the virtual point transformation of measured frequency response functions in dynamic substructuring. In M. Papadrakakis, V. Papadopoulos, V. Plevris (Eds.) *4th ECCOMAS thematic conference on computational methods in structural dynamics and earthquake engineering (COMPDYN)* (pp. 4334–4347). Kos Island, Greece. https://doi.org/10.13140/RG.2.1.2715.3126.

Voormeeren, S., & Rixen, D. (2009). Substructure decoupling techniques - a review and uncertainty propagation analysis. In *IMAC-XXVII: International modal analysis conference, Orlando, FL*. Bethel, CT: Society for Experimental Mechanics.

Voormeeren, S., & Rixen, D. (2012). A family of substructure decoupling techniques based on a dual assembly approach. *Mechanical Systems and Signal Processing*, *27*, 379–396.

Voormeeren, S., de Klerk, D., & Rixen, D. (2010). Uncertainty quantification in experimental frequency based substructuring. *Mechanical Systems and Signal Processing*, *24*(1), 106–118.

Wang, Z. W., & Wang, J. (2011). Inverse substructure method of three-substructures coupled system and its application in product-transport-system. *Journal of Vibration and Control*, *17*(6), 943–951.

Wang, Z., Zhu, P., & Liu, Z. (2018). Relationships between the decoupled and coupled transfer functions: Theoretical studies and experimental validation. *Mechanical Systems and Signal Processing*, *98*, 936–950.

Zhen, J., Lim, T.C., & Lu, G. (2004). Determination of system vibratory response characteristics applying a spectral-based inverse sub-structuring approach. Part i: Analytical formulation. *International Journal of Vehicle Noise and Vibration1*(1), 1–30.

Zhen, J., Lim, T.C., & Lu, G. (2004). Determination of system vibratory response characteristics applying a spectral-based inverse substructuring approach. Part ii: Motor vehicle structures. *International Journal Vehicle Noise and Vibration1*(1/2), 31–67.

Sjövall, P., & Abrahamsson, T. (2008). Substructure system identification from coupled system test data. *Mechanical Systems and Signal Processing*, *22*, 1–2[illegible].

Sjövall, P., & Abrahamsson, T. (2007). Component system identification and state-space model synthesis. *Mechanical Systems and Signal Processing*, *21*(7), 2697–2714. [illegible]

Sjövall, P., McKelvey, T., & Abrahamsson, T. (2006). Constrained state-space system identification with application to structural dynamics. *Automatica*, *42*, 1539–1546.

Tol, Ş., & Özgüven, H. N. (2015). Dynamic characterization of bolted joints using FRF decoupling and optimization. *Mechanical Systems and Signal Processing*, *54*, 124–138.

Urgueira, A. P. V. (1989). *Dynamic analysis of coupled structures using experimental data*. Ph.D. thesis, Imperial College, London.

van Overschee, P., & de Moor, B. (1996). *Subspace identification for linear systems*. Kluwer Academic Publishers.

[illegible]

Chapter 5
Industrial Applications & Related Concepts

Abstract This section highlights some specialized substructuring methods, such as methods for estimating the fixed-interface modes of a substructure from measurements of the free–free structure with a fixture attached at the interface, and also highlights some industrial examples. Transfer path analysis is reviewed, elaborating some of the similarities in the theoretical foundations. Most information about TPA has been obtained from van der Seijs et al. (2016), please refer to the original paper for a more elaborate discussion. — *Chapter Authors: Maarten van der Seijs, Randall Mayes and Matt Allen.*

5.1 Introduction to Transfer Path Analysis

Transfer Path Analysis (TPA) has been a valuable engineering tool for as long as noise and vibrations of products have been of interest. An TPA concerns a product's actively vibrating components (such as engines, gearing systems or turbochargers) and the transmission of these vibrations to the connected passive structures. TPA is particularly useful when the actual vibrating mechanisms are too complex to model or measure directly, as it allows to represent a source by forces and vibrations displayed at the interfaces with the passive side. In this way, the *source excitations* can be separated from the structural/acoustic *transfer characteristics*, allowing to troubleshoot the dominant paths of vibration transmission. The engineer can then anticipate by making changes to either the source itself or the receiving structures that are connected to it.

An TPA often rises from the need to reduce some sort of undesired noise or vibration, for instance, to improve product comfort or lifetime, ensure safety, or preserve stealthiness. Aside from automotive development, applications are also seen in industries such as marine and airplane engineering, building acoustics, and acoustic modeling of musical instruments. An TPA is generally motivated by one of the following desires:

M. S. Allen et al., *Substructuring in Engineering Dynamics*,
CISM International Centre for Mechanical Sciences 594,
https://doi.org/10.1007/978-3-030-25532-9_5

1. *Secrecy*: perhaps, the earliest TPA studies were triggered by the need to reduce the transmission of engine vibrations in military ships and submarines in order to make them stealthy. Many publications in the 1950s and 1960s document on isolation of ship engines by means of absorbers and decoupling mechanisms (Harrison et al. 1952; Ungar and Dietrich 1966; Plunkett 1958; Soliman and Hallam 1968; Snowdon 1979) to minimize the transmission through the interfaces.
2. *Safety*: along with the rapid development of airplanes and spacecraft in the 1960s, TPA concepts started to be of use to study fatigue and stability (flutter) problems due to active or induced vibrations. As sources of vibrations are much more persistent in aeronautics—think of vortex-induced vibrations—focus was on characterizing the passive transfer paths by means of modal analysis (Bisplinghoff et al. 1955; On 1967).
3. *Comfort*: over the past decades, TPA tends to be particularly associated with Noise, Vibration, and Harshness (NVH) engineering as commonly encountered in the automotive industry. The majority of recent developments and commercial solutions have been tailored toward this engineering society or related industries, driven by the increasing customer expectations on acoustic comfort (Penne 2004; Plunt 2005; van der Auweraer et al. 2007; Zeller 2009; Cerrato 2009).

In response to the evolving demands, TPA methods have been under continuous development and their family members have grown numerous. Some designations that found their way into literature include Operational TPA (OTPA), Operational Path Analysis with exogenous inputs (OPAX), blocked-force TPA, Gear Noise Propagation, in situ Source Path Characterization and Virtual Acoustic Prototyping. Very often, those methods are presented from highly case-specific derivations. Not surprisingly, as the underlying physical concepts are similar, some of the abovementioned show strong similarities or are even identical.

An TPA workflow can typically be subdivided into the following steps:

1. operational measurement on the active component;
2. determination of the passive (sub)system characteristics (commonly by means of FRFs);
3. identification of interface loads;
4. calculation of path contributions.

The steps are shown schematically in Fig. 5.1. Depending on the TPA method at hand, some or all of these steps may be performed in arbitrary order. The optimization actions that follow from such an analysis are generally not considered part of the workflow.

The next section presents a unified framework for derivation of a large range of TPA methods. It is chosen to present and classify the methods separate from their typical fields of application, such that the underlying physical concepts are exposed and can be compared. In Sect. 5.2, a general framework for TPA is introduced, starting by depicting the transfer problem using the Dynamic Substructuring paradigm (de Klerk et al. 2008). Hereafter, the TPA methods are derived and classified along three families, namely the *classical* (Sect. 5.3), *component-based* (Sect. 5.4), and *transmissibility-based* (Sect. 5.5) TPA methods, as depicted vertically in Fig. 5.1.

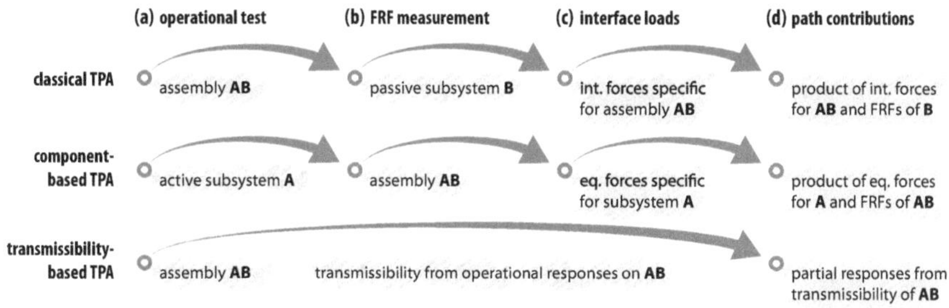

Fig. 5.1 The TPA workflow, depicted stepwise for the three TPA families

5.2 The Transfer Path Problem

Let us consider the dynamic system AB as schematically depicted in Fig. 5.2a. Two subsystems can be distinguished: an active subsystem A containing an excitation at node 1 and a passive subsystem B comprising the responses of interest at node 3. The subsystems are rigidly interconnected at the interface node 2. For simplicity of derivation, the Degrees of Freedom (DoFs) in this example is restricted to three distinct nodes. These may, however, represent a larger set of DoFs, representing, respectively:

(1) *source*: internal DoFs belonging to the active component that causes the operational excitation but are unmeasurable in practice;
(2) *interface*: coupling DoFs residing on the interface between the active and passive component;
(3) *receiver*: response DoFs at locations of interest at the passive component, possibly including acoustic pressures and other physical quantities.

Hence, the example of Fig. 5.2a is illustrative for a wide range of practical problems, provided that the structure of interest can be decomposed into an active and passive part. In what follows, all methods assume that the operational excitation at node 1 is unmeasurable in practice, but transmits vibrations through the interfaces at node 2 to receiving locations at node 3. The responses shall then be built up from a certain description of the vibrations measured at the interface (node $1 \rightarrow 2$) and an appropriate set of transfer functions relating these vibrations to the receiving responses (node $2 \rightarrow 3$). The fundamental choices herein dictate to which TPA family the method is classified.

5.2.1 Transfer Path from Assembled Admittance

Let us first approach the transfer problem top-down for the assembled system AB, see Fig. 5.2a. We are interested in the response spectra at the receiving locations $\mathbf{u}_3(\omega)$

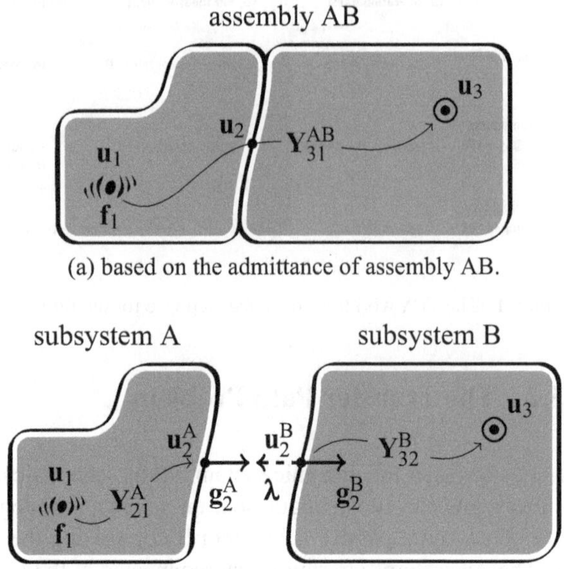

Fig. 5.2 The transfer path problem

(a) based on the admittance of assembly AB.

(b) based on the admittances of subsystems A and B.

for source excitations at node 1, given by the force spectra $\mathbf{f}_1(\omega)$. For the assembled problem, this is simply obtained from a superposition of the individual contributions, i.e. the excitation force spectra multiplied with their respective linear(ized) transfer functions, contained in the columns of admittance FRF matrix $\mathbf{Y}^{AB}(\omega)$.

$$u_i(\omega) = \sum_j Y^{AB}_{ij}(\omega) f_j(\omega) \qquad \Longrightarrow \qquad \mathbf{u}_3(\omega) = \mathbf{Y}^{AB}_{31}(\omega)\,\mathbf{f}_1(\omega). \tag{5.1}$$

In the equations that follow, the explicit frequency dependency (ω) will be omitted to improve readability. Also note that the response set $\mathbf{u}$ can include displacements, velocities, accelerations, or any other quantity, provided that the rows of the FRF matrices are obtained accordingly. Furthermore, in order to keep the derivations brief and understandable, it is chosen to only consider the structure-borne paths. Nevertheless, Eq. (5.1) can easily be extended to include contributions of airborne paths if the application so requires. In that case, $\mathbf{Y}$ and $\mathbf{f}$ need to be augmented with a set of (responses to) acoustic loads such as volume velocities (m^3/s) (van der Seijs et al. 2016).

5.2.2 Transfer Path from Subsystem Admittance

The same transfer function is now derived for an assembly of the individual subsystems, as depicted in Fig. 5.2b. Let us first put the subsystem's FRF matrices $\mathbf{Y}^{A}$ and

$\mathbf{Y}^{\mathrm{B}}$ in a block-diagonal format. The force vector comprising the excitation force is augmented with interface forces $\mathbf{g}_2$ for both sides of node 2, that is yet to be determined. The obtained system of equations resembles the admittance variant of dual assembly, which is a standard form of Dynamic Substructuring (de Klerk et al. 2008).

$$\begin{bmatrix} \mathbf{u}_1 \\ \mathbf{u}_2^{\mathrm{A}} \\ \mathbf{u}_2^{\mathrm{B}} \\ \mathbf{u}_3 \end{bmatrix} = \begin{bmatrix} \mathbf{Y}_{11}^{\mathrm{A}} & \mathbf{Y}_{12}^{\mathrm{A}} & \mathbf{0} & \mathbf{0} \\ \mathbf{Y}_{21}^{\mathrm{A}} & \mathbf{Y}_{22}^{\mathrm{A}} & \mathbf{0} & \mathbf{0} \\ \mathbf{0} & \mathbf{0} & \mathbf{Y}_{22}^{\mathrm{B}} & \mathbf{Y}_{23}^{\mathrm{B}} \\ \mathbf{0} & \mathbf{0} & \mathbf{Y}_{32}^{\mathrm{B}} & \mathbf{Y}_{33}^{\mathrm{B}} \end{bmatrix} \left(\begin{bmatrix} \mathbf{f}_1 \\ \mathbf{0} \\ \mathbf{0} \\ \mathbf{0} \end{bmatrix} + \begin{bmatrix} \mathbf{0} \\ \mathbf{g}_2^{\mathrm{A}} \\ \mathbf{g}_2^{\mathrm{B}} \\ \mathbf{0} \end{bmatrix} \right) \quad \text{or} \quad \mathbf{u} = \mathbf{Y}\left(\mathbf{f} + \mathbf{g}\right). \tag{5.2}$$

The following physical explanation can now be reasoned to solve Eq. (5.2). The excitation force at node 1 induces a motion at node 2 of subsystem A. As subsystem B is not directly affected by forces at A (due to the block-diagonal form of the global FRF matrix), an incompatibility is caused between $\mathbf{u}_2^{\mathrm{A}}$ and $\mathbf{u}_2^{\mathrm{B}}$. This is denoted by the interface "gap" $\boldsymbol{\delta}$, which can conveniently be written using a signed Boolean matrix[1] $\mathbf{B}$ as shown by Eqs. (5.3a) and (5.3c). Next, assuming that no additional mass is present between the subsystems, the *equilibrium condition* is satisfied, requiring the interface forces $\mathbf{g}_2$ on both sides to be equal in magnitude and opposing in sign. The interface forces are expressed by Eqs. (5.3b) and (5.3c), using a Lagrange multiplier $\boldsymbol{\lambda}$ for the magnitude and the transposed Boolean matrix to account for the interface force direction.

$$\boldsymbol{\delta} = \mathbf{u}_2^{\mathrm{B}} - \mathbf{u}_2^{\mathrm{A}} \quad \text{or} \quad \boldsymbol{\delta} \triangleq \mathbf{B}\mathbf{u}, \tag{5.3a}$$

$$\mathbf{g}_2^{\mathrm{A}} = -\mathbf{g}_2^{\mathrm{B}} = \boldsymbol{\lambda} \quad \text{or} \quad \mathbf{g} \triangleq -\mathbf{B}^T\boldsymbol{\lambda}, \tag{5.3b}$$

$$\text{with} \quad \mathbf{B} = \begin{bmatrix} \mathbf{0} & -\mathbf{I} & \mathbf{I} & \mathbf{0} \end{bmatrix}. \tag{5.3c}$$

The interface forces $\boldsymbol{\lambda}$ that pull the two subsystems together can be determined from Eq. (5.3a) by requiring $\boldsymbol{\delta} = \mathbf{0}$, which enforces the *compatibility condition* $\mathbf{u}_2^{\mathrm{A}} = \mathbf{u}_2^{\mathrm{B}}$. Considering Eq. (5.3b), the interface forces that ensure compatibility can be determined by equating the second and third line of Eq. (5.2) and solving for $\boldsymbol{\lambda}$:

$$\begin{aligned} \mathbf{Y}_{21}^{\mathrm{A}}\mathbf{f}_1 + \mathbf{Y}_{22}^{\mathrm{A}}\mathbf{g}_2^{\mathrm{A}} &= \mathbf{Y}_{22}^{\mathrm{B}}\mathbf{g}_2^{\mathrm{B}} \\ \left(\mathbf{Y}_{22}^{\mathrm{A}} + \mathbf{Y}_{22}^{\mathrm{B}}\right)\boldsymbol{\lambda} &= -\mathbf{Y}_{21}^{\mathrm{A}}\mathbf{f}_1 \\ \boldsymbol{\lambda} &= -\left(\mathbf{Y}_{22}^{\mathrm{A}} + \mathbf{Y}_{22}^{\mathrm{B}}\right)^{-1}\mathbf{Y}_{21}^{\mathrm{A}}\mathbf{f}_1 \\ \mathbf{g}_2^{\mathrm{B}} &= \left(\mathbf{Y}_{22}^{\mathrm{A}} + \mathbf{Y}_{22}^{\mathrm{B}}\right)^{-1}\mathbf{Y}_{21}^{\mathrm{A}}\mathbf{f}_1. \end{aligned} \tag{5.4}$$

Equation (5.4) provides the interface forces at the coinciding interface DoFs caused by the operational excitation $\mathbf{f}_1$ inside subsystem A. The response at the receiving side $\mathbf{u}_3^{\mathrm{B}}$ is found by substituting Eq. (5.4) into the last line of Eq. (5.2):

[1]The signed Boolean matrix $\mathbf{B}$ establishes the relations for all interface DoFs of A and B that are vectorially associated, e.g. u_{2x}^{A} and u_{2x}^{B}. Guidelines on the construction and properties of the signed Boolean matrix are found in de Klerk et al. (2008).

$$\mathbf{u}_3 = \mathbf{Y}^{\mathrm{B}}_{32}\mathbf{g}^{\mathrm{B}}_2 = \underbrace{\left[\mathbf{Y}^{\mathrm{B}}_{32}\left(\mathbf{Y}^{\mathrm{A}}_{22} + \mathbf{Y}^{\mathrm{B}}_{22}\right)^{-1}\mathbf{Y}^{\mathrm{A}}_{21}\right]}_{\mathbf{Y}^{\mathrm{AB}}_{31}}\mathbf{f}_1. \tag{5.5}$$

Comparing with Eq. (5.1), it follows that the terms between the brackets indeed represent the admittance of the assembly $\mathbf{Y}^{\mathrm{AB}}_{31}$ by coupling of the subsystems' admittances. It can be verified that this result is in accordance with Lagrange Multiplier Frequency Based Substructuring (LM-FBS) assembly (de Klerk et al. 2008) as detailed in Chap. 3.

So far, it has been assumed that the excitation at node 1 is measurable. In reality, however, it is impossible or impractical to identify the exact force loading. This is solved in TPA by assuming that the dynamics at the interface node 2 due to this excitation are measurable and may very well represent the source excitation. In what follows, different approaches are examined to describe the transmission of vibrations, or rather, the response at the passive subsystem B for a nonmeasurable excitation somewhere inside (or on) the active subsystem A. The notation of Dynamic Substructuring is used to reveal the relations between the different approaches.

5.3 Classical TPA

The family of *Classical TPA* methods is essentially intended to identify transfer path contributions in *existing products*. They have, nowadays, become standard practice to troubleshoot NVH problems in automotive engineering (Plunt 2005; van der Auweraer et al. 2007). A classical TPA performs operational tests on the assembled product AB to obtain interface forces between the active and passive side, namely $\boldsymbol{\lambda}$ in Eq. (5.4). It can be verified from Eqs. (5.2) and (5.5) that these interface forces fully determine the responses at the passive side and are thus representative for the effects of the source vibrations at the receiver locations $\mathbf{u}_3$. To calculate the receiver responses,[2], the passive side interface forces $\mathbf{g}^{\mathrm{B}}_2 = -\boldsymbol{\lambda}$ are applied to the interfaces of subsystem B, as shown in Fig. 5.4a.

$$\mathbf{u}_3 = \mathbf{Y}^{\mathrm{B}}_{32}\mathbf{g}^{\mathrm{B}}_2. \tag{5.6}$$

Both steps pose some challenges in practice. The FRFs of the passive side are typically determined from impact or shaker tests, or in a reciprocal fashion using, for instance, an acoustic source at the receiving location and accelerometers at the interface nodes (Ten Wolde 1973; ten Wolde et al. 1975; Fahy 1995, 2003; ten Wolde 2010). Either way, it requires *dismounting* of the active part(s) from the passive side. With respect to the determination of operational interface forces $\mathbf{g}^{\mathrm{B}}_2$, it could be impractical to

[2]In this framework, the resulting responses $\mathbf{u}_3$ are formulated as a matrix-vector product, namely the sum of the partial responses. Techniques to evaluate the individual transfer paths contributions are discussed in van der Seijs et al. (2016).

mount force sensors between the active and passive part. Therefore, a number of indirect methods have been developed to circumvent direct force measurement.

Hence, the variants of classical TPA are defined according to how $\mathbf{g}_2^{\mathrm{B}}$ is obtained from operational tests, which are discussed next.

5.3.1 Classical TPA: Direct Force

The most straightforward technique to obtain the interface forces is from force transducers mounted directly between the active the passive side, as depicted in Fig. 5.3a. It was demonstrated by Eq. (5.4) that the interface force caused by the operational excitation is given by

$$\boldsymbol{\lambda} = -\left(\mathbf{Y}_{22}^{\mathrm{A}} + \mathbf{Y}_{22}^{\mathrm{B}}\right)^{-1} \mathbf{Y}_{21}^{\mathrm{A}} \mathbf{f}_1 \qquad \Longrightarrow \qquad \mathbf{g}_2^{\mathrm{B}} = -\boldsymbol{\lambda}. \tag{5.7}$$

This is valid under the assumption that $\mathbf{u}_2^{\mathrm{B}} = \mathbf{u}_2^{\mathrm{A}}$, which requires the stiffness of the transducers to be high enough (relative to the stiffness of the actual subsystems) in the frequency range of interest. In fact, the main drawback of the method is the inconvenience of placing the transducers between the active component and the receiving structure. Lack of space, distortion of the original mounting situation, and the incapability to measure all desired degrees of freedom at a connection point render the method impractical, especially for typical automotive applications. In case of large-scale systems such as ship machinery, this method may still be preferred (Verheij 1982).

5.3.2 Classical TPA: Mount Stiffness

An effective way of reducing vibration transmission is by placing resilient mounts between the components instead of rigid fixtures, as illustrated in Fig. 5.3b. By proper tuning of the mount flexibility (stiffness) and absorption (damping) properties, a high level of vibration suppression can be achieved. The mount stiffness method uses these mount properties to determine the interface forces. Assumed that the added mass of the mounts is negligible, the interface force equilibrium condition Eq. (5.3b) is still satisfied. However, the compatibility condition of Eq. (5.3a) is "weakened", hence $\mathbf{u}_2^{\mathrm{B}} - \mathbf{u}_2^{\mathrm{A}}$ is no longer zero. Instead, the m interface forces and coordinate incompatibilities are related by the dynamic stiffnesses of the mounts, denoted by z_{jj}^{mt}, j representing a single interface DoF:

$$\begin{bmatrix} g_j^{\mathrm{A}} \\ g_j^{\mathrm{B}} \end{bmatrix} = -z_{jj}^{\mathrm{mt}} \begin{bmatrix} 1 & -1 \\ -1 & 1 \end{bmatrix} \begin{bmatrix} u_j^{\mathrm{A}} \\ u_j^{\mathrm{B}} \end{bmatrix} \qquad \Longrightarrow \qquad g_j^{\mathrm{B}} = z_{jj}^{\mathrm{mt}}(u_j^{\mathrm{A}} - u_j^{\mathrm{B}}) \qquad j \in 1, \ldots, m.$$

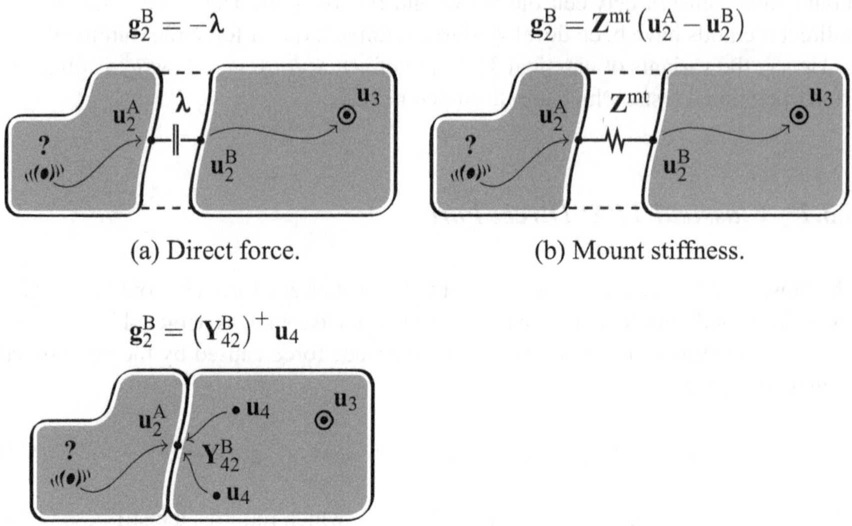

(a) Direct force.

(b) Mount stiffness.

(c) Matrix inverse.

Fig. 5.3 Three approaches to determine the operational interface forces in Classic TPA

A spring-like stiffness matrix can be recognized, however, with a minus sign because the interface forces g_j act on the connected subsystems A and B instead of the mount. Introducing the diagonal mount stiffness matrix $\mathbf{Z}^{\mathrm{mt}}$, the full set of m interface forces $\mathbf{g}_2^{\mathrm{B}}$ can be estimated from the differential interface displacements between the source and receiver side, i.e. the measured displacements at both sides of the mounts:

$$\mathbf{g}_2^{\mathrm{B}} = \mathbf{Z}^{\mathrm{mt}}(\mathbf{u}_2^{\mathrm{A}} - \mathbf{u}_2^{\mathrm{B}}) \quad \text{with} \quad \mathbf{Z}^{\mathrm{mt}} = \mathrm{diag}(z_{11}^{\mathrm{mt}}, \ldots, z_{mm}^{\mathrm{mt}}). \tag{5.8}$$

In most cases, the flexible mounts are already integrated in the design to attenuate the vibration transmission. If they are, however, placed in the system for the mere purpose of TPA, it can be shown that the interface forces and thus the vibrations of the coupled system are altered significantly (Reuss et al. 2012; Barten et al. 2014), namely:

$$\mathbf{g}_2^{\mathrm{B}} = \left(\mathbf{Y}_{22}^{\mathrm{A}} + \mathbf{Y}_{22}^{\mathrm{B}} + \mathbf{Y}^{\mathrm{mt}}\right)^{-1} \mathbf{Y}_{21}^{\mathrm{A}} \mathbf{f}_1 \quad \text{with} \quad \mathbf{Y}^{\mathrm{mt}} = \left(\mathbf{Z}^{\mathrm{mt}}\right)^{-1}.$$

Although the mount stiffness method can be powerful and easy to conduct, the accuracy is highly subject to the terms[3] in $\mathbf{Z}^{\mathrm{mt}}$. Typical absorbers exhibit amplitude-dependent nonlinearities and directional characteristics (Harrison et al. 1952; Ungar and Dietrich 1966; Thompson et al. 1998).

[3]Note that the terms in the dynamic stiffness matrix $\mathbf{Z}^{\mathrm{mt}}$ correspond to differential displacements of the associated interface DoFs $\mathbf{u}_2^{\mathrm{A}} - \mathbf{u}_2^{\mathrm{B}}$ and not the coordinates of both A and B. Some implications for the terms in $\mathbf{Z}^{\mathrm{mt}}$ are discussed in Voormeeren (2012), Barten et al. (2014).

5.3.3 Classical TPA: Matrix Inverse

The third and perhaps most popular classical TPA member is the matrix-inverse method (Powell and Seering 1984; Dobson and Rider 1990; Thite and Thompson 2003a, b). It was observed from Eq. (5.6) that responses at the passive side are found from application of the interface forces to the passive subsystem's FRFs. Recalling these responses from Eq. (5.2):

$$\begin{bmatrix} \mathbf{u}_2^{\mathrm{B}} \\ \mathbf{u}_3 \end{bmatrix} = \begin{bmatrix} \mathbf{Y}_{22}^{\mathrm{B}} \\ \mathbf{Y}_{32}^{\mathrm{B}} \end{bmatrix} \mathbf{g}_2^{\mathrm{B}}.$$

This problem can be inverted if the left-hand side contains sufficient independent responses to describe all m interface forces and moments in $\mathbf{g}_2^{\mathrm{B}}$. The set of receiver responses $\mathbf{u}_3$ is typically too small in number and too distant from the interfaces to be suitable for inversion. Inversion of the first row is theoretically sufficient, but requires complete instrumentation of the assembled structure's interfaces to measure all DoFs $\mathbf{u}_2^{\mathrm{B}}$ associated with $\mathbf{g}_2^{\mathrm{B}}$. In addition, a symmetric FRF matrix $\mathbf{Y}_{22}^{\mathrm{B}}$ would be required for the passive subsystem's interfaces, which is challenging to obtain accurately.

In practice, the passive side is equipped with so-called *indicator responses* $\mathbf{u}_4$ as shown in Fig. 5.3c. An amount of $n \geq m$ responses shall be located in the proximity of the interfaces, such that the full set of m interface forces is properly *observable* from these points (this is addressed below). As these indicator DoFs merely assist in the determination of the interface forces, the sensor type can be chosen rather arbitrarily, although (triaxial) accelerometers are the most common choice.

Two sets of measurements are now required to reconstruct the interface force spectra. First, responses $\mathbf{u}_4$ are measured on the assembled system AB during operational tests. These can be expressed in terms of subsystem admittances, similar to Eq. (5.5):

$$\mathbf{u}_4 = \mathbf{Y}_{41}^{\mathrm{AB}} \mathbf{f}_1 = \mathbf{Y}_{42}^{\mathrm{B}} \underbrace{\left(\mathbf{Y}_{22}^{\mathrm{A}} + \mathbf{Y}_{22}^{\mathrm{B}}\right)^{-1} \mathbf{Y}_{21}^{\mathrm{A}} \mathbf{f}_1}_{\mathbf{g}_2^{\mathrm{B}}}. \tag{5.9}$$

Next, FRFs need to be measured for the passive subsystem B, relating the motion at these indicator points to forces at the interface, namely $\mathbf{Y}_{42}^{\mathrm{B}}$. Note that this requires dismounting of the active components from the assembly. Nevertheless, the FRFs $\mathbf{Y}_{42}^{\mathrm{B}}$ (needed for the matrix-inverse force determination) and $\mathbf{Y}_{32}^{\mathrm{B}}$ (to calculate responses at the target locations, Eq. (5.6)) can be obtained from the same FRF measurement campaign, as it only involves mounting of additional sensors. The operational interface forces can be reconstructed from a pseudoinverse of the indicator response spectra $\mathbf{u}_4$ with the subsystem FRFs:

$$\mathbf{g}_2^{\mathrm{B}} = \left(\mathbf{Y}_{42}^{\mathrm{B}}\right)^{+} \mathbf{u}_4. \tag{5.10}$$

If $\mathbf{Y}^{\mathrm{B}}_{42}$ is full rank, it holds that $\left(\mathbf{Y}^{\mathrm{B}}_{42}\right)^{+}\mathbf{Y}^{\mathrm{B}}_{42} = \mathbf{I}$, such that in theory the correct interface forces from Eq. (5.4) are obtained by subsequent application of Eqs. (5.9) and (5.10). Hence, the conditioning of the FRF matrix is crucial, which should have

1. sufficient rank to describe all interface forces $\mathbf{g}^{\mathrm{B}}_{2}$ independently from the set $\mathbf{u}_4$ and;
2. a reasonably low condition number in order not to amplify measurement errors in the inversion.

As a rule of thumb, it is common to use at least twice as many response DoFs as strictly required to fully determine the interface forces. Much attention has been devoted to improving the conditioning of the inverse problem by means of singular value decomposition, see for instance, the work of Thite and Thompson (2003a, b, 2006), Dobson and Rider (1990), Choi et al. (2007) or an early review of techniques (Mas and Sas 1994). Time domain implementations have also been developed, such as the inverse structural filtering methods described in Powell and Seering (1984), Kammer (1998).

Recently, there has been an interest in the application of strain gages, instead of the commonly used accelerometers at the indicator points. It is argued that strain responses possess a more direct relation to interface forces and are better able to capture the local phenomena of the structure. Consequently, it is expected that strain measurements lead to better conditioning of the matrix to be inverted (Sachse et al. 2013; dos Santos et al. 2014). More research is currently needed to further verify this assumption.

5.4 Component-Based TPA

A fundamentally different class of methods is that of the *component-based TPA*. As shown in Eq. (5.7), the interface forces obtained from a classical TPA are not a characteristic of the source alone but of the assembled dynamics. For that reason, a classical TPA cannot predict the effects of subsystem modification, as one would need to conduct a new operational test for every change in design. Hence, the interface forces measured in an assembly AB are not transferable to an assembly with another receiving side B.

Component-based TPA tries to characterize the source excitation by a set of equivalent forces or velocities that are an inherent property of the active component itself. The responses at the receiving side can be simulated by applying these forces to the FRFs of an assembled system with the active part shut down, as illustrated in Fig. 5.4b. Hence, the dynamic interaction with the passive side is accounted for in a later stage, at least not during operational measurements. This allows defining a testing environment that is ideal for operational measurement on the active component, explaining the denotation *component-based*.

Interestingly, with respect to the origin of component-based TPA theory, literature has been very unambiguous. Some researchers have found inspiration in acoustics or

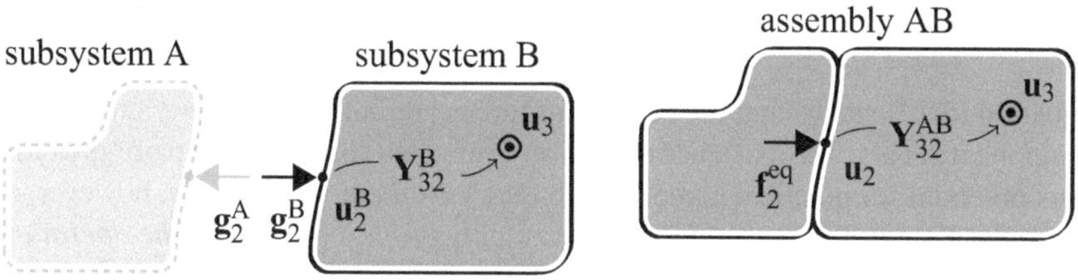

(a) Classic TPA: application of interface forces (measured under operation of assembled system AB) to the passive subsystem B.

(b) Component-based TPA: application of equivalent forces (measured under operation of subsystem A) to the assembled system AB.

Fig. 5.4 Application of forces representing the operational excitation: classical TPA and component-based TPA

electronic network theory (particularly the equivalent source identities of Thévenin and Norton), such as Mondot and Petersson (1987), Petersson and Gibbs (1993), Petersson and Gibbs (2000), Moorhouse (2001), Elliott and Moorhouse (2008), Moorhouse et al. (2009), Bonhoff (2010). Others derived similar theories from a structural mechanical point of view (Janssens et al. 1999; Janssens and Verheij 2000; Gaudin and Beniguel 2012) or dynamic substructuring techniques (de Klerk 2009; de Klerk and Rixen 2010). As a consequence, a wide variety of component-based TPA methods have been derived, largely independent of each other. This section presents a more generic derivation in order to unite all component methods and compare the various concepts.

5.4.1 The Equivalent Source Concept

Approaching the problem top-down, one is looking for a set of equivalent forces $\mathbf{f}_2^{\mathrm{eq}}$ that, applied to the interface of the assembled system AB at rest, yields the correct responses at $\mathbf{u}_3$. This can be directly formulated using the admittance of the assembly $\mathbf{Y}_{32}^{\mathrm{AB}}$, or expanded in terms of its subsystem admittances:

$$\mathbf{u}_3 = \mathbf{Y}_{32}^{\mathrm{AB}} \mathbf{f}_2^{\mathrm{eq}} = \left[\mathbf{Y}_{32}^{\mathrm{B}} \left(\mathbf{Y}_{22}^{\mathrm{A}} + \mathbf{Y}_{22}^{\mathrm{B}} \right)^{-1} \mathbf{Y}_{22}^{\mathrm{A}} \right] \mathbf{f}_2^{\mathrm{eq}}. \tag{5.11}$$

The response $\mathbf{u}_3$, as a result of the equivalent forces, should equal the response obtained when the active component is running in the same assembly, caused by $\mathbf{f}_1$ in Eq. (5.5). Comparing Eqs. (5.11) with (5.5), it follows naturally that the equivalent forces should take the form

$$\mathbf{Y}_{32}^{\mathrm{B}} \left(\mathbf{Y}_{22}^{\mathrm{A}} + \mathbf{Y}_{22}^{\mathrm{B}} \right)^{-1} \mathbf{Y}_{22}^{\mathrm{A}} \mathbf{f}_2^{\mathrm{eq}} = \mathbf{Y}_{32}^{\mathrm{B}} \left(\mathbf{Y}_{22}^{\mathrm{A}} + \mathbf{Y}_{22}^{\mathrm{B}} \right)^{-1} \mathbf{Y}_{21}^{\mathrm{A}} \mathbf{f}_1 \qquad \Longrightarrow$$

$$\mathbf{f}_2^{\text{eq}} = \left(\mathbf{Y}_{22}^{\text{A}}\right)^{-1} \mathbf{Y}_{21}^{\text{A}}\,\mathbf{f}_1. \tag{5.12}$$

Equation (5.12) shows that the equivalent forces are indeed a property on the active component A only. The existence of such an equivalent source problem offers tremendous potential for practical component-based TPA methods. There is, however, one important limitation: *the equivalent forces only properly represent the operational excitations for responses on the receiving structure or on the interface. Responses obtained on the source will be different and therefore of no use.* This limitation was already mentioned in de Klerk (2009), de Klerk and Rixen (2010) and can be understood by examining the system of Eq. (5.2): responses at the passive side B are caused only by forces through or onto the interface, whereas the responses at the source side A are a result of both the direct contribution of $\mathbf{f}_1$ and its reflection through the coupled subsystem B.

In the next section, several approaches are discussed that yield a set of equivalent forces from operational tests, following the definition of Eq. (5.12).

5.4.2 Component TPA: Blocked Force

Consider the case where the boundary of subsystem A is rigidly fixed, as depicted in Fig. 5.5a. The operational excitation $\mathbf{f}_1$ is entirely portrayed by the reaction forces at the "blocked" interface $\mathbf{g}_2^{\text{bl}}$, such that the interface displacements $\mathbf{u}_2^{\text{A}} = \mathbf{0}$. A direct relation is found, leading to the following equivalent force:

$$\begin{bmatrix} \mathbf{u}_1 \\ \mathbf{u}_2 = \mathbf{0} \end{bmatrix} = \begin{bmatrix} \mathbf{Y}_{11}^{\text{A}} & \mathbf{Y}_{12}^{\text{A}} \\ \mathbf{Y}_{21}^{\text{A}} & \mathbf{Y}_{22}^{\text{A}} \end{bmatrix} \begin{bmatrix} \mathbf{f}_1 \\ \mathbf{g}_2^{\text{A}} = -\mathbf{g}_2^{\text{bl}} \end{bmatrix} \quad \Longrightarrow \quad \begin{cases} \mathbf{g}_2^{\text{bl}} = \left(\mathbf{Y}_{22}^{\text{A}}\right)^{-1} \mathbf{Y}_{21}^{\text{A}}\,\mathbf{f}_1 \\ \mathbf{f}_2^{\text{eq}} = \mathbf{g}_2^{\text{bl}}. \end{cases} \tag{5.13}$$

The so-called *blocked-force TPA* is perhaps the most commonly known variant of component-based TPA methods because of its straightforward applicability. It can be seen as an application of the Thévenin equivalent source problem (Thévenin 1883), that found its way into popular mechanics halfway the twentieth century (Hixson 1961; Ungar and Dietrich 1966; Rubin 1967). Mathematically, one can regard the blocked-force method as imposing a Dirichlet boundary condition on the active subsystem's interface.

The blocked-force method assumes the boundary to be infinitely stiff in all directions, which is in practice rarely the case. Hence, the accuracy of the blocked forces is highly subject to the stiffness of the boundary relative to the component at hand (de Klerk 2009; van den Bosch et al. 2014; van der Seijs et al. 2014). An additional difficulty is the measurement of rotational moments, as most commonly used sensors are unable to measure collocated 6-DoF interface loads. As a consequence, the blocked-force method is expected to perform best at the low-frequency end for which the rigid boundary assumption is most valid and rotational effects are in practice least prominent.

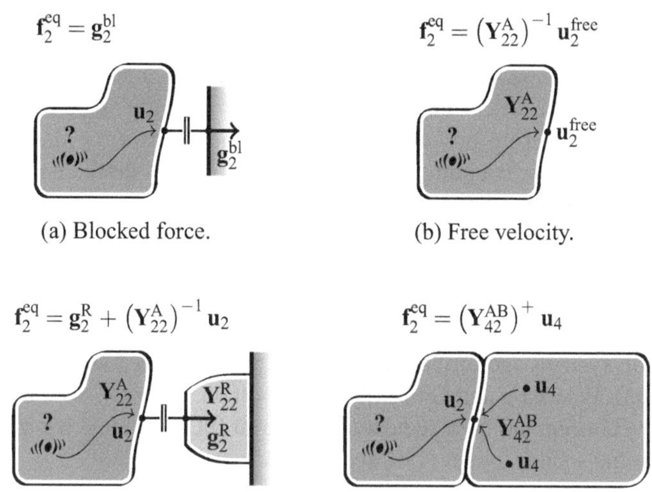

(a) Blocked force. (b) Free velocity.

(c) Hybrid interface on test rig. (d) In-situ in original assembly.

Fig. 5.5 Component-based TPA methods: various approaches to obtain equivalent forces representing the excitation

5.4.3 *Component TPA: Free Velocity*

The direct counterpart of the blocked-force method is the *free velocity TPA* as depicted in Fig. 5.5b. In this case, the component's interfaces are left free, such that all vibrations are seen as "free displacements" $\mathbf{u}_2^{\mathrm{free}}$ at the interface DoFs. From here on, equivalent forces[4] can be calculated by inverting the admittance measured at the free subsystem's interfaces, which can be understood by comparing the free displacements with the blocked force definition of Eq. (5.12):

$$\begin{bmatrix} \mathbf{u}_1 \\ \mathbf{u}_2^{\mathrm{A}} = \mathbf{u}_2^{\mathrm{free}} \end{bmatrix} = \begin{bmatrix} \mathbf{Y}_{11}^{\mathrm{A}} & \mathbf{Y}_{12}^{\mathrm{A}} \\ \mathbf{Y}_{21}^{\mathrm{A}} & \mathbf{Y}_{22}^{\mathrm{A}} \end{bmatrix} \begin{bmatrix} \mathbf{f}_1 \\ \mathbf{g}_2^{\mathrm{A}} = \mathbf{0} \end{bmatrix} \quad \Longrightarrow \quad \begin{cases} \mathbf{u}_2^{\mathrm{free}} = \mathbf{Y}_{21}^{\mathrm{A}} \mathbf{f}_1 \\ \mathbf{f}_2^{\mathrm{eq}} = \left(\mathbf{Y}_{22}^{\mathrm{A}}\right)^{-1} \mathbf{u}_2^{\mathrm{free}}. \end{cases} \tag{5.14}$$

Analogue to the blocked-force TPA, this method can be seen as a strict application of a Neumann boundary condition and is furthermore related to Norton's equivalent source theorem for electric networks (Norton 1926). Again, imposing free boundary conditions can be troublesome as the vibrating system often needs to be mounted at one or more connection points to be able to run in operation. Therefore, the method is in practice expected to perform best for frequencies well above the rigid body modes of the structure. Also note that effects such as friction and damping, which

[4] If the admittances of the subsystems are available separately, one may also apply the free velocities directly, as shown in van Schothorst et al. (2012), Rixen et al. (2015).

would occur at the interfaces of the assembled system, are absent in the operational measurement on the free component.

The fact that the method makes reference to free velocities rather than free displacements has a historical motive, as the former quantity is commonly applied in acoustical engineering in combination with acoustic pressure. For such acoustic problems, dynamic coupling of admittance is normally discarded. This is a fair assumption, considering that the impedance of the radiating source is much larger than the impedance of the receiving surrounding fluid, i.e. air. In structure-borne vibrations, the source and receiving system often exhibit similar dynamics, hence explicit coupling is needed. This was realized for a single-DoF problem by means of a nondimensional coupling function in Mondot and Petersson (1987). Extensions to multi-DoF systems and further discussion of this topic are found in Petersson and Gibbs (1993), Petersson and Gibbs (2000), Moorhouse (2001). As a final note, the free velocity concept has been standardized into an ISO norm for characterization of structure-borne sound sources (ISO Technical Committee 1996).

5.4.4 Component TPA: Hybrid Interface

As both abovementioned methods pose their limitations in practice, one often prefers to conduct operational tests on an appropriate support structure. Such a coupled structure unavoidably displays its own dynamics on the interfaces, which need to be accounted for in order to render the equivalent forces independent of any connected part. Let us, therefore, imagine the active component fixed onto a test bench or test rig R as illustrated in Fig. 5.5c. Denoting the interface admittance of the test rig by $\mathbf{Y}_{22}^{\mathrm{R}}$, we obtain for the extended system of equations:

$$\begin{bmatrix} \mathbf{u}_1^{\mathrm{A}} \\ \mathbf{u}_2^{\mathrm{A}} \\ \mathbf{u}_2^{\mathrm{R}} \end{bmatrix} = \begin{bmatrix} \mathbf{Y}_{11}^{\mathrm{A}} & \mathbf{Y}_{12}^{\mathrm{A}} & \mathbf{0} \\ \mathbf{Y}_{21}^{\mathrm{A}} & \mathbf{Y}_{22}^{\mathrm{A}} & \mathbf{0} \\ \mathbf{0} & \mathbf{0} & \mathbf{Y}_{22}^{\mathrm{R}} \end{bmatrix} \begin{bmatrix} \mathbf{f}_1 \\ \mathbf{g}_2^{\mathrm{A}} \\ \mathbf{g}_2^{\mathrm{R}} \end{bmatrix} \quad \text{with} \quad \begin{cases} \mathbf{u}_2^{\mathrm{A}} = \mathbf{u}_2^{\mathrm{R}} \\ \mathbf{g}_2^{\mathrm{A}} = -\mathbf{g}_2^{\mathrm{R}}. \end{cases} \tag{5.15}$$

After enforcing the compatibility and equilibrium conditions on the interface DoFs, a derivation similar to Sect. 5.2.2 can be followed to find the operational interface forces $\mathbf{g}_2^{\mathrm{R}}$. These forces are now dependent on the properties of both the active component A and the test rig R. Substituting the forces back into the second row of Eq. (5.15), the corresponding interface displacements $\mathbf{u}_2$ are obtained as well (the superscript is dropped because of compatibility):

$$\begin{cases} \mathbf{g}_2^{\mathrm{R}} = \left(\mathbf{Y}_{22}^{\mathrm{A}} + \mathbf{Y}_{22}^{\mathrm{R}}\right)^{-1} \mathbf{Y}_{21}^{\mathrm{A}} \mathbf{f}_1 & \text{(5.16a)} \\ \mathbf{u}_2 = \left[\mathbf{I} - \mathbf{Y}_{22}^{\mathrm{A}} \left(\mathbf{Y}_{22}^{\mathrm{A}} + \mathbf{Y}_{22}^{\mathrm{R}}\right)^{-1}\right] \mathbf{Y}_{21}^{\mathrm{A}} \mathbf{f}_1. & \text{(5.16b)} \end{cases}$$

Equations (5.16a), (5.16b) hold for any $\mathbf{Y}_{22}^{\mathrm{R}}$, as long as no external force is acting on the test rig. The desired set of equivalent forces Eq. (5.12) is obtained by combining Eqs. (5.16a) and (5.16b) in such way that the dynamics of the test rig $\mathbf{Y}_{22}^{\mathrm{R}}$ are eliminated.

$$\mathbf{f}_2^{\text{eq}} = \left(\mathbf{Y}_{22}^{\text{A}}\right)^{-1} \mathbf{Y}_{21}^{\text{A}} \mathbf{f}_1 \qquad \Longrightarrow \qquad \mathbf{f}_2^{\text{eq}} = \mathbf{g}_2^{\text{R}} + \left(\mathbf{Y}_{22}^{\text{A}}\right)^{-1} \mathbf{u}_2. \tag{5.17}$$

As it turns out, Eq. (5.17) forms the sum of contributions of both the blocked force and the free velocity experiment, respectively Eqs. (5.13) and (5.14). One could therefore speak of a *hybrid interface* condition, or Robin boundary condition in a mathematical sense. It can indeed be verified that Eqs. (5.16a)–(5.16b) converge[5] to the blocked forces for $\mathbf{Y}^{\text{R}} \rightarrow \mathbf{0}$ and to free velocities when $\mathbf{Y}^{\text{R}} \rightarrow \infty$.

The hybrid interface approach combining force and motion was originally published in de Klerk and Rixen (2010); the displacement term was regarded in this work as the "non-rigid test bench compensation" to the imperfect blocked forces. Although generally applicable in theory, it should be mentioned that the method can be rather costly and time consuming in practice, as one needs to explicitly measure collocated forces and motion in all directions for every interface node (van den Bosch et al. 2014).

In lack of force measurement, one may substitute the third row of (5.15) for $\mathbf{g}_2^{\text{B}}$. The so obtained approach was suggested in van der Seijs et al. (2014) and takes displacement measurement only:

$$\mathbf{f}_2^{\text{eq}} = \left(\mathbf{Y}_{22}^{\text{R}}\right)^{-1} \mathbf{u}_2 + \left(\mathbf{Y}_{22}^{\text{A}}\right)^{-1} \mathbf{u}_2. \tag{5.18}$$

The price for not having to measure interface forces is that separate FRF measurements should now be conducted, to obtain the subsystem admittances of the active component and the test rig.

5.4.5 Component TPA: In Situ

Looking again at Eq. (5.18), we observe that the two inverted admittance FRF matrices, in fact, represent the dynamic stiffness matrices of respectively component A and R for the same set of collocated interface DoFs. They can be combined by simple impedance addition[6]:

$$\mathbf{f}_2^{\text{eq}} = \left(\mathbf{Z}_{22}^{\text{R}} + \mathbf{Z}_{22}^{\text{A}}\right) \mathbf{u}_2 = \mathbf{Z}_{22}^{\text{AR}} \mathbf{u}_2. \tag{5.19}$$

The result of Eq. (5.19) is indeed indifferent to the dynamics of R or any other mounting structure that component A is connected to. Transforming back to admittance notation, we find that the blocked forces can be calculated inversely from the admittance of the assembly's interface or, in an overdetermined fashion, using a sufficient set of indicator points $\mathbf{u}_4$ on the receiving subsystem:

---

[5]A intuitive presentation of the range between the two limit cases is given in Moorhouse (2001).

[6]This is standard practice for FEM assembly.

$$\mathbf{f}_2^{\text{eq}} = \left(\mathbf{Y}_{22}^{\text{AR}}\right)^{-1} \mathbf{u}_2 \quad \text{or} \quad \mathbf{f}_2^{\text{eq}} = \left(\mathbf{Y}_{42}^{\text{AR}}\right)^{+} \mathbf{u}_4. \tag{5.20}$$

Obtained from a different derivation, the approach of Eq. (5.20) has first been proposed by Moorhouse and Elliott (Elliott and Moorhouse 2008; Moorhouse et al. 2009) as the *in situ* method. As implied by the name, the operational measurement may even be conducted in the target assembly AB, avoiding dismounting of any part, namely:

$$\mathbf{f}_2^{\text{eq}} = \left(\mathbf{Y}_{22}^{\text{AB}}\right)^{-1} \mathbf{u}_2 \quad \text{or} \quad \mathbf{f}_2^{\text{eq}} = \left(\mathbf{Y}_{42}^{\text{AB}}\right)^{+} \mathbf{u}_4. \tag{5.21}$$

The equivalent forces resulting from application of Eq. (5.21) are nonetheless a property of component A only and thus transferable to an assembly with another passive side.

The in situ force determination procedure is illustrated in Fig. 5.5d. Indeed, Eq. (5.21) represents the inverse of Eq. (5.11), yet with an extended set of response DoFs in order to render the system (over)determined and thus invertible. As a consequence, the method shows great resemblance with the classical matrix-inverse method of Eq. (5.10), the difference being the choice for the assembled admittance instead of the passive subsystem's admittance. Similar techniques regarding matrix conditioning (e.g. overdetermination, singular value rejection) apply as well to the in situ method. Most remarkable is that the receiving side can be chosen arbitrarily, as the equivalent forces identified by Eq. (5.20) or (5.21) are theoretically invariant of any subsystem coupled to it. In that respect, two important conditions need to be kept in mind:

1. Operational excitations $\mathbf{f}_1$ may only originate from the domain of component A. Any excitation coming from the passive side will disturb the determination of equivalent forces.
2. Although the responses used for the matrix inversion ($\mathbf{u}_2$ or $\mathbf{u}_4$) can be positioned rather arbitrarily, they are bounded to the domain of the interface and the passive side. This relates back to the remark made after Eq. (5.12): vibrations at the source structure are not only caused by the interface forces but also by the source excitations directly.

Physically, one could interpret the in situ method as follows: knowing the transfer functions from the interface DoFs to several points on the passive side, Eq. (5.20) or (5.21) seeks for a set of equivalent forces $\mathbf{f}_2^{\text{eq}}$ that, applied to the interface DoFs of the assembled structure, generates the same responses $\mathbf{u}_4$ at the passive side. Given that this response set is overdetermined, the equivalent forces are calculated such that they minimize the sum of the squared errors (or l^2-norm) in the extended set[7] $\mathbf{u}_4$.

Several numerical and experimental studies of in situ source characterization have been reported (Elliott 2009; Bonhoff 2010; Elliott et al. 2013; Madsen 2014; van der

[7] Similar ideas are used in the field of experimental substructure decoupling: the identification of the force that decouples the residual substructure can be improved by defining an "extended interface", adding some additional DoFs on the structure of interest distant from the interface (Sjövall and Abrahamsson 2008; D'Ambrogio and Fregolent 2010; Voormeeren and Rixen 2012).

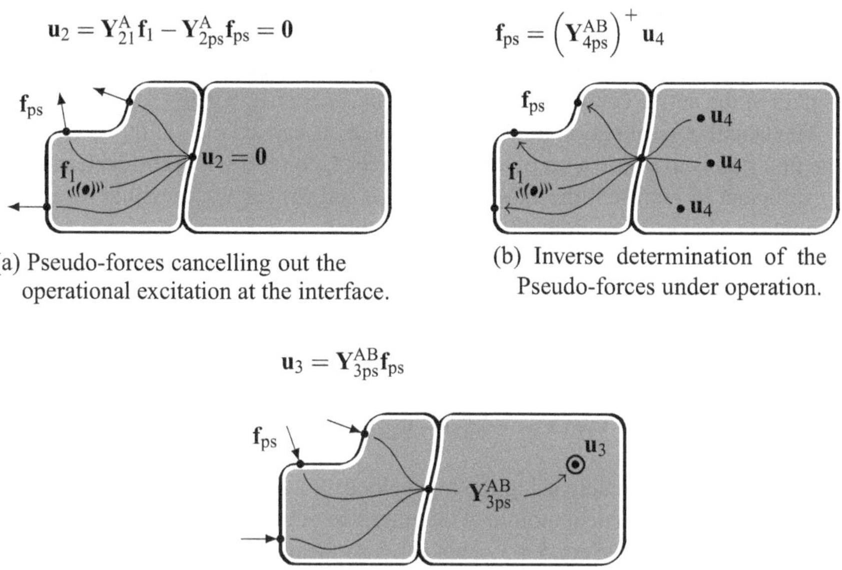

(a) Pseudo-forces cancelling out the operational excitation at the interface.

(b) Inverse determination of the Pseudo-forces under operation.

(c) Application of pseudo-forces to the assembly, causing the same response.

Fig. 5.6 The concept of the pseudo-forces component TPA method

Seijs et al. 2014, 2015). A time domain force reconstruction algorithm was proposed by Sturm et al. (2012, 2013). Further generalization of the in situ concept can be recognized in the *pseudo-forces* method that is discussed next.

5.4.6 Component TPA: Pseudo-forces

The last member of the component-based TPA family to discuss is the *pseudo-forces* method as proposed by Janssens and Verweij in the late 1990s (Janssens et al. 1999; Janssens and Verheij 2000; Janssens et al. 2002). It assumes the existence of a nonunique set of pseudo-forces acting on the outer surface of the active component, canceling out the effect of the operational vibrations transmitted through the interface to the passive side. This is illustrated in Fig. 5.6a. If those forces are now applied in the opposite direction to the assembly with the source shut down, an identical response at the receiving side should be obtained, see Fig. 5.6c. In other words, the pseudo-forces are supposed to represent the source excitation for responses at the passive side.

The fact that it can be regarded as a component TPA method lies in the former assumption: if there exists a set of pseudo-forces that cancels out the operational dynamics at the interface, the responses beyond this interface shall be zero as well, hence, these forces are invariant of any structure attached to it. A similar reasoning may as well be applied to the previously discussed equivalent force methods, yet the locations of the pseudo-forces are not bounded to the interface but extend to the full domain of the active component (see Fig. 5.6).

The actual determination of the pseudo-forces is carried out slightly differently. The first step is to define a set of s pseudo-forces $\mathbf{f}_{\mathrm{ps}}$ on the active component that is (a) sufficient to represent the excitation source and (b) easily accessible for impact hammer or shaker measurement. One could think of a minimum of $s = 6$ forces when the interface behavior can be considered as rigid, or a larger number in case of more intricate connectivity. Second, a set of $n \geq s$ indicator response DoFs has to be chosen on the passive side from which the operational excitation is monitored. These responses are denoted by $\mathbf{u}_4$ and can be written as a result of $\mathbf{f}_1$, similar to Eq. (5.9):

$$\mathbf{u}_4 = \mathbf{Y}^{\mathrm{AB}}_{41}\,\mathbf{f}_1 = \mathbf{Y}^{\mathrm{B}}_{42}\left(\mathbf{Y}^{\mathrm{A}}_{22} + \mathbf{Y}^{\mathrm{B}}_{22}\right)^{-1}\mathbf{u}^{\mathrm{free}}_{2,\mathbf{f}_1} \quad \text{with} \quad \mathbf{u}^{\mathrm{free}}_{2,\mathbf{f}_1} \triangleq \mathbf{Y}^{\mathrm{A}}_{21}\,\mathbf{f}_1. \tag{5.22a}$$

A substitution is made here using the free velocity caused by the excitation $\mathbf{f}_1$, or in other words: the theoretical motion at the interface of component A if the interfaces were left free, see also Sect. 5.4.3. This is by no means a quantity that needs to be measured, but will prove useful for the derivation later on.

The next step is to determine pseudo-forces $\mathbf{f}_{\mathrm{ps}}$ that best recreate the operational responses at the indicator DoFs $\mathbf{u}_4$ when the source excitation is shut down. Note that the assumption is similar to the equivalent forces problem stated in Sect. 5.4.1, yet the pseudo-forces act on the outer surface of the active component rather than its interfaces. Similar to Eq. (5.22a), the response at the indicator DoFs as a result of the pseudo-forces reads:

$$\mathbf{u}_4 = \mathbf{Y}^{\mathrm{AB}}_{4\mathrm{ps}}\mathbf{f}_{\mathrm{ps}} = \mathbf{Y}^{\mathrm{B}}_{42}\left(\mathbf{Y}^{\mathrm{A}}_{22} + \mathbf{Y}^{\mathrm{B}}_{22}\right)^{-1}\mathbf{u}^{\mathrm{free}}_{2,\mathbf{f}_{\mathrm{ps}}} \quad \text{with} \quad \mathbf{u}^{\mathrm{free}}_{2,\mathbf{f}_{\mathrm{ps}}} \triangleq \mathbf{Y}^{\mathrm{A}}_{21}\,\mathbf{f}_{\mathrm{ps}}. \tag{5.22b}$$

Now, the pseudo-forces $\mathbf{f}_{\mathrm{ps}}$ that best describe the operational source excitations are found from solving an overdetermined system with the response set $\mathbf{u}_4$ that was measured under operation (see Fig. 5.6b):

$$\mathbf{f}_{\mathrm{ps}} = \left(\mathbf{Y}^{\mathrm{AB}}_{4\mathrm{ps}}\right)^{+}\mathbf{u}_4. \tag{5.23}$$

It can be shown that these forces, under certain conditions, are a property of component A only. A sufficient condition is that the free interface velocities $\mathbf{u}^{\mathrm{free}}_2$ as a result of the original source excitation can be fully represented by the set of pseudo-forces determined from Eq. (5.23). In that case the definitions of Eqs. (5.22a) and (5.22b) may be equated and one finds that the pseudo-forces are specific for the source component:

$$\mathbf{u}^{\mathrm{free}}_{2,\mathbf{f}_{\mathrm{ps}}} = \mathbf{u}^{\mathrm{free}}_{2,\mathbf{f}_1} \quad \Longrightarrow \quad \mathbf{f}_{\mathrm{ps}} = \left(\mathbf{Y}^{\mathrm{A}}_{2\mathrm{ps}}\right)^{+}\mathbf{Y}^{\mathrm{A}}_{21}\,\mathbf{f}_1. \tag{5.24}$$

Applying these pseudo-forces to the FRFs of an assembled system of interest, i.e. $\mathbf{Y}^{\mathrm{AB}}_{3\mathrm{ps}}$, it can be verified that the correct receiver responses $\mathbf{u}_3$ at the passive side are obtained:

$$\mathbf{u}_3 = \mathbf{Y}^{\mathrm{AB}}_{3\mathrm{ps}}\mathbf{f}_{\mathrm{ps}} = \left[\mathbf{Y}^{\mathrm{B}}_{32}\left(\mathbf{Y}^{\mathrm{A}}_{22} + \mathbf{Y}^{\mathrm{B}}_{22}\right)^{-1}\mathbf{Y}^{\mathrm{A}}_{2\mathrm{ps}}\right]\left(\mathbf{Y}^{\mathrm{A}}_{2\mathrm{ps}}\right)^{+}\mathbf{Y}^{\mathrm{A}}_{21}\,\mathbf{f}_1 = \mathbf{Y}^{\mathrm{AB}}_{31}\mathbf{f}_1. \tag{5.25}$$

The pseudo-forces determined from Eq. (5.23) are indeed transferable to an assembly with another passive side. Equation (5.25) yields the responses for this new assembly, provided that the columns of the FRF matrix $\mathbf{Y}^{\mathrm{AB}}_{3\mathrm{ps}}$ correspond to the same pseudo-forces, i.e. excitation points at the source.

The previously discussed in situ method can be regarded as a special case of the pseudo-force method, namely for the case where the pseudo-forces are located at the interfaces. The pseudo-forces calculated from Eq. (5.23) shall then equal the equivalent (blocked) forces from Eq. (5.12), namely $\mathbf{f}_{\mathrm{ps}} = \mathbf{f}^{\mathrm{eq}}_2$. With regard to the positioning of the indicator DoFs $\mathbf{u}_4$, the same restriction holds as for the in situ method, namely that they must be located at the passive system B or at the interface.[8]

5.5 Transmissibility-Based TPA

The two previously discussed families of TPA have in common that they attempt to model the vibration transmission in a physically correct sense, namely by determining as many forces and moments as required to describe the subsystem connectivity in full. Consequently, both families explicitly require the corresponding FRFs for the interface DoFs to the receiving response locations. It is evident that this approach ultimately reveals a wealth of information on the particular functioning of the active component, force distribution over the interface, resonances in the structure, etc. If, however, the mere purpose of an TPA is to identify the dominant path contributions in the assembled product, the efforts to set up and conduct the full experiment can be fairly disproportionate and costly. This is, especially, the case when multiple incoherent vibration sources are to be investigated.

The last family to discuss avoids the stage of explicit force determination. Instead, the path contributions are determined from so-called "transmissibilities" between sensors, possibly calculated from operational measurements. Various methods are discussed in the following sections that share the following properties:

1. Measurements are conducted on the assembled product only, saving time to dismount the active components. In fact, the interfaces between the active and passive components are no longer of principal interest.
2. Path contributions are determined from well-chosen indicator points around the sources or connections. These indicator points function as inputs to the TPA.
3. The result of the analysis is highly subject to the choice for these indicator points; care should, therefore, be taken to include all transmission paths.

[8]This was erroneously stated in the original work (Janssens and Verheij 2000).

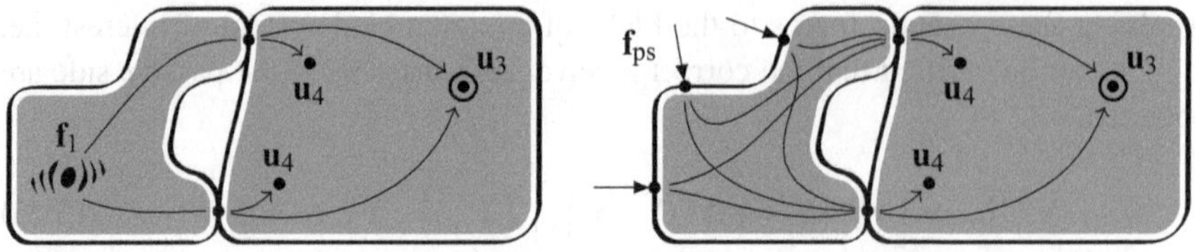

(a) Transmissibility from source FRFs. (b) Transmissibility from alternative FRFs.

Fig. 5.7 The transmissibility concept for an example with two connection points

The family of *transmissibility-based TPA* methods indeed departs from the traditional source–transfer–receiver model that assumes a physically meaningful set of loads, FRFs, and responses. Although potentially less accurate, methodologies derived from this concept tend to be easy to set up, versatile concerning sensor type and particularly effective for ranking contributions from several sources. From a practical point of view, transmissibility-based TPA tries to outrun the physically correct methods by its ability to conduct multiple cycles in a shorter time. Nevertheless, under certain conditions, results of similar (or even equivalent) accuracy can be achieved as compared to classical and component-based TPA methods.

5.5.1 The Transmissibility Concept

To discuss the theoretical concepts behind transmissibility-based methods, consider an assembled system AB with two connection points[9] as shown in Fig. 5.7. The active side contains a vibration mechanism that is characterized by internal forces $\mathbf{f}_1$; the receiver responses at the passive side are denoted by p DoFs in $\mathbf{u}_3$. To monitor the vibrations transmitted across the interfaces, n indicator DoFs $\mathbf{u}_4$ are positioned around the connection points.

Let us first assume that the source excitation $\mathbf{f}_1$ can be described by o forces (or independent force distributions) and that FRFs are measurable for all of the abovementioned DoFs. The equations for the passive side responses then read:

$$\begin{cases} \mathbf{u}_3 = \mathbf{Y}^{AB}_{31}\mathbf{f}_1 & \text{for } p \text{ receiver DoFs,} \quad (5.26a) \\ \mathbf{u}_4 = \mathbf{Y}^{AB}_{41}\mathbf{f}_1 & \text{for } n \text{ indicator DoFs.} \quad (5.26b) \end{cases}$$

Provided that $n \geq o$ and $\mathbf{Y}^{AB}_{41}$ is full rank, all excitation forces $\mathbf{f}_1$ are observable from $\mathbf{u}_4$. Hence Eq. (5.26b) can be inverted and substituted into Eq. (5.26a). The responses $\mathbf{u}_3$ are now expressed in terms of the DoFs $\mathbf{u}_4$ that can be measured under operation:

[9] An example with two connection points was chosen here merely to provide better insight into some important cross-correlation properties. There is no fundamental consequence for the generality of the methods derived.

$$\mathbf{u}_3 = \mathbf{T}^{\mathrm{AB}}_{34,\mathbf{f}_1}\mathbf{u}_4 \quad \text{with} \quad \mathbf{T}^{\mathrm{AB}}_{34,\mathbf{f}_1} \triangleq \mathbf{Y}^{\mathrm{AB}}_{31}\left(\mathbf{Y}^{\mathrm{AB}}_{41}\right)^{+}. \tag{5.27}$$

The so obtained transmissibility matrix $\mathbf{T}^{\mathrm{AB}}_{34,\mathbf{f}_1}$ relates motion at the indicator DoFs $\mathbf{u}_4$ (the inputs) to the receiver DoFs $\mathbf{u}_3$ (the outputs) for excitation forces $\mathbf{f}_1$. Interestingly, the size of the transmissibility matrix has become $p \times n$, obfuscating the o excitations associated with the original FRFs. This raises the question which excitations are really needed to construct the transmissibility matrix and to what extend this matrix is generally valid for the problem ($\mathbf{u}_4 \to \mathbf{u}_3$) under different excitations of the source structure.

To gain more insight in the transmissibility problem stated by Eq. (5.27), let us expand Eqs. (5.26a) and (5.26b) in terms of the subsystems' admittances. As seen in previous derivations, the terms that couple the subsystems are identical for all responses ($\mathbf{u}_3, \mathbf{u}_4$) at the passive side:

$$\begin{bmatrix}\mathbf{u}_3\\ \mathbf{u}_4\end{bmatrix} = \begin{bmatrix}\mathbf{Y}^{\mathrm{AB}}_{31}\\ \mathbf{Y}^{\mathrm{AB}}_{41}\end{bmatrix}\mathbf{f}_1 = \begin{bmatrix}\mathbf{Y}^{\mathrm{B}}_{32}\\ \mathbf{Y}^{\mathrm{B}}_{42}\end{bmatrix}\left(\mathbf{Y}^{\mathrm{A}}_{22} + \mathbf{Y}^{\mathrm{B}}_{22}\right)^{-1}\mathbf{Y}^{\mathrm{A}}_{21}\mathbf{f}_1.$$

Recalling now the expressions for the interface force and free velocity, respectively Eqs. (5.4) and (5.14), the following substitutions can be made:

$$\begin{gathered}\begin{bmatrix}\mathbf{u}_3\\ \mathbf{u}_4\end{bmatrix} = \begin{bmatrix}\mathbf{Y}^{\mathrm{B}}_{32}\\ \mathbf{Y}^{\mathrm{B}}_{42}\end{bmatrix}\mathbf{g}^{\mathrm{B}}_2\\ \text{width} \quad \begin{cases}\mathbf{g}^{\mathrm{B}}_2 = \left(\mathbf{Y}^{\mathrm{A}}_{22} + \mathbf{Y}^{\mathrm{B}}_{22}\right)^{-1}\mathbf{u}^{\mathrm{free}}_2 & \text{for } m \text{ interface forces}\\ \mathbf{u}^{\mathrm{free}}_2 = \mathbf{Y}^{\mathrm{A}}_{21}\mathbf{f}_1 & \text{for } m \text{ free velocities.}\end{cases}\end{gathered} \tag{5.28}$$

Hence, Eq. (5.28) shows that the transmission of vibrations from o forces $\mathbf{f}_1$ to n responses $\mathbf{u}_4$ is limited by the number of DoFs of the interface forces/displacements m. This means that the interface acts as a *bottleneck*: it limits the effective rank of the transmissibility problem to a maximum of m. Furthermore, Eq. (5.28) exposes two interesting properties of the transmissibility concept:

1. Regarding the source excitation, Eq. (5.28) shows a direct relation between the interface forces $\mathbf{g}^{\mathrm{B}}_2$ of the coupled system and the theoretical free velocities $\mathbf{u}^{\mathrm{free}}_2$ at the disconnected interfaces of component A. As understood from the component-based TPA methods, various sets of forces can be defined that equivalently produce these free interface velocities $\mathbf{u}^{\mathrm{free}}_2$, such as the pseudo-forces of Eq. (5.22b). This is illustrated in Fig. 5.7b. In fact, any set of forces on the source that renders the interface fully *controllable* will guarantee convergence to a transmissibility matrix that generally captures the transmission of vibrations caused in component A. This property is used in the operational TPA method as discussed in the next section.
2. Regarding the transmissibility, the problem ($\mathbf{u}_4 \to \mathbf{u}_3$) is in fact *specific to the passive side only*, provided that the interface forces $\mathbf{g}^{\mathrm{B}}_2$ are fully observable from $\mathbf{u}_4$. That means that the second line of Eq. (5.28) can be inverted, which is equivalent to Eq. (5.10) from the classical matrix-inverse method. With respect to

transmissibility-based TPA, it implies that if $\mathbf{u}_4$ is chosen in such a way that all interface forces are *observable*, $\mathbf{g}_2^{\mathrm{B}}$ can be eliminated from Eq. (5.28) and the transmissibility matrix becomes a property of the passive side only, namely:

$$\mathbf{u}_3 = \mathbf{T}_{34}^{\mathrm{B}}\mathbf{u}_4 \quad \text{with} \quad \mathbf{T}_{34}^{\mathrm{B}} \triangleq \mathbf{Y}_{32}^{\mathrm{B}}\left(\mathbf{Y}_{42}^{\mathrm{B}}\right)^{+}. \tag{5.29}$$

An elegant duality can be observed here with the component-based TPA concept that characterizes the excitation as a property of the source component. More properties of the transmissibility matrix are discussed in Guasch et al. (2013), Weijtjens et al. (2014).

Theoretically, it can be stated that the controllability is a property of the source component A, whereas the observability is specific to the receiving structure B. If both conditions are satisfied, Eqs. (5.27) and (5.29) generate the same transmissibility matrix, hence $\mathbf{T}_{34}^{\mathrm{AB}} = \mathbf{T}_{34}^{\mathrm{B}}$ or simply $\mathbf{T}_{34}$. The pseudo-inversion step involved in Eq. (5.27) or (5.29) "decorrelates" the transmissibility columns associated with $\mathbf{u}_4$ into linearly independent contributions to the receiver DoFs $\mathbf{u}_3$, such that an element of $\mathbf{T}_{34}$ is defined as follows:

$$T_{ij} \triangleq \left.\frac{u_i}{u_j}\right|_{u_{k \neq j} = 0} \qquad \begin{cases} u_i \in \mathbf{u}_3 \\ u_j, u_k \in \mathbf{u}_4. \end{cases}$$

This process is often referred to as Cross-Talk Cancelation (CTC). Note that this concerns a spatial or "modal" decorrelation[10] only, just like other methods involving FRF matrix inversion. It should therefore by no means be understood as some special signal processing step.

In practice, to obtain $\mathbf{T}_{34}$ from FRFs as illustrated by Fig. 5.7a or b, one requires sufficient excitations and indicator DoFs to respectively *control* and *observe* the dynamics at the interface. However, the foremost reason for using a transmissibility-based TPA method is to identify source path contributions without the need to conduct a tedious FRF measurement campaign. The next section discusses how to obtain $\mathbf{T}_{34}$ from operational responses only.

5.5.2 *Operational Transfer Path Analysis (OTPA)*

The transmissibility matrix $\mathbf{T}_{34}$ can be estimated statistically from the correlation between $\mathbf{u}_3$ and $\mathbf{u}_4$ when monitored under a variety of operational test conditions. The fundamental assumption is that, by testing the vibration source under different

[10]This decorrelation has been approached in several ways: for instance, based on so-called Global and Direct Transmissibility Functions (GTDT) in the work of Magrans (1981), Guasch and Magrans (2004); Guasch (2009); Guasch et al. (2013) or along conventional FRFs as seen in Varoto and McConnell (1998), Ribeiro et al. (1999, 2000), Maia et al. (2001).

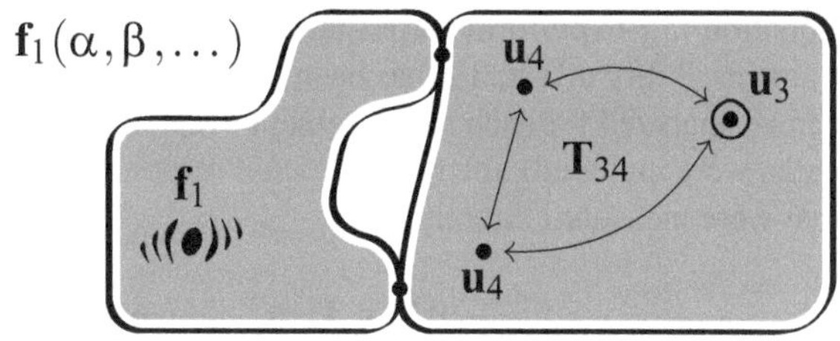

Fig. 5.8 Operational TPA: the source excitation $\mathbf{f}_1$ depends on operational parameters $(\alpha, \beta, \dots)$ such as engine speed or torque

operational conditions (e.g. a motor operating at different speeds or torques), multiple load cases are exercised that are slightly independent. However, as seen in the previous section, the effect of these load cases across the interface are theoretically limited to a maximum of m independent modes of vibration at the passive side, because of the bottleneck effect of the interface. This is further discussed below.

Let us now assume that the operational excitation is able to generate sufficient independent load cases $\mathbf{f}_1(\alpha, \beta \dots)$, where α and β represent independent operational parameters.[11] This is illustrated in Fig. 5.8. Equation (5.27) can be established for r sets of measured responses obtained under different operational conditions, e.g. the sliced time blocks of a run-up measurement. This amounts to stacking the r measured spectra of the receiving DoFs and indicator DoFs columnwise into arrays $\mathbf{U}_3$ and $\mathbf{U}_4$ to construct the following system of equations (the superscript on $\mathbf{u}$ indicates the measurement block number):

$$\underbrace{\left[\mathbf{u}_3^{(1)}\ \mathbf{u}_3^{(2)}\ \cdots\ \mathbf{u}_3^{(r)}\right]}_{p\times r} = \underbrace{\mathbf{T}_{34}}_{p\times n}\ \underbrace{\left[\mathbf{u}_4^{(1)}\ \mathbf{u}_4^{(2)}\ \cdots\ \mathbf{u}_4^{(r)}\right]}_{n\times r} \quad \Longrightarrow \quad \mathbf{U}_3 = \mathbf{T}_{34}\mathbf{U}_4. \tag{5.30}$$

When sufficient measurement data is acquired to ensure $r \geq n$, the problem of Eq. (5.30) can be made invertible by post-multiplying with $\mathbf{U}_4^H$. The transmissibility matrix is then found from solving this system or, more conveniently, from the H_1 estimator[12] of the Cross-Power Spectra (CPS) and Auto-Power Spectra (APS) of the measured DoFs:

$$\begin{aligned} \mathbf{T}_{34} &= \mathbf{U}_3\mathbf{U}_4^H\left(\mathbf{U}_4\mathbf{U}_4^H\right)^{-1} \\ \Longrightarrow \quad \mathbf{T}_{34} &= \mathbf{S}_{34}\mathbf{S}_{44}^{-1} \quad \text{with} \quad \begin{cases} \mathbf{S}_{34} = \frac{1}{r}\,\mathbf{U}_3\mathbf{U}_4^H & \text{(CPS)} \\ \mathbf{S}_{44} = \frac{1}{r}\,\mathbf{U}_4\mathbf{U}_4^H & \text{(APS).} \end{cases} \end{aligned} \tag{5.31}$$

[11] In Roozen and Leclère (2013), it is suggested that external excitation on the source structure, e.g. by use of a non-instrumented hammer, leads to better conditioning of the transmissibility matrix compared to a sequence of operational excitations. This resembles the approach depicted by Fig. 5.7b.

[12] The H_1 estimator is a well-known principle in experimental modal analysis to determine FRFs from a Multiple Input–Output (MIMO) test, see for instance, Bendat and Piersol (1980), Ewins (2000). Alternative ways to obtain the transmissibility matrix have recently been explored, such as the H_2 or H_s estimator to balance the error contributions between the inputs and outputs (Leclère et al. 2012).

Equation (5.31) performs cross-talk cancelation similar to Eq. (5.27). Now that the transmissibility matrix $\mathbf{T}_{34}$ has been estimated globally, the actual path contributions can be analyzed for each measurement cycle. If performed correctly (i.e. all transfer paths were included), the reconstructed response spectra equal the response spectra that were measured directly.

$$\mathbf{u}_3^{(l)} = \mathbf{T}_{34}\mathbf{u}_4^{(l)} \qquad l \in \{1, \ldots, r\}. \tag{5.32}$$

The approach governed by Eqs. (5.30) and (5.32) is generally known as *Operational Transfer Path Analysis* (OTPA) (Noumura and Yoshida 2006). Clearly, the performance of this method is subject to the choice and positioning of the indicator sensors $\mathbf{u}_4$, that function as "inputs" to the analysis. Too few sensors (or too distant from the source connections) could lead to neglecting important transmission paths, resulting in unrealistic prognoses (Gajdatsy et al. 2008). On the other hand, having too many sensors (or too near to each other) complicates the inversion of $\mathbf{S}_{44}$ due to poor conditioning, resulting in amplification of measurement noise (Gajdatsy et al. 2008).

Theoretically, the rank of $\mathbf{U}_4$ is limited by the number of DoFs at the interface m. This implies that $\mathbf{U}_4$ and thus $\mathbf{S}_{44}$ is rank deficient, even if $r > n$. It is therefore common practice to perform a Principal Component Analysis (PCA) in order to identify how many independent modes of vibration are present in the system and transferred across the interfaces. Principal components are calculated from the singular value decomposition of $\mathbf{U}_4$. The left-singular vectors represent the vibration mode shapes that build up to the vibrations in the indicator DoFs, sorted from the largest contribution to the smallest. By controlling the amount of principal components that are used in the computation of the transmissibility by Eq. (5.32), the condition number can be kept low such that one balances between the completeness of path descriptions and attenuation of measurement noise. This is not discussed in detail here, guidelines can be found in specific literature on OTPA (Noumura and Yoshida 2006; de Klerk and Ossipov 2010; Putner et al. 2012).

Another strong advantage of OTPA is the ease of combining various types of sensors, both for the input (indicator) and output (receiver) DoFs. Quantities such as accelerations, velocities, sound pressures and even forces and strains can be used in a mixed fashion, as long as proper scaling (unit normalization) is taken into account (Noumura and Yoshida 2006). This makes OTPA a suitable method to quickly investigate the proportion of structure-borne and airborne contributions (de Klerk et al. 2009). Other related extensions or applications include Response Modification Analysis (RMA) for providing structural sensitivities to the target responses (Scheuren and Lohrmann 2014), pass-by analysis of road vehicle noise (Putner et al. 2013; ISO Technical Committee 2015) and time domain auralization of OTPA, sometimes referred to as Transfer Path Synthesis (TPS) (Vorländer and Dietrich 2008).

5.6 Substructuring to Estimate Fixed-Interface Modes

Suppose that a structure of interest is attached to a flexible fixture. Suppose that measurements can be performed on this assembly to measure the displacements at several DOF on the substructure itself, denoted $\mathbf{u}_s$, and on the fixture $\mathbf{u}_s$. Similarly, the fixture could be removed from the assembly and measurements were taken at the same DOF on the isolated fixture $\mathbf{u}^{\text{fixt}_f}$, where the subscript f could be omitted since now all measurements must be on the fixture.

A modal test could be performed to obtain N modes of vibration of the assembly, resulting in the modal parameters for the fixture+structure, ω_r, ζ_r and $[\phi_{fr}^{\mathrm{T}}, \phi_{sr}^{\mathrm{T}}]^{\mathrm{T}}$, where ω_r is the rth natural frequency of the structure, ζ_r is the corresponding damping ratio and the vectors ϕ_{fr} and ϕ_{sr} contain the mass-normalized mode vector of the rth mode at the measurement points on the fixture and substructure, respectively. The index r ranges over all of the measured modes, $r = 1...N$. The fixture is a dynamic system itself, although it is meant to approximate a rigid boundary condition.

One can also obtain the modal parameters of the fixture alone (without the structure of interest attached) either through test or analysis, and its modal parameters are denoted: ω_r^{fixt}, ζ_r^{fixt} and ϕ_{fr}^{fixt}. The natural frequencies and damping ratios of the fixture are not needed in any of the following, only the mass-normalized mode shapes of the fixture.

If the fixture were truly rigid and perfectly fixed to ground, then the motion of the structure when it is attached to the fixture, $\mathbf{u}_f$, would be zero. In practice this will not be the case. The most straightforward remedy would be to estimate both the displacement and rotation at the points where the structure joins the fixture and then to apply constraints to force that motion to be zero. However, that method is not reliable because, as will be illustrated in the examples presented in the lecture. Instead, the following constraints are applied.

5.6.1 Modal Constraints

The measured fixture motion, $\mathbf{u}^{fixt}$, can be approximated as follows in terms of N_{fixt} modes of the fixture:

$$\mathbf{u}^{\text{fixt}} \approx \boldsymbol{\phi}_f^{\text{fixt}} \mathbf{q}^{\text{fixt}}, \tag{5.33}$$

where $\mathbf{q}^{\text{fixt}}$ denotes the modal coordinates of the fixture. This expression can be inverted to find the participation of each fixture mode, as $\mathbf{q}^{\text{fixt}} \approx \left(\boldsymbol{\phi}_f^{\text{fixt}}\right)^+ \mathbf{u}^{\text{fixt}}$, where $()^+$ denotes the Moore–Penrose pseudoinverse of the matrix. The estimate of the fixture modal amplitudes is only meaningful here if one has at least as many measurement points as modes of interest and if the measurement locations on the fixture are chosen such that $\boldsymbol{\phi}_f^{\text{fixt}}$ has full column rank.

If the modal basis of the fixture is sufficiently rich, then it will be adequate to span the space of the motion of the fixture even after the structure is attached (i.e., of the fixture+structure). If this is the case, then the fixture modal motions can be estimated from the measured motions, $\mathbf{u}^{\text{fixt}}$, as follows.

$$\mathbf{q}^{\text{fixt}} = \left(\boldsymbol{\phi}_f^{\text{fixt}}\right)^+ \mathbf{u}_f = \left(\boldsymbol{\phi}_f^{\text{fixt}}\right)^+ \boldsymbol{\phi}_f \mathbf{q}, \tag{5.34}$$

where $\mathbf{q}$ denotes the modal coordinates of the assembled fixture+structure. Our desire is to estimate the modal parameters that the structure would have if it were attached to a truly rigid fixture. One way to do this is to apply the constraints,

$$\mathbf{q}^{\text{fixt}} = 0 \tag{5.35}$$

to the modes of the fixture+structure using the methods outlined in this course. In terms of the modal coordinates of the fixture+structure, the constraint equations are

$$\left(\boldsymbol{\phi}_f^{\text{fixt}}\right)^+ \mathbf{u}_f = 0 \tag{5.36}$$

or

$$\left(\boldsymbol{\phi}_f^{\text{fixt}}\right)^+ \boldsymbol{\phi}_f \mathbf{q} = 0 \tag{5.37}$$

either of which constitutes a set of N_{fixt} constraint equations. The matrix multiplying q is the matrix B in this course and in the review by De Klerk, Rixen and Voormeeren (2008) or [**a**] in the text by Ginsberg (2001). The procedure described in either of those works can be used to enforce these constraints and hence to estimate the modes of the fixture+structure with the fixture motion nullified. The "cms.m" MATLAB®routine, which is freely available on the MATLAB Central File Exchange (search under Matthew S. Allen) or in the materials provided with this course, implements the method in Ginsberg (2001) and was used to perform the calculations described in the lecture.

It is important to note that the constraints above only enforce zero motion at the fixture measurement points if the number of measurement points equals the number of fixture modes. In practice, one should use more measurement points on the fixture than there are active modes in order to average out noise and measurement errors. However, when this is done the motions of the physical measurement points may not be exactly zero after applying the constraints. In the best case, the residual motion would be due only to measurement noise, but there might also be residual motion in the fixture that is physical, since one is seeking to constrain an infinite dimensional system with a finite number of constraints. Fortunately, one can readily observe the fixture motions after applying the constraints to see whether the constraints were effective in enforcing a rigid boundary condition. This provides a valuable way to check whether enough modes were used in Eq. 5.38.

5.6.2 Singular Vector Constraints

In some cases, it may not be feasible to perform a modal test on the fixture alone in order to estimate $\boldsymbol{\phi}_f^{\text{fixt}}$. There also may be situations in which the free modes of the fixture do not form an efficient basis for the motion of the fixture+structure, so some other basis must be considered. In these cases, one alternative is to use the dominant singular vectors of the fixture mode shape matrix to form the constraints. The first step is to perform a Singular Value Decomposition (SVD) of the fixture motions observed on the fixture+structure

$$\boldsymbol{\phi}_f = \mathbf{USV}^{\text{T}} \tag{5.38}$$

where $\mathbf{U}$ is a matrix of singular vectors, $\mathbf{S}$ a diagonal matrix of corresponding singular values and $\mathbf{V}$ reveal how each singular vector participates in each of the modes. Both $\mathbf{U}$ and $\mathbf{V}$ are orthonormal. If the motion of the fixture is sufficiently simple, then it is likely to be captured in only a few, N_{SVD}, dominant vectors, so one can approximate the fixture motion as follows:

$$\mathbf{u}_f = \mathbf{U}^{N_{SVD}}\mathbf{S}^{N_{SVD}}(\mathbf{V}^{N_{SVD}})^{\text{T}}\mathbf{q}, \tag{5.39}$$

where the matrices now contain only the first N_{SVD} columns or rows or both. The following constraint will set this approximation to the fixture's motion to zero,

$$(\mathbf{U}^{N_{SVD}})^{\text{T}}\mathbf{u}_f = 0. \tag{5.40}$$

This is an alternative to the constraints in Eq. 5.38 and avoids having to measure the modes of the fixture alone.

5.7 Payload and Component Simulations Using Effective Mass and Modal Craig–Bampton Form Modal Models

The methods in the prior section can be extended and used to estimate Hurty/Craig–Bampton models of subcomponents from measurements. In particular, for small components a quantity of interest is the effective mass. In Sedaghati et al. (2003), Mayes (2015), Mayes et al. (2013) and Mayes and Linehan (2014), a methodology was developed and validated on several structures.

Consider a component mounted on a larger structure. If the structure is large, we can consider the component to be subjected to a base acceleration that is independent of the component's motion. Then, the HCB model allows one to compute the effective mass, which describes how much each of the fixed-base modes of the structure is excited by the base acceleration. For example, if one considers a clamped–clamped beam on a rigid foundation, then each symmetric mode would be excited by a uniform motion of the base but asymmetric modes would not be. Hence, the asymmetric modes would have zero effective mass and the symmetric modes' effective mass would vary depending on their propensity to be excited; typically, the lower modes have higher effective mass. Such a model can be effective when seeking to determine if a subcomponent survives a given environment.

Effective mass models are usually derived from finite element models, but methods were recently presented that can derive them experimentally (Mayes et al. 2013). To do so, a test is performed with the component on a transmission simulator, with the assembly free–free. Then the substructuring method in the prior section can be used to constrain the transmission simulator to ground and estimate the fixed-interface modes of the test article. Further processing steps yield the effective mass. See Sedaghati et al. (2003), Mayes (2015), Mayes et al. (2013) and Mayes and Linehan (2014) for further details.

5.8 Calibration of an SUV Rear Subframe

In the automotive industry, finalized products are commonly made up of thousands of components, which can, in turn, be made out of smaller components. Finite element modeling is used to increasing extent to study these structures. This is because of the flexibility the computer models bring in terms of analyzing structural modifications. One modeling approach that is often taken is to divide the structure into parts and use substructuring techniques. For an accurate modeling, it is then important to understand the physics of every substructure and not to the least the physical behavior of the interfacing parts through which the substructures are physically connected. Many different connection types exist, where rubber bushings are commonly used when vibration isolation is needed such as in the isolation of subframes from the body in white. It is well known that the stiffness of rubber materials is temperature and frequency dependent and can exhibit large variation from one unit to the next. Furthermore, good model characteristics for damping can be crucial for the full model's prediction quality. The damping is usually not modeled based on first principles due to its complexity, but rather by calibration of models to experimentally acquired data. Therefore, there is an interest in updating FE models toward experimentally obtained data from vibration tests. A study, reported in Bylin et al. (2018), Bylin (2018) and Gibanica and Abrahamsson (2017), has been conducted using a rear subframe of a passenger car, see Fig. 5.9. It primarily concerned the estimation of the bushing rubber material parameters in a mass- loaded boundary configuration, but also to calibrate the rear subframe model in general. The model calibration MATLAB application FEMcali (Abrahamsson 2019) was used as a tool for calibrating the model parameters. By additive mass loading of the bushings in an experimental modal analysis setting, the bushings are considerably activated and bring local bushing modes down in frequency where it is possible to study them with modal analysis. Furthermore, the additive mass loading provides a more realistic boundary condition for the bushings. In Fig. 5.9, the two different bushing types are

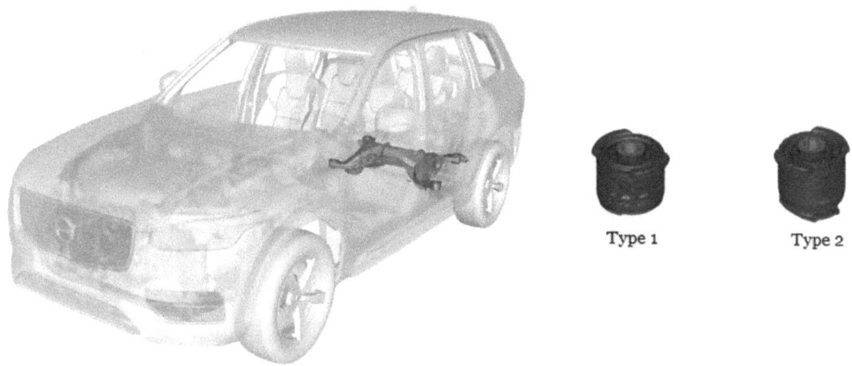

Fig. 5.9 Rear subframe location in the car and two bushing types of particular interest

shown and the mass loading component can be seen in Fig. 5.10 together with a small region of an FE representation of the mass-loaded bushing.

Two finite element models are used, one with and the other without mass-loaded bushings. The FE models of both configurations each consist of more than half a million degrees of freedom. The calibration procedure minimizes the deviation between the FE model's frequency response functions and the experimentally identified model's ditto. Equalized damping is imposed on both models to circumvent the mode pairing problem. A bootstrapping cross-validation is further used to quantify the parameter uncertainties. A SIMO stepped-sine testing procedure was used to obtain test data. The FE model with mass loading was calibrated first to obtain good parameter estimation of the rubber's Young's modulus, i.e., all other model parameters were then fixed at their nominal values. The configuration without mass loading was then calibrated with fixed calibrated rubber parameters for an overall good parameter estimation of the rear subframe FE model. In the final step, a subset of all parameters was updated toward a higher frequency region for the configuration without mass loading to gain further physical understanding of the behavior of the component.

Experimental modal analysis test setup. The vibration tests were performed in the Vibration Lab at the Chalmers University of Technology. Three nominally identical rear subframes of a Volvo XC90 with and without mass loading were used to obtain data for the spread between individuals. The testing procedure was the same for all components. The rear subframe with and without mass loading is illustrated in Fig. 5.11. The components were hung in long thin lines (red and thickened for visibility in Fig. 5.10) attached to a support structure via soft springs. The support modes were all below 5 Hz and were thereby well below the first flexible eigenfrequency that appeared above 40 Hz in both configurations.

With the target to obtain particularly good test data for the first 20 flexible body modes, the position of 20 uniaxial sensor locations was found from a set of 170 candidate positions with a sensor placement pretest planning method (Kammer 2005;

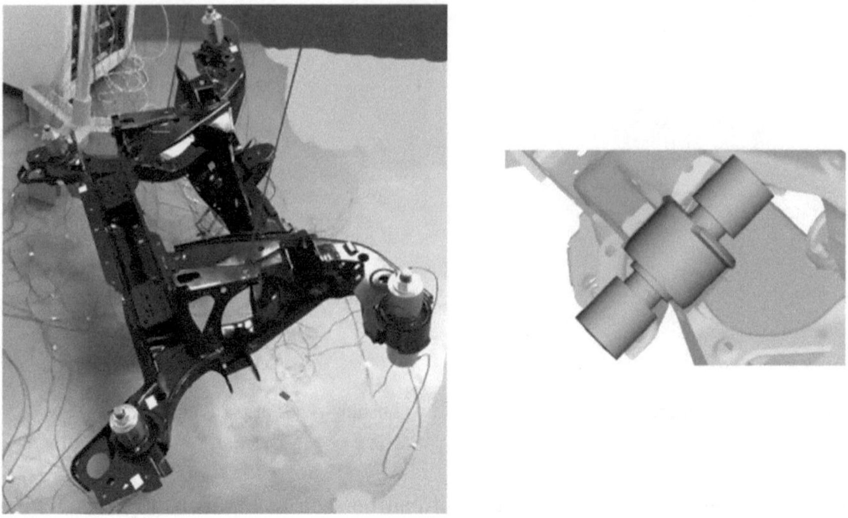

Fig. 5.10 Rear subframe with additional weights fitted to four bushings in instrumented test configuration (left). CAD illustration of additional weight (right)

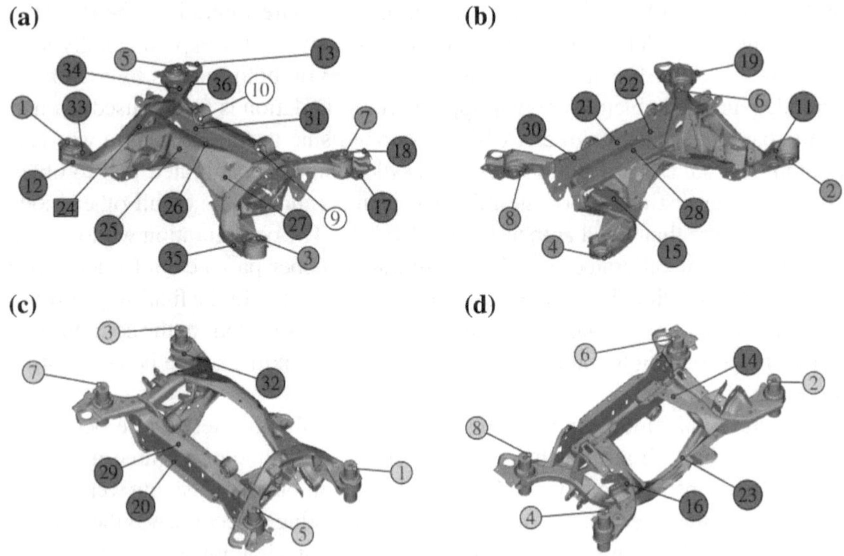

Fig. 5.11 Top view of configuration without additional weights in **a** and bottom view in **b**. Top view of mass-loaded configuration in **c** and bottom view in **d**. Circles mark accelerometer locations and a rectangle marks the excitation position. Excitation was normal to the surface, with a direct accelerance force and accelerometer sensor configuration. Green (uniaxial) and yellow (triaxial) circles represent accelerometer locations common for both configurations. Cyan (without weights) and red (with weights) markings indicate triaxial sensor positions that differ between the configurations

Gibanica et al. 2016). Six more uniaxial accelerometer positions were added for visualization purpose. Ten additional triaxial accelerometers were placed on the bushings. All sensor placements can be seen in Fig. 5.11. In the colocated and colinear force/acceleration sensor configuration, the accelerometer and force transducer was placed on the opposite side of a thin sheet metal part of the subframe. Accelerometers 9–36 (numbering; see Fig. 5.11) were placed identically on both configurations while accelerometers 1–8 were placed directly on the bushings in the configuration without added weights and these were instead placed on the additional weights in the mass-loaded configuration. The triaxial accelerometers were PCB 356A03 (1 gram, glued) and the uniaxial were PCB 352C22 (0.5 gram, bees wax fastened). The accelerometer masses were included in the FE models. The shaker was an LDS V201 attached to the subframe via a metallic stinger of approximately 5 mm length. The excitation force was measured with a Brüel&Kjær 8203 force sensor together with an B&K IEPE converter 2647B, attached to the component through a glued stinger attachment plate with 0.2 grams weight. The force transducer and stinger attachment plate masses were not included in the FE models.

Two excitation methods were used for both configurations. Periodic chirp tests at various amplitudes were performed to assess the linearity of the systems to obtain proper excitation loads. Most test data used for calibration were collected with multi-sine tests for diminishing the influence of noise in the model updating procedure. For the mass-loaded configuration, a frequency range (40,500) Hz with 2000 discrete frequencies was deemed sufficient due to the relatively strong damping. In the configuration without added weights, a frequency range (60,500) Hz with 3000 discrete frequency were used instead. The discrete frequencies were selected based on the method specified in Vakilzadeh et al. (2015) to minimize system identification model bias.

The calibration procedure used gives a parameter setting that minimizes the deviation between the FE model and an identified state-space system model, and thus indirectly from raw test data through the identified model. It is therefore crucial for the system identification model to be of high quality. For the configuration without mass loading, all accelerometers were used in the identification, from 60 to 350 Hz with 20 states. Low- and high-frequency residual modes were added to the obtained system and the state-space model's **B** and **C** matrices were augmented with additional data from chirp measurements for the low-frequency region from 20 to 60 Hz and stepped-sine data for the high-frequency region from 350 to 470 Hz. This produced a system that accurately predicted the system's resonance and antiresonance behavior.

For the mass-loaded configuration, only the bushing accelerometers, accelerometers 1 through 8 in Fig. 5.11, were used initially for the system identification due to the many local bushing modes identified from the measurements, not visible in other channels. Thus an identified system with 40 states from 40 to 225 Hz was identified for 24 channels (8 first triaxial accelerometers). A low- and high-frequency residual system pole was added to the system and additional data from a chirp test from 20 to 40 Hz and stepped-sine test from 225 to 250 Hz was used to reestimate the system's **B** and **C** matrices by the procedure given in Sect. 4.9.1. In addition, all 36 accelerometers, or 56 channels, throughout the frequency region from 20 to 250 Hz were used in the reestimation for a system with more system outputs.

The raw data from the channels for which the system identification model gave the poorest fit were used as validation data in a validation procedure. The calibration channels selected from the 56 available channels for the mass-loaded configuration were 1x, 1z, 2x, 2z, 3x, 3z, 4x, 4z, 5x, 5z, 6x, 6y, 6z, 7z, 8x, 8z, 9x, 11, 12, 16, 17, 18, 20, 21, 22, 25, 26, 28, 29, 33, 34, 35, 36, see Fig. 5.11. For the configuration without mass loading the following set was selected: 1x, 1y, 1z, 2x, 2y, 2z, 3x, 3y, 3z, 4x, 4y, 4z, 5y, 5z, 6y, 6z, 8z, 10x, 11, 12, 13, 15, 16, 17, 18, 19, 20, 21, 22, 23, 25, 26, 27, 28, 29, 30, 31, 32, 33, 34, 35, 36. Approximately, 60% (with mass loading) and 75% (without mass loading) of the channels were used for calibration.

The finite element model. The rear subframe FE model consists mainly of shell elements and rubber bushings modeled with an isotropic linear material model with solid elements. It holds over 600,000 degrees of freedom. The same FE model was used in the mass- loaded boundary configuration where four additional rigid components was attached to the bushings to represent the additional weights. The additional weights were designed to be simple to model with high accuracy. They consist of cylinders with two different outer radii dimensions with a through hole to allow for a through bolt. They were made of steel and each weight was of 1 kg total mass and were attached with bolts to each bushing of interest (four of the subframe's six rubber bushings) of the subframe. MSC Nastran was used to establish the mass and stiffness matrices of the FE models.

Calibration parameter selection. In the mass-loaded configuration, only the rubber stiffness parameters were used in the calibration and all other model parameters were fixed to their nominal values. The parametrization for the four bushings can be seen in Fig. 5.12. For the configuration without mass loading, 10 additional free parameters were selected which are shown in Fig. 5.12b through d. All other FE model data were fixed to their nominal values. Two parameter types were used, material stiffness and shell element thickness. The inverse Fisher Information Matrix (FIM) (Kay 1993) was used in an identifiability analysis to fix parameters that would render the calibration problem unidentifiable. From this study, it was found that parameter p_{12} was only marginally identifiable and could be fixed. In the deterministic calibration, most parameters were bounded to stay within 25% from their nominal setting.

Calibration results. The nominal and updated parameters from the deterministic calibration, and the mean value and Coefficient of Variation (COV %) from a cross-validation are shown in Table 5.1. The table column order illustrates the calibration order, i.e., mass-loaded configuration first followed by the configuration without mass loading up to 265 and 350 Hz in the second and final step, respectively. Thus, three calibrations were performed in total. Fields with (–) or (←) denote parameters that were fixed in that particular calibration.

In an initial step, the mass-loaded configuration was calibrated from 40 to 225 Hz so that the bushings rubber stiffness could be estimated, i.e., parameters p_1 to p_4. All

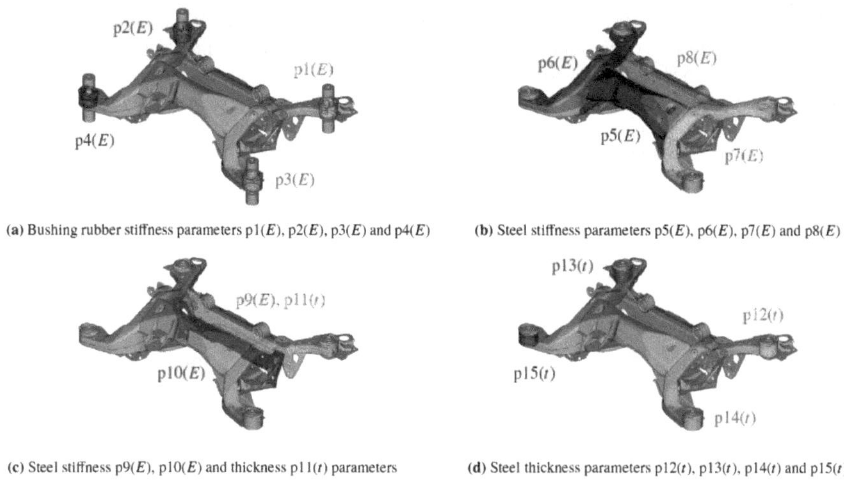

(a) Bushing rubber stiffness parameters p1(*E*), p2(*E*), p3(*E*) and p4(*E*)

(b) Steel stiffness parameters p5(*E*), p6(*E*), p7(*E*) and p8(*E*)

(c) Steel stiffness p9(*E*), p10(*E*) and thickness p11(*t*) parameters

(d) Steel thickness parameters p12(*t*), p13(*t*), p14(*t*) and p15(*t*)

Fig. 5.12 Parameterization of the FE models with E representing stiffness and t thickness, in **a** for the mass loaded configuration and in **b**, **c**, and **d** for the configuration without mass loading. Parameters $p1$ through $p4$ were also parameterized in the configuration without mass loading

other parameters were at this stage fixed to their nominal values, where parameters p_{12} to p_{14} were set to 1.20 mm. These four parameters were manually updated to a nominal setting of 1.40 mm in the calibration of the configuration without mass loading.

The configuration without mass loading was first calibrated from 60 to 265 Hz, with updated bushing parameters, by updating parameters p_5 to p_{11}. Updated parameters p_5 to p_{11} were then used as nominal parameters in the final calibration from 60 to 350 Hz where parameters p_1, p_2, p_{10}, p_{13}, p_{14} and p_{15} were updated. Parameter p_{12} was kept fixed in all three calibrations.

In Table 5.1, it can be seen that the mean value of the cross-validation results, based on 20 data splits with raw test data and unconstrained minimization, differ somewhat from the deterministic calibrated estimates. There are significant noise levels in the raw test data, especially noticable around the low-frequency antiresonances for some channels (see Figs. 5.13 and 5.14) that might affect the cross-validation results. Furthermore, the identified model used in the calibration does not perfectly fit test data in all channels, especially around the antiresonances (see again Fig. 5.13), and a difference between calibration and cross-validation results is therefore to be expected. The parameters show little variation, as indicated by the COV in Table 5.1, with parameters p_9, p_{10}, and p_{11} for calibration FE265 varying more than than the others, approximately from 3 to 6%.

It can be seen in Table 5.1 that the two calibrated stiffness parameters p_1 and p_2 in the mass-loaded configuration, (see Fig. 5.12a), are considerably higher than p_3 and p_4. This is a consequence of making the FE model fit experimental data so that the FE modes as shown in Fig. 5.15 fit the eigenfrequency order of that was

Table 5.1 Nominal parameters (Nom) along with three calibration results. Calibrated parameters for the mass-loaded configuration denoted FE_{mass}, configuration without mass loading with calibration up to 265 Hz denoted FE_{265} and up to 350 Hz denoted FE_{350}. The mean value and coefficient of variation (%) of the cross-validation is denoted with superscripts μ and COV. Empty fields (–) represent fixed parameters in that particular calibration and left arrows (←) point to numerical data used in that calibration. E represent Young's modulus and t denote plate thickness

#	Param	Nom	FE_{mass}	FE_{265}	FE_{350}	FE^{μ}_{mass}	FE^{COV}_{mass}	FE^{μ}_{265}	FE^{COV}_{265}	FE^{COV}_{350}	FE^{COV}_{350}
p_1	E (MPa)	5.00	4.01	←	4.62	3.67	0.72	–	–	4.98	0.39
p_2	E	5.00	3.82	←	7.01	3.48	0.85	–	–	5.25	0.11
p_3	E	5.00	2.83	←	5.83	2.69	0.94	–	–	–	–
p_4	E	5.00	2.41	←	5.41	2.94	1.19	–	–	–	–
p_5	E (GPa)	210.	←	192.3	←	–	–	196.4	0.99	–	–
p_6	E	210.	←	208.0	←	–	–	198.9	0.13	–	–
p_7	E	210.	←	240.7	←	–	–	228.6	0.09	–	–
p_8	E	210.	←	222.4	←	–	–	249.1	1.73	–	–
p_9	E	210.	←	250.8	←	–	–	264.4	3.86	–	–
p_{10}	E	210.	←	166.6	←	–	–	162.6	3.10	189.2	0.10
p_{11}	t (mm)	1.80	←	1.46	←	–	–	1.35	5.81	–	–
p_{12}	t	1.40	1.20	1.40	←	–	–	–	–	–	–
p_{13}	t	1.40	1.20	1.40	3.67	–	–	–	–	2.56	0.71
p_{14}	t	1.40	1.20	1.40	1.81	–	–	–	–	1.75	1.12
p_{15}	t	1.40	1.20	1.40	1.61	–	–	–	–	1.80	0.74

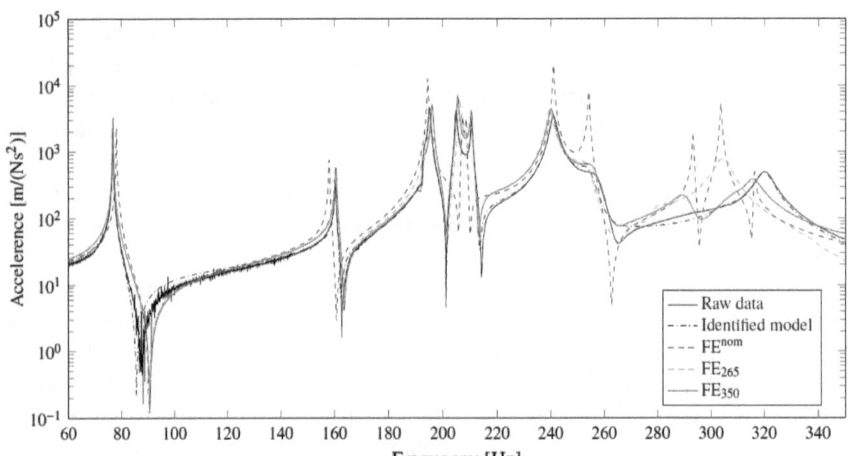

Fig. 5.13 Cross-accelerance for input at position 24 and output at 4x. Note that raw test data and data from identified model almost are almost coincident

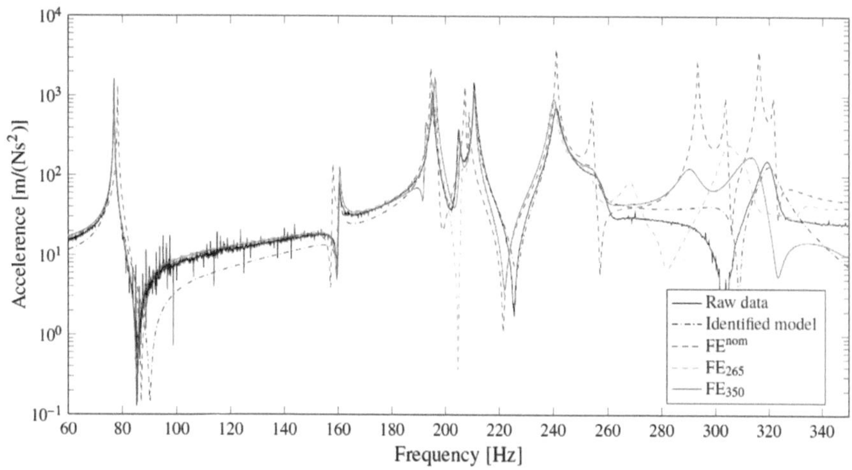

Fig. 5.14 Cross-accelerance for input at position 24 and output at 30

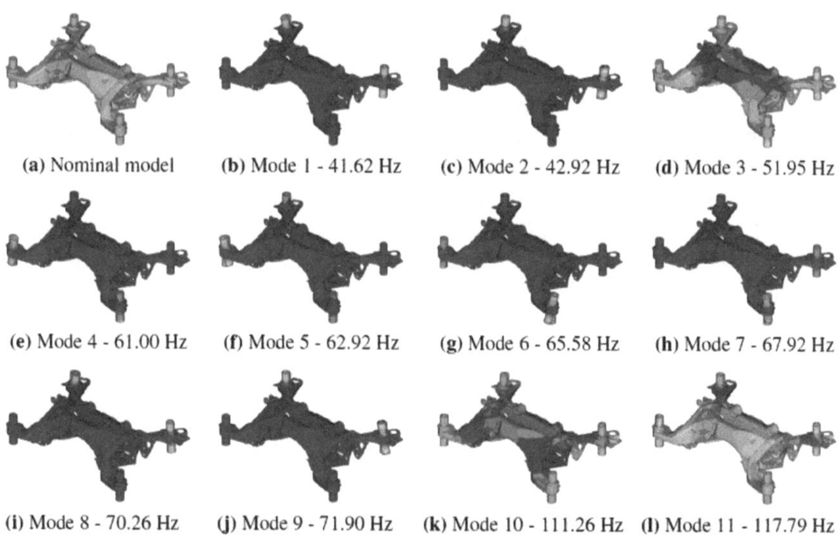

Fig. 5.15 The nominal model with mass-loaded bushings (**a**) along with the 11 first calibrated flexible modes, from mode 1 (**b**) to mode 11 (**l**). Blue indicate small motion and red large motion

found in experimental data. This is possibly caused by a nominal model inadequacy given that the bushings are different for these two sets of parameters, i.e., p_1 and p_2 correspond to bushing of Type 2 shown in Fig. 5.9 while p_3 and p_4 correspond to bushing of Type 1. The calibration results indicate that bushings of Type 2 are less stiff than bushings of Type 1.

Table 5.2 Mass-loaded and massless configurations' experimental eigenfrequencies denoted Exp_{mass} and Exp, respectively. Nominal FE models for the two configurations denoted with superscript *nom*. Other notations as in Table 5.1. Δf_{mass} (%) represent the relative difference between FE_{mass} and Exp_{mass}, and Δf_{265} and Δf_{350} the relative differences between Exp and FE_{265} and FE_{350}, respectively. Total mass of the different models are shown in bottom row

Mode	Exp_{mass}	$FE^{\text{nom}}_{\text{mass}}$	FE_{mass}	Δf_{mass}	Exp	FE^{nom}	FE_{265}	Δf_{265}	FE_{350}	Δf_{350}
1	40.95	46.45	41.62	1.64	76.82	78.20	76.86	0.05	76.90	0.10
2	–	48.16	42.92	–	160.54	158.18	160.53	0.01	160.63	0.06
3	52.07	53.92	51.95	0.23	193.08	194.63	193.13	0.03	192.62	0.24
4	59.12	78.60	61.00	3.18	195.25	196.97	195.40	0.08	196.24	0.51
5	63.21	80.03	62.92	0.46	204.94	207.37	206.29	0.66	205.59	0.32
6	66.07	82.50	65.58	0.74	210.84	208.84	210.84	0.00	210.71	0.06
7	–	84.96	67.92	–	240.93	241.10	239.62	0.54	240.13	0.33
8	71.84	86.42	70.26	2.20	254.36	254.27	253.34	0.40	255.99	0.64
9	74.64	88.03	71.90	3.67	–	293.76	268.52	–	290.45	–
10	110.59	116.27	111.26	0.61	306.89	304.23	278.47	9.26	310.18	1.07
11	118.19	126.17	117.79	0.34	320.09	316.98	304.01	5.02	315.46	1.45
Mass	37.18	37.01	37.01		26.80	26.73	26.51		26.87	

In Table 5.2, the eigenfrequencies of the first eleven flexible modes are shown for the configuration with and without mass loading, along with their masses. It can be seen that for the mass-loaded configuration the nominal results are very different from what is experimentally identified. It should be noted that mode 2 is very close to mode 1 in the experimental data and could not be identified, as indicated by the FE modes show in Fig. 5.15b, c. The same holds for modes 6 and 7, shown in Fig. 5.15g and h. FE results indicate that modes 4 and 5, and also modes 8 and 9 almost coalesce, as seen from their mode shapes shown in Fig. 5.15. The calibrated FE_{mass} has a much better correspondence to experimental data compared to the nominal model FE_{nom} mass which is also indicated by the deviation metric. The deviation metric between the experimental data Exp_{mass} and the nominal FE model FE_{nom} is 0.90 and 1.02 for the calibration and cross-validation models, respectively. The calibrated model gives a deviation metric of 0.44 and 0.50 for the calibration and cross-validation data, respectively. Thus an improvement of 51% is seen in both the calibration and cross-validation metrics.

Table 5.2 indicate that eigenfrequencies of the nominal configuration without mass loading, FE^{nom}, correlates rather well with those of the experimental model. However, the modal assurance criterion (MAC, see Ref. Allemang and Brown 1982) analysis results reported in Fig. 5.17 show that there are mode switches between modes 3 and 4, and modes 5 and 6. Also, some modes show a very low MAC correlation and the last two modes have a high cross-correlation. The updated model FE_{350}, for which the MAC analysis is shown in Fig. 5.17b, shows a much better MAC correlation, even for the modes above 300 Hz. It can be noted that the FE analysis indicates that there is a mode, not identified in the experiments, around 290 Hz. It is shown in Fig. 5.16j where it can be seen that in that mode the motion is strongly

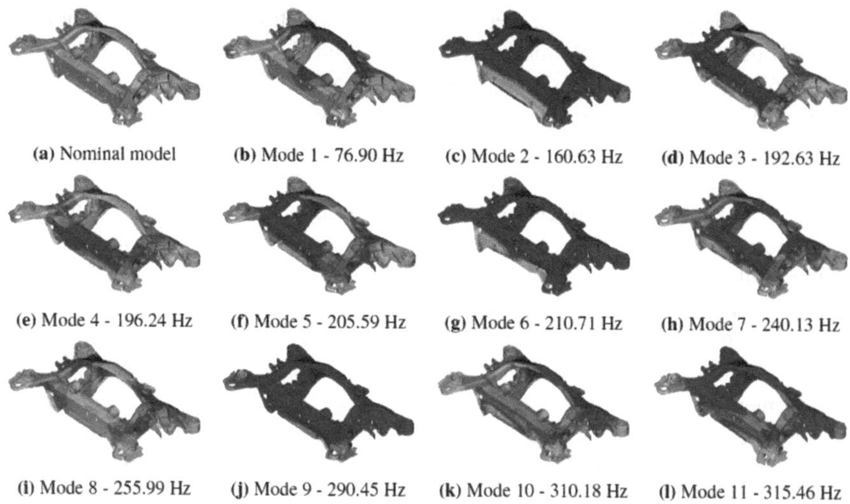

Fig. 5.16 The nominal model without mass-loaded bushings (**a**) along with the 11 first calibrated flexible modes, from mode 1 (**b**) to mode 11 (**l**)

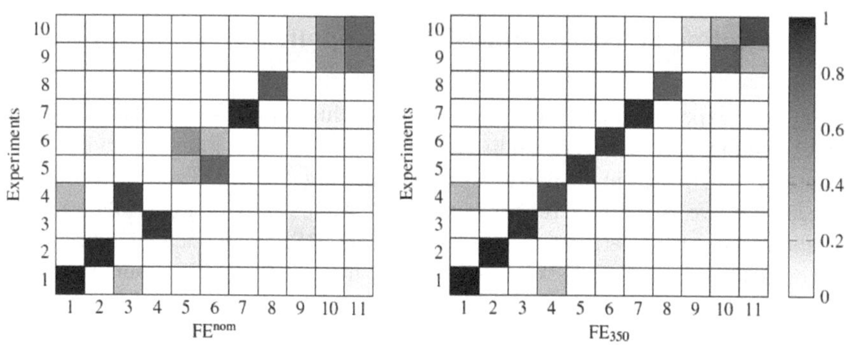

Fig. 5.17 MAC correlation between the experimentally identified system and the nominal FE model FE^{nom} (left) and identified system and the calibrated FE model FE_{350} (right)

localized to one bushing (upper left). It is therefore just marginally controllable from the input position far away from the bushing. Also, due to strong damping in the bushing modes, it would be very hard to identify it from the raw frequency response data of the used sensor set. Further in Fig. 5.16, it can be seen that in the two modes above 300 Hz that the bushings are considerably activated. In order to capture the behavior of the these modes, the stiffness parameters p_3 and p_4 of the rubber bushings were again used as free parameters in the last calibration step, FE_{350}. A justification for this is that rubber has frequency dependent stiffness properties, i.e., its stiffness increases with frequency (see Ref. Jones 2001). In the mass-loaded configuration, the calibration focus is more on the lower frequency spectrum as compared to that of the calibration without mass loading.

The intermediate calibration step FE_{265} does not show a good correlation for modes above the 8:th flexible modes, as seen in Table 5.2. This is due to that modes above 265 Hz are not included in the model calibration.

The deviation metric for the intermediate model FE_{265} was improved by 60% (from 0.91 to 0.38) after the calibration. The validation metric was improved with 23% (from 1.25 to 0.97). It should be noted that the model from which FE_{265} was calibrated started from a model with updated bushing parameters from the previous calibration step, and that this is reflected in the deviation metric. The deviation metric for the final model calibration FE_{350} was improved with 71% (from 0.88 to 0.26) for the calibration. The calibration of this model started from the parameter settings of model FE_{265}, with exception of parameters p_3 and p_4 that was set according to above. The validation metric was improved with 33% (from 1.20 to 0.81). The improvement in the validation metric is lower compared to that of the calibration metric. One reason for this is likely that poorer frequency response test data were used as validation data. In Fig. 5.13, one of the better channels is shown for raw test data, the identified model, the nominal FE model, and the calibrated FE model up to 265 and 350 Hz. Figure 5.14 depicts one of the poorer channel estimates.

5.9 Coupling of Two Major SUV Components

In a study of experimental substructure coupling, a high-modal-count Body in White (BiW) (see Fig. 5.18) was coupled to a moderate-modal-count Rear Subframe (RSF) of same type as treated in the previous chapter (see Fig. 5.10). The physical coupling between the two is via four rubber bushings that provide vibration isolation. The vibration testing was performed with the BiW and RSF component stand-alone to get component data. Testing was also done of the full BiW-RSF assembly to get validation data. The coupling procedure was to

- Make vibration testing of the two components.
- Make system identification to create state-space models of the two.
- Keep the state-space model of the BiW as-is as attempts to calibrate a BiW FE model was found to require more resources than were available.
- Do model calibration of the finite element model using the state-space model of the RSF and take that to state-space form.
- Do a transformation of the state-space models to coupling form and do the coupling.
- Validate the coupling outcome by comparing with test results of the BiW-RSF assembly.

These steps are detailed below.

Vibration testing of the components. Testing of the BiW was made with dummy adapters at the four coupling points (see Fig. 5.19). These adapters were aluminum cylinders with similar geometry as the vulcanized aluminum part of the rubber bushings mounted to the RSF. The adapters were fitted with three triaxial accelerometers

Fig. 5.18 Body in White chassis of a Volvo XC90 resting on rubber cushions (at yellow arrows) ready for vibration testing. Loose rear subframe resting on lab floor in front

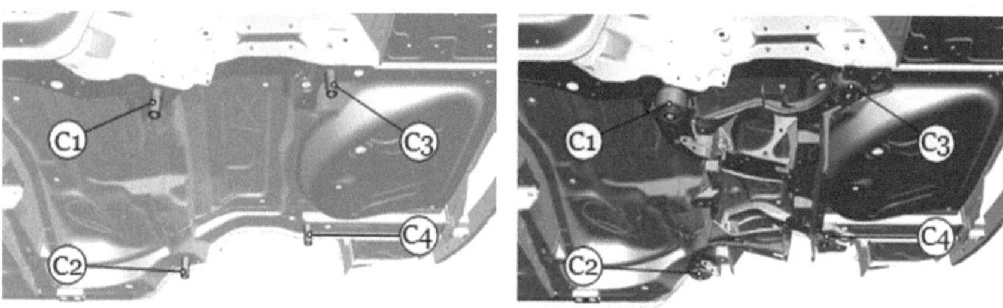

Fig. 5.19 Location of dummy adapters C1 through C4 under rear part of BiW (left). Location of virtual points C1 through C4 on RSF fitted to BiW (right)

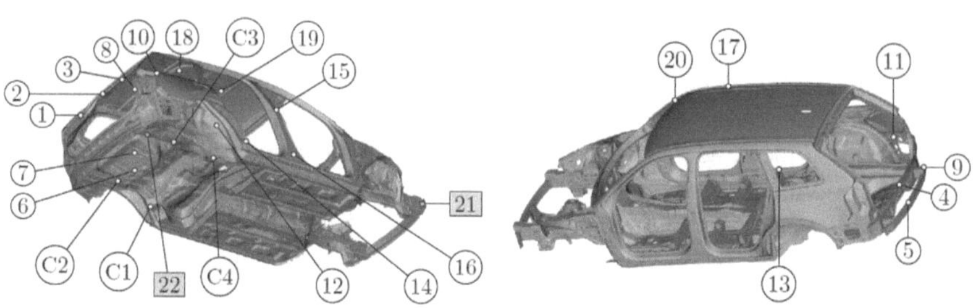

Fig. 5.20 Location of BiW-mounted accelerometers 1 through 22. Coupling adapters C1 through C4 have three triaxial accelerometers and are subjected to three successive SIMO stimuli tests each. Accelerometer numbers in rectangles also indicate location of validation stimuli

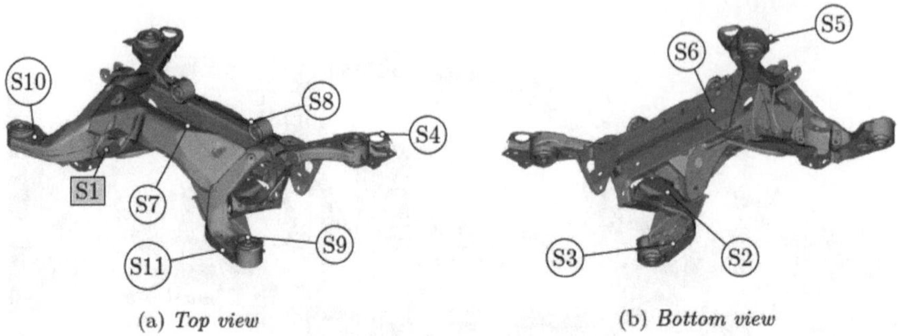

Fig. 5.21 Subframe with positions of accelerometers marked. Circle markings are for triaxial accelerometers and the rectangular marking is for one uniaxial accelerometer and the input force position

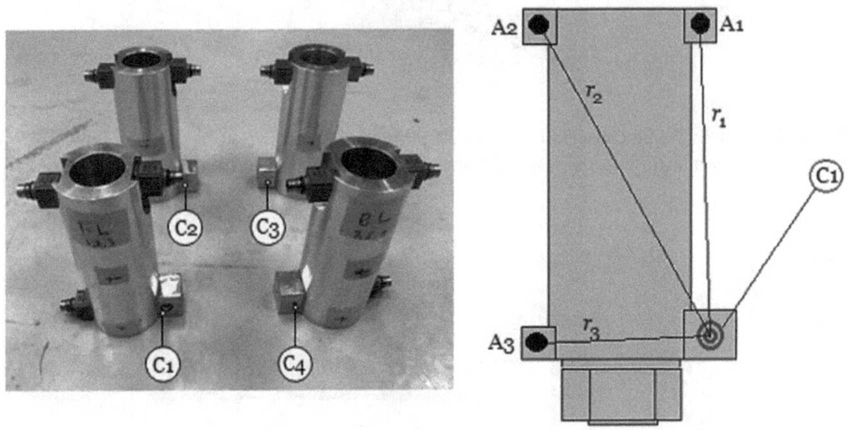

Fig. 5.22 Adapters C1 through C4 equipped with triax accelerometers (left). Location vectors r_1 through r_3 indicate kinematical transformation to virtual point C1 (right)

each to provide sufficient information about the motion of the stiff adapters in a stepped-sine vibration testing. Besides the $4 \times 3 \times 3 = 36$ sensor channels of the triaxial accelerometers, an additional 22 uni-axis accelerometers were spread out over the chassis (see Fig. 5.20). The signals of the triaxial accelerometers were transformed to provide estimates of the translational and rotational motion of so-called virtual points associated with each adapter (see Fig. 5.22). Its three translation and three rotations were obtained by least squares fitting and kinematic transformations on the nine signals from its sensors. The system stimulus was provided with a roving shaker that provided $4 \times 3 = 12$ force inputs to the adapters in 12 repeated SIMO experiments to give frequency response data. Two additional excitation points were used for validation purpose. The chassis was put on soft rubber cushions to mimic free–free boundary conditions during testing. Test data indicated at least 136 flexible body modes below 200 Hz for the BiW.

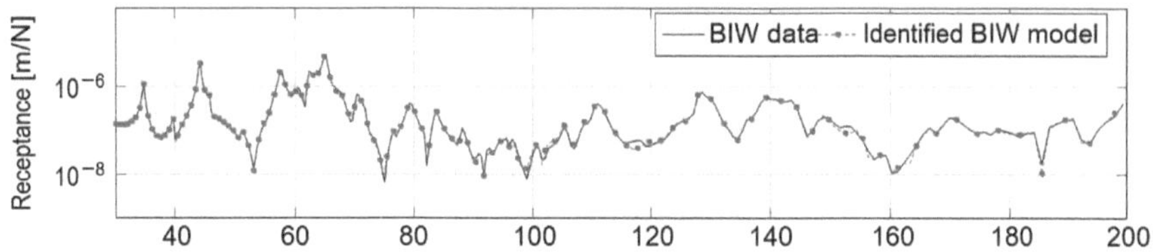

Fig. 5.23 Experimental FRF and ditto synthesized from identified state-space model of BiW

The testing procedure outlined in Sect. 5.8 was used to get data for the RSF in two configurations, one with free bushings and one with mass-loaded bushings. Testing with mass-loaded rubber bushings was crucial for obtaining good calibration for the rubber material properties. Since the main purpose of the bushings is to provide good vibration isolation it is safe to say that the material parameters of the rubber are of utmost importance for a successful coupling procedure. Since the rubber bushings are parts of the RSF, the translational and rotational dynamic stiffnesses of these parts are accounted for. The virtual points for coupling are included as nodes in the finite element model and provide input and output information of the FE-based state-space model of this component. Test data indicated about 40 flexible body modes below 200 Hz for the RSF.

Make state-space system identification of BiW. Preliminary identification results indicated that direct accelerance FRF data were particularly troublesome in the state-space model fitting. As the direct dynamical stiffness associated to the coupling degrees of freedom plays an important role in the coupling, increased effort was put in obtaining a good state-space model for these. A procedure given in Ref. Gibanica et al. (2018) was followed. The procedure is based on observation from direct accelerance measurements. These observations indicate that strong outside-band eigenmodes strongly affects the frequency response within the frequency band of interest in the testing. In comparison with more traditional approaches that compensate for outside-band low frequency modes with a residual mass model and high frequency modes with a residual stiffness model, the procedure is to add virtual system poles (residual states) just outside the frequency range of interest to compensate for the effect of outside-band modes just above the upper frequency of interest. The effect of residual states was identified by linear regression to test data. The influence of these states was first removed from test data before a traditional State-space Subspace Identification method (N4SID) was applied and then superimposed to the identified model to get a higher order state-space model with a very good fit to data, see Fig. 5.23 for one specific transfer function. An assessment of all transfer functions is reported in Fig. 5.24 which utilizes the FRAC correlation metric defined as

$$\text{FRAC}_{ij} = |\sum_{k=1}^{K} H_{ij}^{A}(\omega_k)(H_{ij}^{X}(\omega_k))^H|^2 / (\sum_{k=1}^{K} |H_{ij}^{A}(\omega_k)|^2 \sum_{k=1}^{K} |H_{ij}^{X}(\omega_k)|^2) \quad (5.41)$$

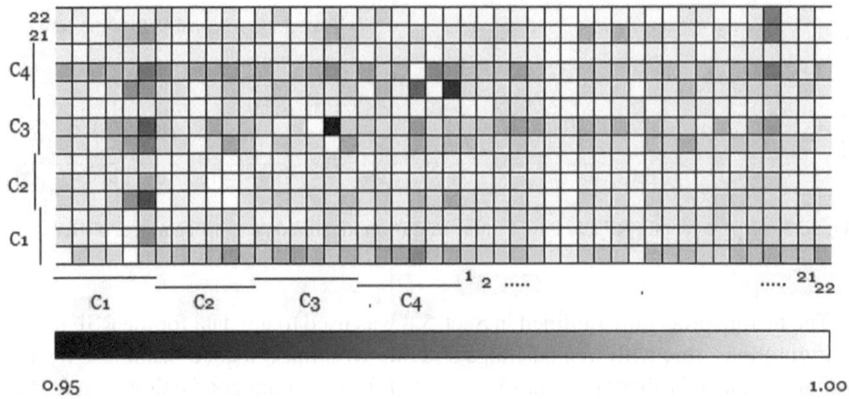

Fig. 5.24 Model fit (by FRAC) of BiW state-space model in range (30,200) Hz. Stimuli locations are three forces at each cylinder C1 through C4 and at locations 21 and 22. Responses stem from same locations but also from accelerometers 1 through 20

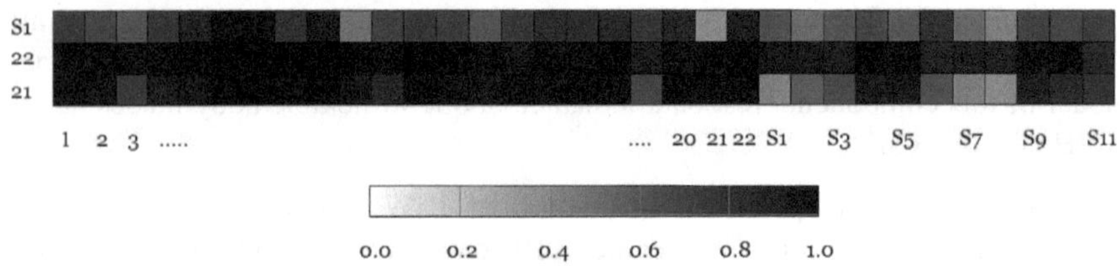

Fig. 5.25 Model fit (by FRAC) of coupled model in range (30,200) Hz

for which a perfect fit between test data from the tested assembly and the synthesized system would render a value of 1 (one).

Physically motivated constraints on reciprocity were imposed by averaging off-diagonal elements of the raw test data transfer function matrices. Also, in order to fully couple components that share the same translation and rotational motion at their mutual interfaces, all translation and rotation outputs and corresponding force and moment inputs must be available at the coupling points. Since the test lacks moment input, the transfer functions related to moment input have to be estimated by through a modal system expansion, see paper E of Liljerehn's doctoral thesis (Liljerehn 2016). For comparison, the coupling was performed with and without the rotational couplings, see Fig. 5.26.

Make state-space system identification of RSf and calibrate its FE model. The system identification from RSf data and the proceeding calibration of the RSf finite element model were conducted along the lines described in Sect. 5.8. In particular, the stiffness parameters of the rubber bushings were much reduced in comparison to the nominal parameter configuration.

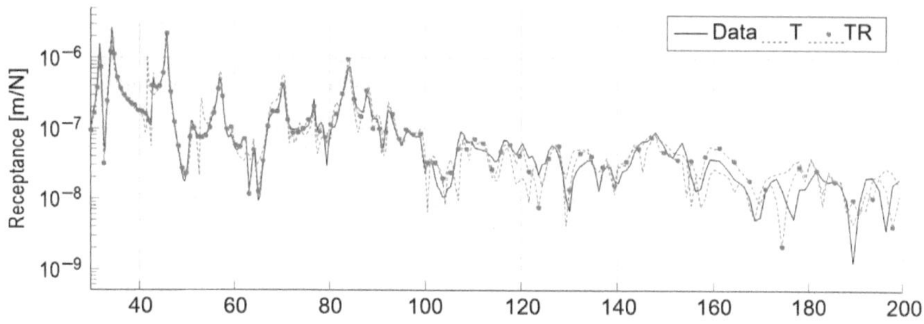

Fig. 5.26 An example of an FRF of the coupled system; the receptance from input at 22 to response at 12. TR denotes results obtained with full coupling (translations and rotations) while T denotes results obtained with translational coupling only

Do transformation of state-space models. The numerical linear algebra operations given by Eqs. (4.42) and (4.43) were performed to transform the state-space models of the two components into coupling form.

Validate the coupling by comparing coupled model with test results. The overall quality of the synthetic frequency response function is evaluated with FRAC correlation. The results for all transfer functions can be seen in Fig. 5.25. It can be noted that the synthesized FRF:s associated with stimulus on the subframe (see Fig. 5.21 at location S1), in general, render somewhat poorer correlations than the others. An example of a transfer function is given in Fig. 5.26.

References

Abrahamsson, T. J. S. (2019). FEMcali (Finite element model calibration). https://www.mathworks.com/matlabcentral/fileexchange/

Allemang, R., Brown, D. L. (1982). A correlation coefficient for modal vector analysis.

Barten, E., van der Seijs, M.V., de Klerk, D. (2014). A complex power approach to characterise joints in experimental dynamic substructuring. In *Dynamics of Coupled Structures, Proceedings of the 32nd IMAC, A Conference and Exposition on Structural Dynamics* (Vol. 1, pp. 281–296), Chap. 27. New York: Springer. https://doi.org/10.1007/978-3-319-04501-6_27.

Bendat, J. S., & Piersol, A. G. (1980). *Engineering applications of correlation and spectral analysis*. New York: Wiley Inc.

Bisplinghoff, R. L., Ashley, H., & Halfman, R. L. (1955). *Aeroelasticity*. Cambridge: Addison-Wesley Publishing Company.

Bonhoff, H. A. (2010). The influence and significance of cross-order terms in interface mobilities for structure-borne sound source characterization. Ph.D. thesis, Technische Universität Berlin

Bylin, A. (2018). Dynamic substructuring using experimental-analytical state-space models of automotive components. Master's thesis, Chalmers University of Technology, Dept Mechanics and Maritime Sciences, report 2018:69.

Bylin, A., Gibanica, M., & Abrahamsson, T. J. S. (2018). Experimental-analytical state-space synthesis of a body-in-white and rear subframe.

Cerrato, G. (2009). Automotive sound quality – powertrain, road and wind noise. *Sound and Vibration*, *43*(4), 16–24. http://www.sandv.com/downloads/0904cerr.pdf.

Choi, H. G., Thite, A. N., & Thompson, D. J. (2007). Comparison of methods for parameter selection in Tikhonov regularization with application to inverse force determination. *Journal of Sound and Vibration*, *304*(3), 894–917. https://doi.org/10.1016/j.jsv.2007.03.040.

D'Ambrogio, W., & Fregolent, A. (2010). The role of interface DoFs in decoupling of substructures based on the dual domain decomposition. *Mechanical Systems and Signal Processing*, *24*(7), 2035–2048. https://doi.org/10.1016/j.ymssp.2010.05.007.

de Klerk, D., & Ossipov, A. (2010). Operational transfer path analysis: Theory, guidelines and tire noise application. *Mechanical Systems and Signal Processing*, *24*(7), 1950–1962. https://doi.org/10.1016/j.ymssp.2010.05.009. Special Issue: ISMA 2010.

de Klerk, D., & Rixen, D. J. (2010). Component transfer path analysis method with compensation for test bench dynamics. *Mechanical Systems and Signal Processing*, *24*(6), 1693–1710. https://doi.org/10.1016/j.ymssp.2010.01.006.

de Klerk, D., Rixen, D. J., & Voormeeren, S. N. (2008). General framework for dynamic substructuring: History, review and classifcation of techniques. *AIAA Journal*, *46*(8), 1169–1181. https://doi.org/10.2514/1.33274.

de Klerk, D. (2009). Dynamic response characterization of complex systems through operational identification and dynamic substructuring. Ph.D. thesis, Delft University of Technology, The Netherlands. http://resolver.tudelft.nl/uuid:2e4dd63e-68d5-41a4-8201-1e784f2f752c

de Klerk, D., Lohrmann, M., Quickert, M., & Foken, W. (2009). Application of operational transfer path analysis on a classic car. In *DAGA*. Rotterdam.

Dobson, B. J., & Rider, E. (1990). A review of the indirect calculation of excitation forces from measured structural response data. *Proceedings of the Institution of Mechanical Engineers, Part C: Journal of Mechanical Engineering Science*, *204*(2), 69–75. https://doi.org/10.1243/PIME_PROC_1990_204_080_02.

dos Santos, F. L. M., Peeters, B., Lau, J., & Desmet, W., Góes, L. C. S. (2014). An overview of experimental strain-based modal analysis methods. In *Proceedings of the International Conference on Noise and Vibration Engineering (ISMA)*. Leuven, Belgium

Elliott, A. S. (2009). Characterisation of structure borne sound source in-situ. Ph.D. thesis, University of Salford. http://usir.salford.ac.uk/id/eprint/26661.

Elliott, A. S., & Moorhouse, A. T. (2008). Characterisation of structure borne sound sources from measurement in-situ. *Journal of the Acoustical Society of America*, *123*(5), 3176. https://doi.org/10.1121/1.2933261.

Elliott, A. S., Moorhouse, A. T., Huntley, T., & Tate, S. (2013). In-situ source path contribution analysis of structure borne road noise. *Journal of Sound and Vibration*, *332*(24), 6276–6295. https://doi.org/10.1016/j.jsv.2013.05.031.

Ewins, D. J. (2000). *Modal testing: Theory, practice and application* (2nd ed.). New York: Wiley Inc.

Fahy, F. J. (1995). The vibro-acoustic reciprocity principle and applications to noise control. *Acta Acustica United with Acustica*, *81*(6), 544–558.

Fahy, F. J. (2003). Some applications of the reciprocity principle in experimental vibroacoustics. *Acoustical Physics*, *49*(2), 217–229. https://doi.org/10.1134/1.1560385.

Gajdatsy, P., Janssens, K., Gielen, L., Mas, P., & Van der Auweraer, H. (2008). Critical assessment of operational path analysis: effect of neglected paths. In *Proceedings of the XV International Congress on Sound and Vibration (ICSV)* (pp. 1090–1097). Daejeon, Korea.

Gajdatsy, P., Janssens, K., Gielen, L., Mas, P., & Van Der Auweraer, H. (2008). Critical assessment of operational path analysis: Effect of coupling between path inputs. *Journal of the Acoustical Society of America*, *123*(5), 3876. https://doi.org/10.1121/1.2935777.

Gaudin, A., & Beniguel, J. F. (2012). Low frequency road noise decomposition at wheel center on an roller bench. In *Acoustics 2012*. Société Française d'Acoustique, Nantes, France. https://hal.archives-ouvertes.fr/hal-00811236.

Gibanica, M., & Abrahamsson, T. J. S. (2017). Parameter estimation and uncertainty quantification of a subframe with mass loaded bushings. In R. B. (Ed.), *Proceedings of 32nd IEEE Conference on Decision and Control* (Vol. 3, pp. 61–76). Berlin: Springer International Publishing.

Gibanica, M., Abrahamsson, T. J. S., & Allemang, R. J. (2018). Residual states for modal models indentified from accelerance data. In: M. Maines & B. Dilworth (Eds.), *Topics in modal analysis and testing* (Vol. 9). Berlin: Springer International Publishing.

Gibanica, M., Abrahamsson, T. J. S., Kammer, D. C. (2016). Redundant information rejection in sensor localisation using system gramians. In: M. Mains (ed.) *Topics in modal analysis and testing* (Vol. 10). Berlin: Springer International Publishing.

Ginsberg, J. H. (2001). *Mechanical and structural vibrations* (1st ed.). New York: Wiley.

Guasch, O. (2009). Direct transfer functions and path blocking in a discrete mechanical system. *Journal of Sound and Vibration, 321*(3), 854–874. https://doi.org/10.1016/j.jsv.2008.10.006.

Guasch, O., & Magrans, F. X. (2004). The global transfer direct transfer method applied to a finite simply supported elastic beam. *Journal of Sound and Vibration, 276*(1), 335–359. https://doi.org/10.1016/j.jsv.2003.07.032.

Guasch, O., García, C., Jové, J., & Artís, P. (2013). Experimental validation of the direct transmissibility approach to classical transfer path analysis on a mechanical setup. *Mechanical Systems and Signal Processing, 37*(1), 353–369. https://doi.org/10.1016/j.ymssp.2013.01.006.

Harrison, M., Sykes, A. O., & Martin, M. (1952). Wave effects in isolation mounts. *Journal of the Acoustical Society of America, 24*(1), 62–71. https://doi.org/10.1121/1.1906850.

Hixson, E. L. (1961). Mechanical impedance and mobility. In: C. M. Harris & C. E. Crede (Eds.), *Shock and vibration handbook* (1st ed., pp. 1–59), Chap. 10. New York: McGraw-Hill.

ISO Technical Committee 43/Subcommittee 1/Workgroup 22 (ISO/TC43/SC1/WG22) (2015). Measurement of noise emitted by accelerating road vehicles – Engineering method – Part 1: M and N categories. ISO 362-1, International Standards Organisation.

ISO Technical Committee 43/Subcommittee 1/Workgroup 22 (ISO/TC43/SC1/WG22). (1996). Acoustics – Characterization of sources of structure-borne sound with respect to sound radiation from connected structures – Measurement of velocity at the contact points of machinery when resiliently mounted. ISO 9611, International Standards Organisation.

Janssens, M. H. A., & Verheij, J. W. (2000). A pseudo-forces methodology to be used in characterization of structure-borne sound sources. *Applied Acoustics, 61*(3), 285–308. https://doi.org/10.1016/S0003-682X(00)00035-9.

Janssens, M. H. A., Verheij, J. W., & Thompson, D. J. (1999). The use of an equivalent forces method for the experimental quantification of structural sound transmission in ships. *Journal of Sound and Vibration, 226*(2), 305–328. https://doi.org/10.1006/jsvi.1999.2303.

Janssens, M. H. A., Verheij, J. W., & Loyau, T. (2002). Experimental example of the pseudo-forces method used in characterisation of a structure-borne sound source. *Applied Acoustics, 63*(1), 9–34. https://doi.org/10.1016/S0003-682X(01)00023-8.

Jones, D. I. G. (2001). *Handbook of viscoelastic vibration damping*. Chichester: Wiley.

Kammer, D. C. (1998). Input force reconstruction using a time domain technique. *Journal of Vibration and Acoustics, 120*(4), 868–874. https://doi.org/10.1115/1.2893913.

Kammer, D. C. (2005). Sensor set expansion for modal vibration testing. *Mechanical Systems and Signal Processing, 19*, 700–713.

Kay, S. M. (1993). *Fundamentals of statistical signal processing* (Vol. I: Estimation Theory, 1st ed.). Englewood Cliffs: Prentice Hall.

Klerk, D. D., Rixen, D. J., & Voormeeren, S. N. (2008). General framework for dynamic substructuring: History, review and classification of techniques. *AIAA Journal, 46*(5), 1169–1181.

Leclère, Q., Roozen, N. B., & Sandier, C. (2012). Experimental estimation of transmissibility matrices. In *Proceedings of the International Conference on Noise and Vibration Engineering (ISMA)*. Leuven, Belgium.

Liljerehn, A.: Machine tool dynamics - a constrained state-space substructuring approach. Ph.D. thesis, Chalmers University of Technology, Dept Applied Mechanics (2016)

Madsen, M. B. (2014). Electrical power assisted steering – dynamic source strength characteristic and vehicle NVH prediction. Master's thesis, University of Southern Denmark.

Magrans, F. X. (1981). Method of measuring transmission paths. *Journal of Sound and Vibration*, *74*(3), 321–330. https://doi.org/10.1016/0022-460X(81)90302-3.

Maia, N. M. M., Silva, J. M. M., & Ribeiro, A. M. R. (2001). The transmissibility concept in multi-degree systems. *Mechanical Systems and Signal Processing*, *15*(1), 129–137. https://doi.org/10.1006/mssp.2000.1356.

Mas, P., Sas, P. (1994). Indirect force identification based upon impedance matrix inversion: A study on statistical and deterministical accuracy. In: *Proceedings of the International Conference on Noise and Vibration Engineering (ISMA)*. Leuven, Belgium.

Mayes, R. (2015). A modal Craig-Bampton substructure for experiments, analysis, control and specifications.

Mayes, R., & Linehan, D. (2014). Measuring effective mass of a circuit board.

Mayes, R., Schoenherr, T., Blecke, J., & Rohe, D. (2013). Efficient method of measuring effective mass of a system.

Mondot, J. M., & Petersson, B. A. T. (1987). Characterization of structure-borne sound sources: The source descriptor and the coupling function. *Journal of Sound and Vibration*, *114*(3), 507–518. https://doi.org/10.1016/S0022-460X(87)80020-2.

Moorhouse, A. T., Elliott, A. S., & Evans, T. A. (2009). In situ measurement of the blocked force of structure-borne sound sources. *Journal of Sound and Vibration*, *325*(4–5), 679–685. https://doi.org/10.1016/j.jsv.2009.04.035. http://www.sciencedirect.com/science/article/pii/S0022460X09003794.

Moorhouse, A. T. (2001). On the characteristic power of structure-borne sound sources. *Journal of Sound and Vibration*, *248*(3), 441–459. https://doi.org/10.1006/jsvi.2001.3797.

Norton, E. L. (1926). Design of finite networks for uniform frequency characteristic. Technical Report TM26–0–1860, Bell Laboratories.

Noumura, K., & Yoshida, J. (2006). A method of transfer path analysis for vehicle interior sound with no excitation experiment. In *Proceedings of FISITA 2006 World Automotive Congress, F2006D183. JSAE*. http://www.fisita.com/publications/papers?id=4748

On, F. J. (1967). *Mechanical impedance analysis for lumped parameter multi-degree of freedom/multi-dimensional systems*. Technical report, NASA Technical Note D-3865 (1967)

Penne, F. (2004). Shaping the sound of the next-generation BMW. In *Proceedings of the International Conference on Noise and Vibration Engineering (ISMA)* (pp. 25–39). Leuven, Belgium.

Petersson, B. A. T., & Gibbs, B. M. (1993). Use of the source descriptor concept in studies of multi-point and multi-directional vibrational sources. *Journal of Sound and Vibration*, *168*(1), 157–176. https://doi.org/10.1006/jsvi.1993.1367.

Petersson, B. A. T., & Gibbs, B. M. (2000). Towards a structure-borne sound source characterization. *Applied Acoustics*, *61*(3), 325–343.

Plunkett, R. (1958). Interaction between a vibratory machine and its foundation. *Noise Control*, *4*(1), 18–22.

Plunt, J. (2005). Finding and fixing vehicle NVH problems with transfer path analysis. *Sound and Vibration*, *39*(11), 12–16.

Powell, R. E., & Seering, W. (1984). Multichannel structural inverse filtering. *Journal of Vibration and Acoustics*, *106*(1), 22–28. https://doi.org/10.1115/1.3269147.

Putner, J., Fastl, H., Lohrmann, M., Kaltenhauser, A., & Ullrich, F. (2012). Operational transfer path analysis predicting contributions to the vehicle interior noise for different excitations from the same sound source. In InterNoise12, no. 3 in INTER-NOISE and NOISE-CON Congress and Conference Proceedings (pp. 2336–2347). New York City, NY. http://www.ingentaconnect.com/content/ince/incecp/2012/00002012/00000009/art00034.

Putner, J., Lohrmann, M., & Fastl, H. (2013). Contribution analysis of vehicle exterior noise with operational transfer path analysis. *Journal of the Acoustical Society of America*, *133*(5), 3323. https://doi.org/10.1121/1.4805558.

Reuss, P., Zeumer, B., Herrmann, J., & Gaul, L. (2012). Consideration of interface damping in dynamic substructuring. In *Topics in Experimental Dynamics Substructuring and Wind Turbine Dynamics, Proceedings of the 30th IMAC, A Conference on Structural Dynamics* (Vol. 2, pp. 81–88) Chap. 10. New York: Springer. https://doi.org/10.1007/978-1-4614-2422-2_10.

Ribeiro, A. M. R., Maia, N. M. M., & Silva, J. M. M. (1999). Experimental evaluation of the transmissibility matrix. In *Proceedings of the XVII International Modal Analysis Conference (IMAC)*. Kissimmee, FL

Ribeiro, A. M. R., Silva, J. M. M., & Maia, N. M. M. (2000). On the generalisation of the transmissibility concept. *Mechanical Systems and Signal Processing*, *14*(1), 29–35. https://doi.org/10.1006/mssp.1999.1268.

Rixen, D. J., Boogaard, A., van der Seijs, M. V., van Schothorst, G., & van der Poel, T. (2015). Vibration source description in substructuring: A theoretical depiction. *Mechanical Systems and Signal Processing*, *60–61*, 498–511. https://doi.org/10.1016/j.ymssp.2015.01.024.

Roozen, N. B., & Leclère, Q. (2013). On the use of artificial excitation in operational transfer path analysis. *Applied Acoustics*, *74*(10), 1167–1174. https://doi.org/10.1016/j.apacoust.2013.04.011.

Rubin, S. (1967). Mechanical immittance- and transmission-matrix concepts. *Journal of the Acoustical Society of America*, *41*(5), 1171–1179. https://doi.org/10.1121/1.1910455.

Sachse, D., Geluk, T., van Wayenberge, T. (2013). Accuracy of inverse load identification techniques for transfer path analysis. In *Automotive Acoustics Conference 2013, 2nd International ATZ Conference. ATZlive*

Scheuren, J., & Lohrmann, M. (2014). Transfer path analysis – experiences, expectations and perspectives. In *International Noise and Vibration Colloquium*. SAE Brazil.

Sedaghati, R., Soucy, Y., & Etienne, N. (2003). Experimental estimation of effective mass for structural dynamics and vibration applications.

Sjövall, P., & Abrahamsson, T. (2008). Substructure system identification from coupled system test data. *Mechanical Systems and Signal Processing*, *22*(1), 15–33. https://doi.org/10.1016/j.ymssp.2007.06.003.

Snowdon, J. C. (1979). Vibration isolation: use and characterization. *Journal of the Acoustical Society of America*, *66*(5), 1245–1274. https://doi.org/10.1121/1.383546.

Soliman, J., & Hallam, M. (1968). Vibration isolation between non-rigid machines and non-rigid foundations. *Journal of Sound & Vibration*, *8*(2), 329–351. https://doi.org/10.1016/0022-460X(68)90236-8.

Sturm, M., Moorhouse, A. T., Alber, T., & Li, F. F. (2012). Force reconstruction using an adaptive algorithm in time domain. In *Proceedings of the International Conference on Noise and Vibration Engineering (ISMA)* (, pp. 17–19). Leuven, Belgium.

Sturm, M., Moorhouse, A. T., Kropp, W., & Alber, T. (2013). Robust calculation of simultaneous multi-channel blocked force signatures from measurements made in-situ using an adaptive algorithm in time domain. In *Proceedings of the XX International Congress on Sound and Vibration (ICSV)* (, vol. 2, pp. 1610–1617). Bangkok, Thailand.

Ten Wolde, T. (1973). Reciprocity experiments on the transmission of sound in ships. Ph.D. thesis, Delft University of Technology, The Netherlands. http://resolver.tudelft.nl/uuid:6d3d9f26-e631-4043-a3ab-04bd45edde98

Ten Wolde, T. (2010). Reciprocity measurements in acoustical and mechano-acoustical systems. Review of theory and applications. *Acta Acustica United with Acustica*, *96*(1), 1–13. https://doi.org/10.3813/AAA.918250

Ten Wolde, T., Verheij, J. W., & Steenhoek, H. F. (1975). Reciprocity method for the measurement of mechano-acoustical transfer functions. *Journal of Sound and Vibration*, *42*(1), 49–55. https://doi.org/10.1016/0022-460X(75)90301-6.

Thévenin, M. L. (1883). Sur un nouveau théoreme d'électricité dynamique [On a new theorem of dynamic electricity]. *Comptes Rendus hebdomadaires des des Séances de l'Académie des Sciences*, *97*, 159–161.

Thite, A. N., Thompson, D. J. (2003a). The quantification of structure-borne transmission paths by inverse methods. Part 1: Improved singular value rejection methods. *Journal of Sound and Vibration*, *264*(2), 411–431. https://doi.org/10.1016/S0022-460X(02)01202-6

Thite, A. N., Thompson, D. J. (2003b). The quantification of structure-borne transmission paths by inverse methods. Part 2: Use of regularization techniques. *Journal of Sound and Vibration*, *264*(2), 433–451. https://doi.org/10.1016/S0022-460X(02)01203-8

Thite, A. N., & Thompson, D. J. (2006). Selection of response measurement locations to improve inverse force determination. *Applied Acoustics*, *67*(8), 797–818. https://doi.org/10.1016/j.apacoust.2006.01.001.

Thompson, D. J., van Vliet, W. J., & Verheij, J. W. (1998). Developments of the indirect method for measuring the high frequency dynamic stiffness of resilient elements. *Journal of Sound and Vibration*, *213*(1), 169–188. https://doi.org/10.1006/jsvi.1998.1492.

Ungar, E., & Dietrich, C. (1966). High-frequency vibration isolation. *Journal of Sound & Vibration*, *4*(2), 224–241. https://doi.org/10.1016/0022-460X(66)90123-4.

Vakilzadeh, M. K., Yaghoubi, V., McKelvey, T., Abrahamsson, T. J., & Ljung, L. (2015). *Experiment design for improved frequency domain subspace system identification of continuous-time systems* (pp. 886–891). IFAC-PapersOnLine.

van der Seijs, M. V., de Klerk, D., & Rixen, D. J. (2016). General framework for transfer path analysis: History, theory and classification of techniques. *Mechanical Systems & Signal Processing*, *68–69*, 217–244. https://doi.org/10.1016/j.ymssp.2015.08.004.

van Schothorst, G., Boogaard, M. A., van der Poel, G. W., & Rixen, D. J. (2012). Analysis of ground vibration transmission in high precision equipment by frequency based substructuring. In *Proceedings of the International Conference on Noise and Vibration Engineering (ISMA)* (pp. 3501–3514). Leuven, Belgium.

van den Bosch, D. D., van der Seijs, M. V., & de Klerk, D. (2014). Validation of blocked-force transfer path analysis with compensation for test bench dynamics. In *Dynamics of Coupled Structures, Proceedings of the 32nd IMAC, A Conference and Exposition on Structural Dynamics*, Chap. 4 (Vol. 1, pp. 37–49). New York: Springer. https://doi.org/10.1007/978-3-319-04501-6_4.

van der Auweraer, H., Mas, P., Dom, S., Vecchio, A., Janssens, K., van de Ponseele, P. (2007). Transfer path analysis in the critical path of vehicle refinement: The role of fast, hybrid and operational path analysis. Technical Report 2007-01-2352, SAE Technical Paper. https://doi.org/10.4271/2007-01-2352

van der Seijs, M. V., Pasma, E. A., de Klerk, D., & Rixen, D. J. (2014). A robust transfer path analysis method for steering gear vibrations on a test bench. In *Proceedings of the International Conference on Noise and Vibration Engineering (ISMA)*. Leuven, Belgium. https://doi.org/10.13140/RG.2.1.4297.8647.

van der Seijs, M. V., Pasma, E. A., de Klerk, D., & Rixen, D. J. (2015). A comparison of two component TPA approaches for steering gear noise prediction. In *Dynamics of Coupled Structures, Proceedings of the 33rd IMAC, A Conference and Exposition on Structural Dynamics*, Chap. 7 (vol. 4, pp. 71–79). New York: Springer. https://doi.org/10.1007/978-3-319-15209-7_7.

Varoto, P. S., & McConnell, K. G. (1998). Single point versus multi point acceleration transmissibility concepts in vibration testing. In *Proceedings of the XVI International Modal Analysis Conference (IMAC), Santa Barbara, CA* (vol. 1, pp. 83–90). Bethel, CT: Society for Experimental Mechanics.

Verheij, J. W. (1982). Multi-path sound transfer from resiliently mounted shipboard machinery: Experimental methods for analyzing and improving noise control. Ph.D. thesis, Delft University of Technology, The Netherlands. http://resolver.tudelft.nl/uuid:30d02e94-2fce-4b88-ab8f-6a58a08f0c06

Voormeeren, S.N. (2012). Dynamic substructuring methodologies for integrated dynamic analysis of wind turbines. Ph.D. thesis, Delft University of Technology, The Netherlands. https://doi.org/10.4233/uuid:f45f0548-d5ec-46aa-be7e-7f1c2b57590d

Voormeeren, S., & Rixen, D. (2012). A family of substructure decoupling techniques based on a dual assembly approach. *Mechanical Systems and Signal Processing*, *27*(18), 379. https://doi.org/10.1016/j.ymssp.2011.07.028.

Vorländer, M., & Dietrich, P. (2008). Transfer path analysis and synthesis for auralization. In *39th Congreso Espanol de Acústica 2008*. Coimbra, Portugal.

Weijtjens, W., De Sitter, G., Devriendt, C., & Guillaume, P. (2014). Operational modal parameter estimation of mimo systems using transmissibility functions. *Automatica*, *50*(2), 559–564. https://doi.org/10.1016/j.automatica.2013.11.021.

Zeller, P. (2009). Handbuch Fahrzeugakustik. *Springer*,. https://doi.org/10.1007/978-3-8348-8657-6.

Chapter 6
Model Reduction Concepts and Substructuring Approaches for Nonlinear Systems

Abstract This chapter reviews common nonlinearities that are encountered in engineering structures, with a particular emphasis on geometric nonlinearity. Popular ways to construct reduced order models for geometrically nonlinear problems are discussed. The concept of nonlinear normal modes is presented to help understand the dynamics of these structures, and some recently presented substructuring methods are reviewed. —*Chapter Authors: Matt Allen & Paolo Tiso*.

6.1 Geometric Nonlinearities

Up to here, all the developed theory considered systems in their linearized form—that is, the interest focused on small dynamic deviations from the equilibrium. In fact, any engineering system is inherently nonlinear, and is often just *designed* to operate in the linear regime. In other cases, one might want to exploit nonlinear responses to achieve a certain performance. In either case, the nonlinearities need to be properly assessed with the aid of numerical models. Nonlinearities can be classified into three broad categories, namely

1. *Boundary condition nonlinearities*, arising from changing constraints to the system. Typically, this involves contact between subcomponents of the same system, between two or more different systems, or the system and an external boundary.
2. *Material nonlinearities*, which are due to any nonlinearity present in the constitutive law, which relates strain and stress rates. Typical examples are plasticity, hyperelasticity, progressive damage, etc.
3. *Geometric nonlinearities*, arising from the redirection of the internal stresses due to finite rotations. This type of nonlinearity is particularly present in thin-walled, lightweight constructions, where geometric nonlinearities cause coupling between bending and stretching deformations. Likewise, geometric nonlinearities arise in multibody systems, where the components are subjected to holonomic constraints that establish nonlinear mappings between displacements and generalized coordinates.

M. S. Allen et al., *Substructuring in Engineering Dynamics*,
CISM International Centre for Mechanical Sciences 594,
https://doi.org/10.1007/978-3-030-25532-9_6

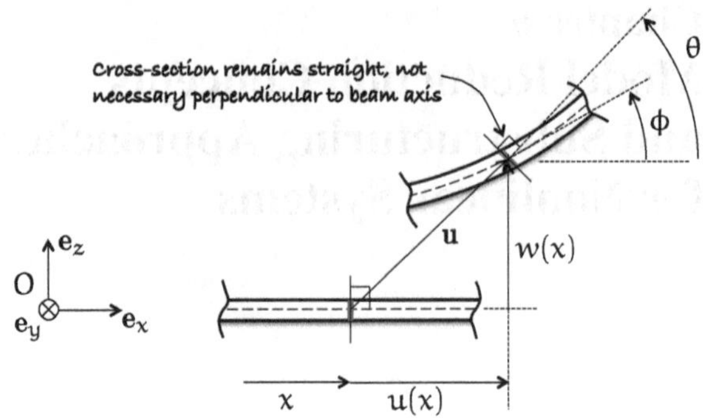

Fig. 6.1 2D beam kinematics

Being the realm of nonlinearities so vast, one cannot devise a general method for the efficient reduction of any given nonlinear system. Here, we limit ourselves to the relevant case of geometric nonlinearities. Apart from the richness in dynamic behavior, one could expect from a geometrically nonlinear system, it is important to highlight here that geometric nonlinearities are *smooth* with respect to the generalized displacements. This important property is a key aspect for many of the reduction methods discussed in this chapter.

6.1.1 2D Beam

In this section, we briefly review the nonlinear beam theory in 2D, and discuss a widely used approximation, namely the von Karman kinematic model (i.e. see Figs. 6.1 and 6.2). We start by assuming a straight, two-dimensional beam, with a rigid cross section, able to undergo arbitrary large displacements and rotations. The material is assumed linear and elastic, and the elastic strains small. The interested reader could refer, for instance, to Woinowski-Krieger (1950).

The displacement field is indicated by $\mathbf{U} = [u; w]$, where u and w are axial and transversal displacements, respectively. Upon an arbitrary deformation of the beam, we can write

$$\mathbf{U} = u\mathbf{e}_x + w\mathbf{e}_z + [\mathbf{R}(\theta) - \mathbf{I}]z\mathbf{e}_z = \frac{1}{2}\left(\mathbf{F}^T\mathbf{F} - \mathbf{I}\right), \tag{6.1}$$

where $\mathbf{R}(\theta)$ is the rotation matrix of the cross section, $\mathbf{I}$ is the identity matrix, and $\mathbf{F}$ is the deformation gradient. The Green–Lagrange strain tensor is then written as

$$\epsilon = \frac{1}{2}\left(\nabla\mathbf{U} + \nabla^T\mathbf{U} + \nabla^T\mathbf{U}\cdot\nabla\mathbf{U}\right). \tag{6.2}$$

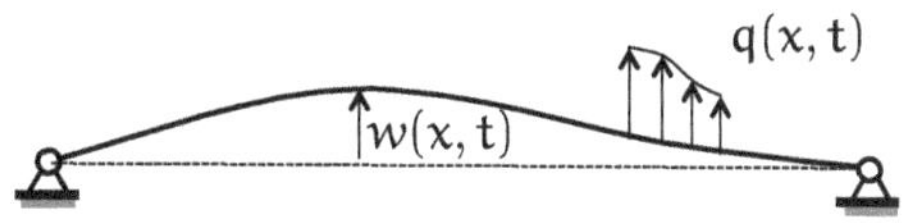

Fig. 6.2 von Karman beam in deformed configuration under distributed load

The polar decomposition theorem allows to split the deformation gradient $\mathbf{F}$ into rotations and deformations as

$$\mathbf{F} = \mathbf{R}(\theta)(\mathbf{I} + \mathbf{L}) \Rightarrow \boldsymbol{\epsilon} = \frac{1}{2}\left(\mathbf{L} + \mathbf{L}^T + \mathbf{L}^T\mathbf{L}\right). \tag{6.3}$$

Linearizing the strains leads to

$$\boldsymbol{\epsilon} \approx \frac{1}{2}\left(\mathbf{L} + \mathbf{L}^T\right) = \begin{bmatrix} e - zk & \gamma \\ \gamma & 0 \end{bmatrix}, \tag{6.4}$$

where the axial strain e, the shear strain γ and the curvature k are, respectively, given by

$$e = (1 + u')\cos\theta + w'\sin\theta - 1 \tag{6.5}$$
$$\gamma = -(1 + u')\sin\theta + w'\cos\theta \tag{6.6}$$
$$k = \theta', \tag{6.7}$$

where $\star' = \frac{\partial \star}{\partial x}$. By adopting a linear, elastic model as

$$\boldsymbol{\epsilon} = \frac{1+\nu}{E}\boldsymbol{\sigma} - \frac{\nu}{E}\mathrm{tr}(\boldsymbol{\sigma})\mathbf{I} \tag{6.8}$$

the resulting cross sectional forces are

$$N = EAe, \tag{6.9}$$
$$T = GA\gamma, \tag{6.10}$$
$$M = EIk\theta', \tag{6.11}$$

where N, T, and M are the axial force, shear force, and bending moment, respectively. The Young's modulus is indicated by E, the shear modulus by G, the cross-sectional area by A and the bending moment of inertia by I.

By using the virtual work principle, the equations of motion are found as

$$\rho A\ddot{u} = (N\cos\theta - T\sin\theta)' \tag{6.12}$$
$$\rho A\ddot{w} = (N\sin\theta + T\cos\theta)' + q \tag{6.13}$$
$$\rho I\ddot{\theta} = T(1 + e) - N\gamma + M' \tag{6.14}$$
$$N = EA[(1 + u')\cos\theta + w'\sin\theta - 1] \tag{6.15}$$

$$T = GA[-(1+u')\sin\theta + w'\cos\theta] \tag{6.16}$$
$$M = EI\theta', \tag{6.17}$$

where q is the distributed transversal load per unit length. Note that these equations are valid for any arbitrary cross section rotation θ. This set of equations is highly nonlinear and therefore of difficult numerical solution. In many applications, the system operates in the mild nonlinear range—that is to say, that the order of the nonlinear terms is the same as that of the linear terms. If this holds, the above model can be expanded and only the significant terms of the expansion need to be retained. This is discussed in the next section.

6.1.2 von Karman Model

We further adopt here the Euler–Bernoulli assumptions, namely

- negligible shear deformation ($\gamma \approx 0$);
- no axial and rotary inertia are considered;
- moderate cross section rotations, i.e., $\cos\theta \approx 1 - \frac{\theta^2}{2}$, $\sin\theta \approx \theta$.

The strains then become

$$e = (1+u')\cos\theta + w'\sin\theta - 1 \;\Rightarrow\; e \approx u' + \frac{1}{2}w'^2 \tag{6.18}$$
$$\gamma = -(1+u')\sin\theta + w'\cos\theta \;\Rightarrow\; \theta \approx w'. \tag{6.19}$$

This is to say that the axial strain e is not only determined by the derivative of the axial displacement u but also by the square of the rotation θ of the cross section. Note also that by assuming $\ddot{u} \approx 0$, (6.12) yields

$$N' = 0, \tag{6.20}$$

that is to say that the axial force is constant along the length of the beam, and given by

$$N = EA\left(u' + \frac{1}{2}w'^2\right). \tag{6.21}$$

The equations of motion then become

$$N' = 0 \tag{6.22}$$
$$\rho A\ddot{w} - EIw'''' - Nw'' = q \tag{6.23}$$
$$T = -M' \tag{6.24}$$
$$M = EIw'' \tag{6.25}$$

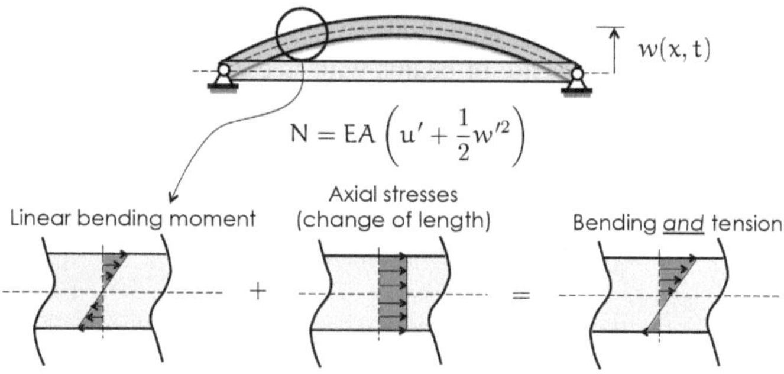

Fig. 6.3 The bending–stretching coupling effect resulting from finite bending displacements

$$N = EA\left(u' + \frac{1}{2}w'^2\right), \tag{6.26}$$

equipped with boundary conditions. Note that the presence of N creates a nonlinear restoring term that depends both on u and w. This effect is known as *string effect*, and it is illustrated in Fig. 6.3. Here we can distinguish between two situations, namely:

1. axially unmovable ends: $u(0) = u(L) = 0$;
2. free ends: $N(0) = N(L)$ prescribed.

Unmovable Ends

In this case, the axial displacement can be integrated along the length of the beam, to give the axial force $N(t)$ as

$$N(t) = \frac{EA}{2L}w'' \int_0^L w'^2 dx \tag{6.27}$$

when plugged into the governing equation for the transverse displacement w, one obtains

$$\rho A\ddot{w} - EIw'''' - \frac{EA}{2L}w'' \int_0^L w'^2 dx = q. \tag{6.28}$$

Note that the nonlinear term is a cubic function of w', and represents the string effect. This term is responsible for the *hardening* of *softening* behavior of the structure when subjected to external loads.

Free Ends

In this case, the value of N is prescribed, and therefore the additional stiffness term in the governing equation is linear:

$$\rho A\ddot{w} - EIw'''' - Nw'' = q. \tag{6.29}$$

As a consequence, no bending–stretching coupling occurs, and the von Karman model is linear. Thus, in the case of beams with free axial ends, a model including higher order terms in rotations ("large rotation" model) is required.

6.2 Finite Element Discretization

Consider the 2-node, 2D beam element sketched in Fig. 6.4. The nodal degrees of freedom can be partitioned into axial and out-of-plane as $\mathbf{u} = [u_1, y_2]^T$ and $\mathbf{u} = [w_1, \theta_1, w_2, \theta_1]^T$. The value of u and w within the element are given by the shape functions as

$$u = \mathbf{h}_u^T \mathbf{u}, \quad w = \mathbf{h}_w^T \mathbf{w}, \tag{6.30}$$

respectively, where the shape functions are given by

$$\mathbf{h}_u^T = \frac{1}{2} \left[1 - \xi,\ 1 + \xi\right], \tag{6.31}$$

$$\begin{aligned} \mathbf{h}_w^T = \frac{1}{8} \big[&4 - 6\xi + 2\xi^3,\ L(\xi^2 - 1)(\xi - 1), \ldots \\ &4 + 6\xi - 2\xi^3,\ L(\xi^2 - 1)(\xi + 1)\big], \end{aligned} \tag{6.32}$$

and $\xi = \frac{2x}{L} - 1$. The axial force N is then given by

$$N = EAe = EA \left(\mathbf{b}_u^T \mathbf{u} + \frac{1}{2} (\mathbf{b}_w^T \mathbf{w})^2 \right), \tag{6.33}$$

where $\mathbf{b}_u = \frac{d\mathbf{h}_u}{d\xi}$ and $\mathbf{b}_w = \frac{d\mathbf{h}_w}{d\xi}$. Likewise, the bending moment M is given by

$$M = EIk = EI\mathbf{c}^T \mathbf{w}, \tag{6.34}$$

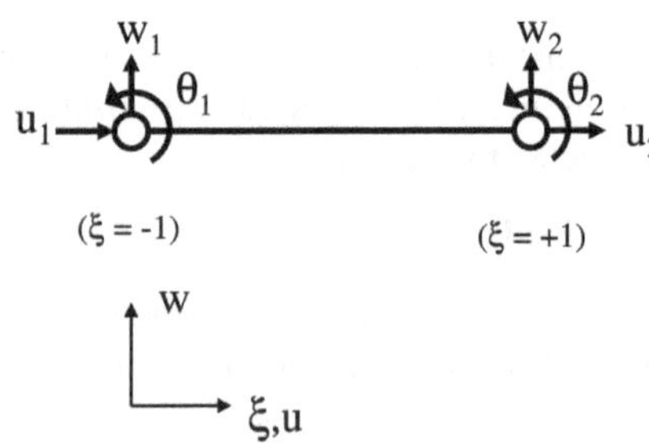

Fig. 6.4 A 2D planar beam element

where $\mathbf{c} = \frac{d^2\mathbf{h}_w}{d\xi^2}$. By using the virtual work principle, the nodal internal forces can be found. The axial contribution $\mathbf{U}_i$ is given by

$$\mathbf{U}_i = \int_0^L N\mathbf{b}_u dx = \int_{-1}^1 N\mathbf{b}_u \frac{L}{2} d\xi, \tag{6.35}$$

while the bending contribution $\mathbf{W}_i$ is given by

$$\mathbf{W}_i = \int_0^L \left(N(\mathbf{b}_w^T\mathbf{w})\mathbf{b}_w + M\mathbf{c}\right) dx = \int_{-1}^1 \left(N(\mathbf{b}_w^T\mathbf{w})\mathbf{b}_w + M\mathbf{c}\right) \frac{L}{2} d\xi. \tag{6.36}$$

Note that the axial forces $\mathbf{U}_i$ is quadratic and depends both on in-plane and out-of-plane deformations, while $\mathbf{W}_i$ is cubic. The tangent stiffness matrix is readily obtained by differentiating the internal nodal forces with respect to the nodal dofs, as

$$\mathbf{K} = \begin{bmatrix} \frac{\partial \mathbf{U}_i}{\partial \mathbf{u}} & \frac{\partial \mathbf{U}_i}{\partial \mathbf{w}} \\ \frac{\partial \mathbf{W}_i}{\partial \mathbf{u}} & \frac{\partial \mathbf{W}_i}{\partial \mathbf{w}} \end{bmatrix} = \begin{bmatrix} \mathbf{K}_{uu} & \mathbf{K}_{uw} \\ \mathbf{K}_{wu} & \mathbf{K}_{ww}, \end{bmatrix} \tag{6.37}$$

where

$$\mathbf{K}_{uu} = \int_0^L EA\mathbf{b}_u\mathbf{b}_u^T dx \tag{6.38}$$

$$\mathbf{K}_{uw} = \mathbf{K}_{wu}^T = \int_0^L EA(\mathbf{b}_w^T\mathbf{w})\mathbf{b}_u\mathbf{b}_w^T dx \tag{6.39}$$

$$\mathbf{K}_{ww} = \int_0^L EI(\mathbf{c}\mathbf{c}^T + EA\mathbf{b}_w^T\mathbf{w})^2\mathbf{b}_w\mathbf{b}_w^T + N\mathbf{b}_w\mathbf{b}_w^T dx. \tag{6.40}$$

6.3 Galerkin Projection

After FE discretization, the motion of the system is governed by a system of second-order Ordinary Differential Equations (ODEs), together with the initial conditions for generalized displacements and velocities, which reads

$$\begin{aligned} \mathbf{M}\ddot{\mathbf{u}}(t) + \mathbf{C}\dot{\mathbf{u}}(t) + \mathbf{f}(\mathbf{u}(t)) &= \mathbf{g}(t)\,, \\ \mathbf{u}(t_0) = \mathbf{u}_0,\ \dot{\mathbf{u}}(t_0) &= \mathbf{v}_0, \end{aligned} \tag{6.41}$$

where the solution $\mathbf{u}(t) \in \mathbb{R}^n$ is a high-dimensional generalized displacement vector, $\mathbf{M} \in \mathbb{R}^{n\times n}$ is the mass matrix, $\mathbf{C} \in \mathbb{R}^{n\times n}$ is the damping matrix, $\mathbf{f} : \mathbb{R}^n \mapsto \mathbb{R}^n$ gives the nonlinear elastic internal force as a function of of the displacement $\mathbf{u}$ of the structure, and $\mathbf{g}(t) \in \mathbb{R}^n$ is the time-dependent external load vector. In our case, the nonlinear term $\mathbf{f}(\mathbf{u})$ models the effect of geometric nonlinearities, as discussed on the previous section. The system (6.41) is usually referred to as the High Fidelity Model (HFM), which solution can be extremely time consuming to compute if the dimension n of the system is large. The classical notion of model reduction aims to reduce this dimensionality by introducing a linear mapping on to a suitable low-dimensional subspace $\mathscr{V}$ as

$$\mathbf{u}(t) \approx \mathbf{V}\mathbf{q}(t)\,, \qquad \mathbf{V} \in \mathbb{R}^{n \times m}, \tag{6.42}$$

where $\mathbf{q}(t) \in \mathbb{R}^m$ $(m \ll n)$ is the low-dimensional vector of reduced variables, and $\mathbf{V}$ is known as the reduction basis. Upon substitution of (6.42) in (6.41), one obtains

$$\mathbf{M}\mathbf{V}\ddot{\mathbf{q}}(t) + \mathbf{C}\mathbf{V}\dot{\mathbf{q}}(t) + \mathbf{f}(\mathbf{V}\mathbf{q}(t)) = \mathbf{g}(t) + \mathbf{r}(t), \tag{6.43}$$

where $\mathbf{r}(t)$ is a residual vector. The ROM is then obtained using Galerkin projection, which requires the residual to be orthogonal to the subspace span $\mathscr{V}$ spanned by $\mathbf{V}$, i.e., posing $\mathbf{V}^T\mathbf{r} = \mathbf{0}$. This leads to

$$\underbrace{\mathbf{V}^T\mathbf{M}\mathbf{V}}_{\tilde{\mathbf{M}}}\,\ddot{\mathbf{q}}(t) + \underbrace{\mathbf{V}^T\mathbf{C}\mathbf{V}}_{\tilde{\mathbf{C}}}\,\dot{\mathbf{q}}(t) + \mathbf{V}^T\mathbf{f}(\mathbf{V}\mathbf{q}(t)) = \mathbf{V}^T\mathbf{g}(t),$$

where $\tilde{\mathbf{M}}, \tilde{\mathbf{C}} \in \mathbb{R}^{m \times m}$ are the reduced mass and damping matrices, respectively. Often, the internal force can be split into its linear and nonlinear contributions as $\mathbf{f}(\mathbf{u}) = \mathbf{K}\mathbf{u} + \mathbf{f}^{nl}(\mathbf{u})$, to obtain a reduced stiffness matrix as well as

$$\tilde{\mathbf{M}}\ddot{\mathbf{q}}(t) + \tilde{\mathbf{C}}\dot{\mathbf{q}}(t) + \underbrace{\mathbf{V}^T\mathbf{K}\mathbf{V}}_{\tilde{\mathbf{K}}}\,\mathbf{q}(t) + \underbrace{\mathbf{V}^T\mathbf{f}^{nl}(\mathbf{V}\mathbf{q}(t))}_{\tilde{\mathbf{f}}(\mathbf{q}(t))} = \mathbf{V}^T\mathbf{g}(t). \tag{6.44}$$

As for the case of linear models, the reduced matrices $\tilde{\mathbf{M}}, \tilde{\mathbf{C}}$ and $\tilde{\mathbf{K}}$ can be pre-computed in the offline stage prior to time integration. In other words, during time integration (online phase), the computational cost associated to the evaluation of the linear terms in (6.44) scales only with the number of reduced variables m. Unfortunately, this is not the case for the computation of the nonlinear term $\tilde{\mathbf{f}}(\mathbf{q}(t))$.

For FE applications, this nonlinear forces are usually evaluated as follows:

$$\tilde{\mathbf{f}}(\mathbf{q}) = \mathbf{V}^T\mathbf{f}^{nl}(\mathbf{V}\mathbf{q}) = \sum_{e=1}^{n_e} \mathbf{V}_e^T\mathbf{f}_e(\mathbf{V}_e\mathbf{q}), \tag{6.45}$$

where $\mathbf{f}_e(\mathbf{u_e}) \in \mathbb{R}^{N_e}$ is the contribution of the element e toward the vector $\mathbf{f}^{nl}(\mathbf{u})$ (N_e being the number of DOFs for the element e), $\mathbf{V}_e$ is the restriction of $\mathbf{V}$ to the rows indexed by the DOFs corresponding to e, and n_e is the total number of elements in the mesh. Since the reduced nonlinear term $\tilde{\mathbf{f}}(\mathbf{q})$ is evaluated in the space of full variables, the computational cost associated to its evaluation does not scale with m

alone. Indeed, (6.45) shows that this cost scales linearly with the number of elements in the structure, and can hence be high for large systems. Thus, despite the reduction in dimensionality achieved in (6.44), the evaluation of the reduced nonlinear term $\tilde{\mathbf{f}}(\mathbf{q})$ hinders any fast prediction of system response using the ROM. Different strategies are available to overcome this problem, and will be addressed in the following section in details. Here, it is sufficient to realize the following:

- In the case of polynomial nonlinearities, as the one arising from von Karman kinematics of linear elastic bodies, it is possible to write the nonlinear reduced force vector $\tilde{\mathbf{f}}(\mathbf{q})$ directly in modal coordinates by constructing reduced tensors that contract on the modal coordinate directly, see for instance,(Touzé et al. 2014).
- When the nonlinearity is not polynomial, one is facing essentially two possibilities:

    1. *assume* a certain form of $\tilde{\mathbf{f}}(\mathbf{q})$ and identify its parameters or
    2. try to compute $\tilde{\mathbf{f}}(\mathbf{q})$ affordably, but still querying the underlying HFM mesh. This strategy is addressed by the so-called *hyper-reduction* methods, which are discussed, at least to some extent, in Sect. 6.8.

Clearly, any effort in efficiently computing the reduced nonlinear terms is useless in the reduction basis $\mathbf{V}$ does not properly span the sought solution. In summary, a successful ROM for (6.41) requires a suitable basis $\mathbf{V}$ *and* an efficient strategy to scale the evaluation cost of $\tilde{\mathbf{f}}$ to the size of the ROM. These aspects will be addressed in the following sections.

6.4 Nonlinear Normal Modes

When a structure's nonlinear behavior is not negligible, the traditional linear modal framework, which is fundamental to our insight into the behavior of linear systems, is no longer valid. We also lose the convenience of linear superposition and the associated computational savings. The Nonlinear Normal Mode (NNM) offers a different theoretical definition of a vibration mode that can be used to understand the behavior of a system, while maintaining a conceptual similarity to linear vibration modes. A structure's nonlinear normal modes can be very useful as a metric to evaluate a candidate Reduced Order Model (ROM) and to compare various ROMs. The NNMs decompose the structure's response into a set of modes, which can each be described by an amplitude (or energy)-dependent natural frequency and mode shape. However, they usually do not form a basis that can be used to uncouple the equations of motion and solve for the response.

The nonlinear normal mode was originally defined by Rosenberg (1960) as a synchronous, periodic motion, or a vibration in unison, of the nonlinear equations of motion. More recently, Vakakis, Kerschen, and others (Vakakis 1997; Kerschen et al. 2009) have extended this definition to include periodic motions where modal interactions may also occur by relaxing the restriction to synchronous motion. According to their definition, which is the one used throughout this paper, a nonlinear normal mode is a *not necessarily synchronous periodic response of the undamped nonlinear equations of motion*. The term normal is retained here to be consistent with the terminology used throughout the literature, even though the properties of modal superposition and orthogonality do not apply. Each NNM describes how the resonant frequency and response of the structure changes as the response amplitude (or energy) increases. They provide a wealth of insight into the structure's behavior since they form the backbone of the systems nonlinear forced response. In other words, each NNM gives an estimate of the deformation of the structure when it is excited near resonance over a range of forcing (or response) amplitudes. For lightly damped structures, the damped invariant manifold can be approximated by the undamped invariant manifold (i.e., the NNM), meaning that the freely decaying response tends to follow an undamped NNM. Nonlinearities in a physical model can come in a variety of forms such as of large deformations, jointed connections, buckling, material constitutive laws, and contact. These nonlinearities introduce behavior such as frequency–energy dependence, localization, modal interactions, and bifurcations. The nonlinear normal mode accounts for such salient nonlinear behavior, providing many insights into the system response.

6.4.1 Periodic Motions of an Undamped System: Nonlinear Normal Modes

The N-Degree of Greedom (DOF) Equations of Motion (EOM) for a nonlinear finite element model generally can be written as follows, and all terms were defined previously.

$$\mathbf{M}\ddot{\mathbf{x}} + \mathbf{C}\dot{\mathbf{x}} + \mathbf{K}\mathbf{x} + \mathbf{f}_{NL}(\mathbf{x}) = \mathbf{f}(t). \tag{6.46}$$

The external loads are applied through the $N \times 1$ force vector $\mathbf{f}(t)$. The $N \times 1$ nonlinear restoring force vector, $\mathbf{f}_{NL}(\mathbf{x})$, accounts for the nonlinearity in the physical system.

The undamped NNM definition used here comes from the works of Kerschen et al. (2009), who defined an NNM as *a not necessarily synchronous periodic motions of the conservative equations of motion*. By this definition, the periodic motions that occur when two or more modes interact are still considered NNMs since they relaxed the restriction of synchronous motion defined originally by Rosenberg (1960). Many new features emerge with this definition of a vibration mode that cannot be described with linear modal analysis, such as frequency–energy dependence, bifurcations, localization, and modal interactions. Typically, the NNMs are represented on the frequency–energy plane, showing how the fundamental frequency of the periodic motion changes as the energy of the system (which is conserved over the period of the response) evolves.

Mathematically, the NNMs are solutions to the undamped, unforced equation of motion given below.

$$\mathbf{M}\ddot{\mathbf{x}} + \mathbf{K}\mathbf{x} + \mathbf{f}_{NL}(\mathbf{x}) = 0. \tag{6.47}$$

A variety of algorithms exist for computing NNMs, for example, the shooting algorithm in Kerschen et al. (2009) has proven very effective. Typically, the algorithm is initialized with the linear natural frequency and mode shape as an initial guess, and continuation methods are used to trace the solution as amplitude increases.

If we consider again the forced, damped system, we can see that the NNM is a solution to the forced, damped system only if

$$\mathbf{C}\dot{\mathbf{x}} = \mathbf{f}(t). \tag{6.48}$$

This corresponds to the resonant case for the system, because only the damping forces are able to resist the applied force. Hence, the NNM can be thought of as a collection of worstcase forced responses for the system.

► As an example, consider the curved beam shown in Fig. 6.5. The beam is 3D printed from PLA plastic onto a stiff, C-shaped backing structure that is constructed from the same material. The structure is excited at its base by a slowly sweeping sinusoidal input (so that it nearly reaches steady state at any instant) over a range of amplitudes.

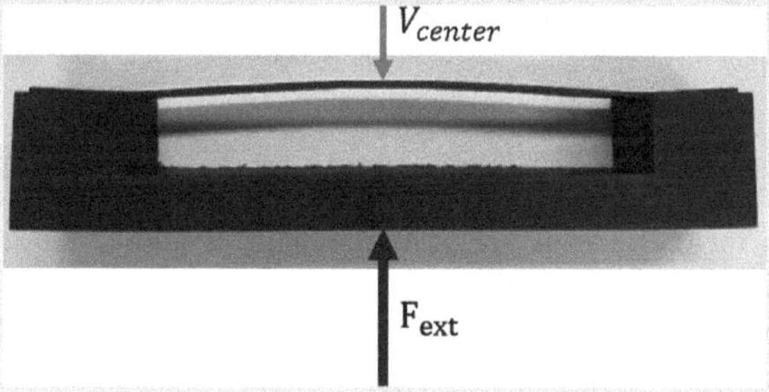

Fig. 6.5 A curved beam is an example of a structure that can exhibit geometric nonlinearity. As the vibration amplitude increases, stresses that build up in the beam start to cancel the bending stiffness, so the effective stiffness and hence effective vibration frequency goes down. At large amplitudes, the structure stiffens since the beam must stretch axially to accommodate higher amplitude motion

The response of the beam is shown in Fig. 6.6. As the amplitude of the excitation increases, the frequency at which the peak response occurs first decreases from about 145 Hz to about 138 Hz. The NNM is also shown, as measured experimentally over a range of amplitude. The NNM correlates with the peak response at each forcing amplitude. At even higher amplitudes, effective resonance frequency increases again, as is predicted well by the NNM.

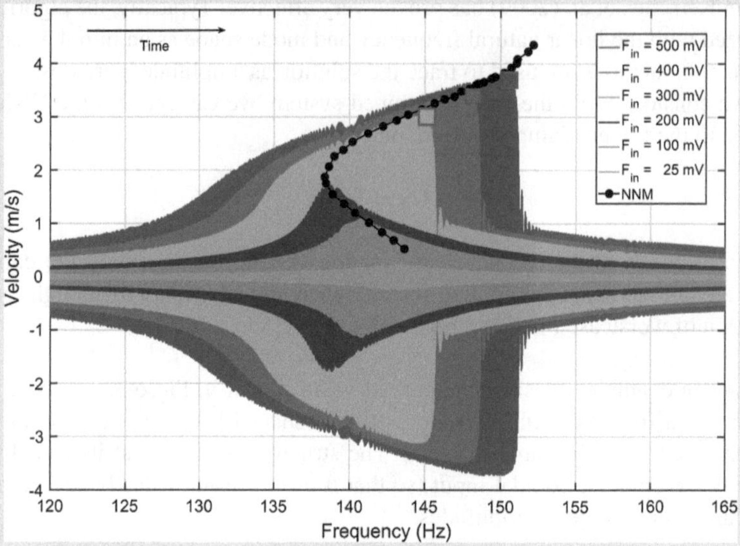

Fig. 6.6 Response of the curved beam to swept-sine input forces at various amplitudes. The NNM is overlaid, expresses as the velocity of the center point of the beam versus frequency

◀

Considering again Eq. 6.48, and noting that the damping forces are generally distributed over the entire structure, it seems that it would be rare that a structure would be excited into an NNM. However, one can use force balance to show that a force may be distributed spatially in a variety of ways, including a simple concentrated force at a point, and still excite an NNM.

Specifically, one can use the energy balance technique in Hill et al. (2014, 2015) to show that the total energy dissipated by the nonlinear system over one cycle is

$$E_{diss/cyc} = \int_0^T P_{diss}\, dt. \tag{6.49}$$

If the system is excited by an arbitrary forcing function $\mathbf{f}(t)$, the energy input into the system as

$$E_{in/cyc} = \int_0^T \dot{\mathbf{x}}(t)^{\mathrm{T}} \mathbf{f}(t)\, dt. \tag{6.50}$$

At steady state, the energy dissipated by the damping forces must match the total energy input to the system over the period T. The balance is enforced by setting $E_{diss/cyc} = E_{in/cyc}$. Then, if the response, $\dot{\mathbf{x}}(t)$ is assumed to take the shape of the NNM that is nearest in frequency, and the spatial distribution of the force is known, then one can solve for the force amplitude. For example, if a single-point co-sinusoidal force is applied with complex amplitude $A + i0$, the scaling on A can be computed by satisfying

$$A \int_0^T \dot{\mathbf{x}}(t)^{\mathrm{T}} \left(\mathbf{e}_n e^{i\omega t}\right) dt = \int_0^T \dot{\mathbf{x}}(t)^{\mathrm{T}} \mathbf{C} \dot{\mathbf{x}}(t)\, dt \tag{6.51}$$

where $\mathbf{e}_n$ is a vector of zeros with a value of one at the location n, which is the point at which the force is applied. If both the NNMs $\mathbf{x}(t)$ and the damping $\mathbf{C}$ in the system are known, Eq. (6.51) can be readily used to estimate the forcing amplitude A that would excite the system at resonance in the associated NNM $\mathbf{x}(t)$. In Kuether et al. (2015), this approach was demonstrated and found to be very effective for predicting the forcing required to drive a system near a known NNM. Furthermore, it also proved useful in providing the initial guesses needed to compute isolated resonance branches.

6.5 Nonintrusive Reduced Order Modeling (ROM) for Geometrically Nonlinear Structures: ICE and ED

By *nonintrusive* methods, we mean procedures that are able to construct the ROM (6.44) *without* requiring access and manipulation of elemental quantities, as for instance internal forces and tangential stiffness matrices. As such, the ROM is formed by properly probing the model by use of general purpose FE programs (e.g., ABAQUS, NASTRAN) and match an a priori determined form for the ROM to the obtained data.

The discretized system of equations for an N Degree of Freedom (DOF) linear elastic, geometrically nonlinear finite element model can be written as

$$\mathbf{M}\ddot{\mathbf{x}} + \mathbf{K}\mathbf{x} + \mathbf{f}_{NL}(\mathbf{x}) = \mathbf{f}(t), \tag{6.52}$$

where $\mathbf{M}$ is the $N \times N$ mass matrix, $\mathbf{K}$ is the $N \times N$ linear stiffness matrix, $\mathbf{f}_{NL}(\mathbf{x})$ is the $N \times 1$ nonlinear restoring force vector, and $\mathbf{f(t)}$ is the $N \times 1$ external force vector. The $N \times 1$ vectors $\mathbf{X}$ and $\ddot{\mathbf{x}}$ are the displacement and acceleration, respectively. The NLROM procedures seek to reduce these undamped, nonlinear equations using the linear vibration modes as the reduction basis, which are found by solving the eigenvalue problem $(\mathbf{K} - \omega_r^2\mathbf{M})\varphi_r$. A small set of mass normalized mode shapes approximate the kinematics of the EOM as

$$\mathbf{x}(t) = \varphi_m \mathbf{q}(t). \tag{6.53}$$

Each column in the $N \times M$ mode shape matrix, φ_m, is a mass normalized mode shape vector, φ, and $\mathbf{q}$ is an vector of $m \times q$ time-dependent modal displacements.

The reduced coordinate space $\mathbf{q}$ is significantly smaller than the physical coordinate space $\mathbf{X}$ (i.e., $m \ll N$).

In order to reduce the full order EOMs, Eq. (6.53) is substituted into Eq. (6.52) and premultiplied by φ_m^T to assure that the residual error is orthogonal to the reduction basis (note that $(\ )^T$ is the transpose operator). The reduced, nonlinear modal form of the rth equation becomes

$$\ddot{q}_r + c_r\dot{q}_r + \omega_r^2 + \theta_r(q_1, q_2, \ldots, q_m) = \varphi_r^T \mathbf{f}(\mathbf{t}), \tag{6.54}$$

where ω_r is the linear natural frequency, and q_r is the rth modal displacement. The nonlinear modal restoring force is given as

$$\theta_r(\mathbf{q}) = \varphi_r^T \mathbf{f}_{NL}(\varphi\mathbf{q}). \tag{6.55}$$

The nonlinear restoring forces are a nonlinear function, $\theta_r(\mathbf{q})$, of the modal displacements. If the finite element model is linear elastic with geometric nonlinearities derived using quadratic strain–displacement relationships, the function $\theta_r(\mathbf{q})$ becomes a quadratic and cubic polynomial, and is given as

$$\theta_r(q_1, q_2, \ldots q_m) = \sum_{i=1}^{m}\sum_{j=1}^{m} B_r(i, j)q_i, q_j + \sum_{i=1}^{m}\sum_{j=1}^{m}\sum_{k=1}^{m} A_r(i, j, k)q_i, q_j, q_k. \tag{6.56}$$

The coefficients and are the quadratic and cubic nonlinear stiffness terms, respectively, for the rth nonlinear modal equation. If the full order EOM in Eq. (6.52) are known in closed form, then the nonlinear coefficients in Eq. (6.56) can be directly computed (referred as direct evaluation in Hollkamp and Spottswood 2005). However, in cases where the closed form equations are not explicitly available (e.g., within commercial finite element packages), an indirect approach must be used. The coefficients are approximated using either the ED or ICE methods using a series of nonlinear static solutions. Once these nonlinear stiffness terms are identified, the NLROM equations are given by Eqs. (6.54) and (6.56), providing a significant computational savings for response prediction compared to direct time integration.

Typically, the linear term is set to ω_r^2 and treated as a known value during the identification of $\mathbf{B}_r$ and $\mathbf{A}_r$.

6.5.1 Enforced Displacements Procedure

The Enforced Displacements Procedure uses a set of prescribed displacements in the shape(s) of the linear modes in the basis in Eq. (6.53). In general, a multimodal displacement can be written as

$$\mathbf{X}_c = \varphi_1\widehat{\mathbf{q}}_1 + \varphi_2\widehat{\mathbf{q}}_2 + \cdots + \varphi_m\widehat{\mathbf{q}}_m, \tag{6.57}$$

where $\mathbf{X}_c$ is an $N \times 1$ displacement vector and $\widehat{\mathbf{q}}_r$ is the scaling factor for the rth mode in the basis. In order to accurately estimate the ROM, the scaling factors should scale the deformation shape of each mode to a level that exercises the geometric nonlinearity in the structure. Using a commercial finite element software package, the resulting reaction forces, $\mathbf{F}_c$, are computed in response to the given displacement field. Then, the quasi-static representation of the NLROM for the rth modal equation becomes

$$\omega_r^2 \widehat{q}_r + \sum_{i=1}^{m} \sum_{j=1}^{m} B_r(i, j) q_i, q_j + \sum_{i=1}^{m} \sum_{j=1}^{m} \sum_{k=1}^{m} A_r(i, j, k) q_i, q_j, q_k = \varphi_r^T \mathbf{F}_c. \quad (6.58)$$

The nonlinear stiffness coefficients are then solved for using the approach presented in Muravyov and Rizzi (2003) with a set of displacement load cases that are in a combination of one, two or three modes in the basis set. The number of nonlinear static solutions, for $m > 3$, required by the ED method (Mignolet and Spottswood 2013) is

$$2m + \frac{3m!}{2(m-2)!} + \frac{m!}{6(m-3)!}. \quad (6.59)$$

6.5.2 *Applied Loads Procedure or Implicit Condensation and Expansion*

The applied loads procedure uses a set of static forces in the shape of the linear modes in order to find the nonlinear stiffness coefficients of the NLROM. For example, a multimode static force can be a combination of those forces that would excite any of the modes in the reduced basis, given as

$$\mathbf{F_c} = \mathbf{M}(\varphi_1 \widehat{f_1} + \varphi_2 \widehat{f_2} + \cdots + \varphi_m \widehat{f_m}), \quad (6.60)$$

where $\mathbf{F}_c$ is the vector of applied forces and $\widehat{f_r}$ is the force scaling factor for the rth mode. Note that the previous works on the applied loads method did not mention the use of the mass matrix when computing the force vector, but it was used here and is needed to obtain a force that exactly isolates a single mode for a linear system. The force scaling factors can be varied to exercise the desired amount of geometric nonlinearity in the structure, as will be discussed further in Section III. The resulting static deformation, $\mathbf{X}_c$, can then be projected onto the linear modal coordinates as follows.

$$q_r = \varphi_r \mathbf{M}\mathbf{x}_c. \quad (6.61)$$

Using the quasi-static force–displacement relationship in Eq. (6.58), one can then solve for the nonlinear stiffness coefficients with the constrained approach (Gordon and Hollkamp 2011) using a set of these applied loads exercising the different modes in the basis. With the applied loads approach, the number of load permutations (for

$m > 3$) required to generate the static response data is given as

$$2m + 2\frac{m!}{(m-2)!} + \frac{4m!}{3(m-3)!}. \tag{6.62}$$

The number of loads differs from that of the Enforced Displacement in Eq. (6.59) due to differences in how the fits are performed; with the ED method sets of two and three load cases can be used to solve algebraic equations for the coefficients, whereas with ICE a larger least squares problem is solved.

6.6 Intrusive Methods

As opposed to nonintrusive methods, *intrusive methods* requires access to the specific finite element formulation. This disadvantage is compensated by the possibility for a more rigorous definition of the ingredients forming the reduction basis.

6.6.1 Modal Derivatives

Vibration Modes

The system in (6.41) can be linearized around its static equilibrium position. Without loss of generality, we assume the equilibrium configuration is $\mathbf{u}_{eq} = \mathbf{0}$ to obtain

$$\mathbf{M}\ddot{\mathbf{u}} + \mathbf{C}\dot{\mathbf{u}} + \mathbf{K}|_{eq}\mathbf{u} = \mathbf{g}(t), \tag{6.63}$$

where

$$\mathbf{K}|_{eq} = \left.\frac{\partial \mathbf{f}(\mathbf{u})}{\partial \mathbf{u}}\right|_{\mathbf{u}=\mathbf{0}}. \tag{6.64}$$

This linearized system would be a good approximation to the original system in (6.41) for small enough displacements from the linearization point. It is well established that for such system, the system response can be written as a linear combination of constant eigenmodes (or Vibration Modes (VMs) in the structural dynamics context) which form a basis of $\mathbb{R}^n$ as follows:

$$\mathbf{u}(t) = \sum_{i=1}^{n} \boldsymbol{\phi}_i q_i(t), \tag{6.65}$$

where the eigenmodes $\boldsymbol{\phi}_i \in \mathbb{R}^n$ are the solution of the generalized eigenvalue problem[1]

$$(\mathbf{K}|_{eq} - \omega_i^2 \mathbf{M})\boldsymbol{\phi}_i = \mathbf{0}. \tag{6.66}$$

(ω_i^2 is the eigenvalue or the eigenfrequency squared). This approach of expressing the solution $\mathbf{u}(t)$ in terms of a basis of eigenvectors is referred to as the principle of *linear modal superposition*. However, if one is considering the slowly varying dynamics of the system, then it can be shown that the response can be very accurately approximated by few modes associated to low eigenfrequencies, and a modal truncation can be obtained:

$$\mathbf{u}(t) \approx \sum_{i=1}^{m} \boldsymbol{\phi}_i \eta_i(t) = \boldsymbol{\Phi}\mathbf{q}(t), \tag{6.67}$$

where $\boldsymbol{\Phi} \in \mathbb{R}^{n \times m}$, $\mathbf{q}(t) = [q_1(t)\ q_2(t)\ \cdots q_m(t)]^T \in \mathbb{R}^m, m \ll n$. Note that this is equivalent to performing a Galerkin projection as shown in Sect. 3.1 for a linearized system, where $\boldsymbol{\Phi}$ is the reduction basis and $\mathbf{q}$ are the corresponding reduced unknowns.

Modal Derivatives

When the deviation from the linearization point increases, the response of (6.63) can no longer be considered as a good approximation for the original nonlinear counterpart. One might still think of using the VMs obtained from the linearized model to form a reduction basis for the reduction of the nonlinear set of equations. A basis composed of a few dominant VMs, however, is typically not sufficient for reduction since it does not feature the dominant coupling effects (e.g., membrane-bending), typical of geometrically nonlinear structures.

Earlier work in Idelsohn and Cardona (1985a, b) and more recent work in Weeger et al. (2016); Witteveen and Pichler (2014) discuss the use of the so-called Modal Derivatives (MDs) to capture the response of the nonlinear system upon departure from the linear behavior. If (6.66) is differentiated with respect to the modal amplitude q_j (assuming $\mathbf{M}$ to be a constant mass matrix) and then evaluated at equilibrium, one obtains

$$(\mathbf{K}|_{eq} - \omega_i^2|_{eq}\mathbf{M}) \left.\frac{\partial \boldsymbol{\phi}_i}{\partial q_j}\right|_{eq} + \left(\left.\frac{\partial \mathbf{K}}{\partial q_j}\right|_{eq} - \left.\frac{\partial \omega_i^2}{\partial q_j}\right|_{eq} \mathbf{M} \right) \boldsymbol{\phi}_i|_{eq} = \mathbf{0}, \tag{6.68}$$

where the MD $\frac{\partial \boldsymbol{\phi}_i}{\partial q_j}$ denotes the derivative of the ith mode in the jth modal direction. The tangent stiffness matrix derivative with respect to η_j is obtained by imposing t a displacement in the direction of $\boldsymbol{\phi}_j|_{eq}$ to the system, i.e.,

[1] Here, we neglect the damping contribution in eigenvalue problem to avoid complex eigenvalues and vectors. Note that for damped linear systems with low damping or *modal/Rayleigh* damping, the eigenvectors for an undamped system are a good approximation for the damped counterpart and still form a good basis for linear modal superposition.

$$\left.\frac{\partial \mathbf{K}}{\partial q_j}\right|_{eq} = \left.\frac{\partial \mathbf{K}(\mathbf{u} = \eta_j \boldsymbol{\phi}_j|_{eq}))}{\partial q_j}\right|_{q_j=0} . \tag{6.69}$$

Physically, an MD represents the sensitivity of VM $\boldsymbol{\phi}_i$ corresponding to a displacement given in the direction of VM $\boldsymbol{\phi}_j$. These MDs can be used as efficient enrichment to the reduction basis to capture the dominant nonlinear behavior effectively.

Calculation of Modal Derivatives

The MDs $\left.\frac{\partial \boldsymbol{\phi}_i}{\partial q_j}\right|_{eq}$ cannot be easily obtained from (6.68) since the coefficient matrix is singular, cf. (6.66). This singularity can be tackled by imposing a normalization condition for the eigenmodes. More details on different solution techniques can be found in Ref. Siddhi (2005). The popular mass normalization is adopted here, i.e.,

$$\boldsymbol{\phi}_i^T \mathbf{M} \boldsymbol{\phi}_i = 1 \qquad \forall i \in \{1, 2, \dots, m\}. \tag{6.70}$$

By differentiating the equation above with respect to the modal amplitude one obtains

$$\boldsymbol{\phi}_i^T \mathbf{M} \frac{\partial \boldsymbol{\phi}_i}{\partial q_j} + \boldsymbol{\phi}_i^T \mathbf{M}^T \frac{\partial \boldsymbol{\phi}_i}{\partial q_j} = 0 \qquad \forall i, j \in \{1, 2, \dots, m\}. \tag{6.71}$$

Exploiting the symmetry of $\mathbf{M}$ and subsequent evaluation at the equilibrium position results in the following relation:

$$\boldsymbol{\phi}_i^T|_{eq} \mathbf{M} \left.\frac{\partial \boldsymbol{\phi}_i}{\partial q_j}\right|_{eq} = 0 \qquad \forall i, j \in \{1, 2, \dots, m\}. \tag{6.72}$$

The following *direct* approach to calculate the MDs can then be formulated using (6.68), (6.72):

$$\begin{bmatrix} [\mathbf{K}|_{eq} - \omega_i^2|_{eq}\mathbf{M}]_{n\times n} & -\left[\mathbf{M}\boldsymbol{\phi}_i|_{eq}\right]_{n\times 1} \\ -\left[\mathbf{M}\boldsymbol{\phi}_i|_{eq}\right]_{1\times n}^T & 0_{1\times 1} \end{bmatrix} \begin{bmatrix} [1] \left.\frac{\partial \boldsymbol{\phi}_i}{\partial q_j}\right|_{eq} \\ \left.\frac{\partial \omega_i^2}{\partial q_j}\right|_{eq} \end{bmatrix} = \begin{bmatrix} -\left.\frac{\partial \mathbf{K}}{\partial q_j}\right|_{eq} \boldsymbol{\phi}_i|_{eq} \\ 0 \end{bmatrix} .$$

The above non-singular system can be used to solve for MDs. This method, however, destroys the band structure of the original system. Nonetheless, it is rigorous and accurate. Apart from this direct approach, the pseudoinverse technique (cf. G'eradin and Rixen 1997) and the Nelson's method (Nelson 1976) are some techniques that preserve the band structure of the matrices.

Regardless of the method adopted to solve (6.68), a high-dimensional matrix needs to be factorized for each ω_i. In Idelsohn and Cardona (1985b), a method to approximate the problem (6.68) is discussed. They propose neglecting the inertial contribution as

$$\mathbf{K}|_{eq} \left.\frac{\partial \boldsymbol{\phi}_i}{\partial q_j}\right|_{eq}^{s} = -\left.\frac{\partial \mathbf{K}}{\partial q_j}\right|_{eq} \boldsymbol{\phi}_i|_{eq} . \tag{6.73}$$

We call the obtained MDs as *Static* MDs (SMDs). The superscript s in $\frac{\partial \boldsymbol{\phi}_i}{\partial q_j}\Big|_{eq}^{s}$ stands for static.

Note that the Static Modal Derivatives given in (6.73) are symmetric, i.e., $\frac{\partial \boldsymbol{\phi}_i}{\partial q_j}\Big|_{eq}^{s} = \frac{\partial \boldsymbol{\phi}_j}{\partial q_i}\Big|_{eq}^{s}$. In fact, it is easy to see that the stiffness matrix derivative given by (6.69) can be written as

$$\frac{\partial \mathbf{K}}{\partial q_j}\bigg|_{eq} = \frac{\partial \mathbf{K}(\mathbf{u})}{\partial \mathbf{u}}\bigg|_{\mathbf{u}=\mathbf{0}} \cdot \boldsymbol{\phi}_j|_{eq} = \frac{\partial^2 \mathbf{f}(\mathbf{u})}{\partial \mathbf{u} \partial \mathbf{u}}\bigg|_{\mathbf{u}=\mathbf{0}} \cdot \boldsymbol{\phi}_j|_{eq} \, . \tag{6.74}$$

Substituting this into (6.73), we obtain

$$\mathbf{K}|_{eq} \frac{\partial \boldsymbol{\phi}_i}{\partial \eta_j}\bigg|_{eq}^{s} = -\left(\frac{\partial^2 \mathbf{f}(\mathbf{u})}{\partial \mathbf{u} \partial \mathbf{u}}\bigg|_{\mathbf{u}=\mathbf{0}} \right) : (\boldsymbol{\phi}_j|_{eq} \otimes \boldsymbol{\phi}_i|_{eq}) \, . \tag{6.75}$$

The third-order tensor $\left(\frac{\partial^2 \mathbf{f}(\mathbf{u})}{\partial \mathbf{u} \partial \mathbf{u}}\right)$ which contains the second-order partial derivatives is symmetric by Shwarz' theorem ($\mathbf{f}(\mathbf{u}) \in C^2(\mathbb{R}^n, \mathbb{R}^n)$), i.e., $\left(\frac{\partial^2 \mathbf{f}(\mathbf{u})}{\partial \mathbf{u} \partial \mathbf{u}}\right)_{Iij} = \left(\frac{\partial^2 \mathbf{f}(\mathbf{u})}{\partial \mathbf{u} \partial \mathbf{u}}\right)_{Iji}$. Therefore, we get

$$\frac{\partial \boldsymbol{\phi}_i}{\partial \eta_j}\bigg|_{eq}^{s} = -(\mathbf{K}|_{eq})^{-1} \left[\frac{\partial^2 \mathbf{f}(\mathbf{u})}{\partial \mathbf{u} \partial \mathbf{u}}\bigg|_{\mathbf{u}=\mathbf{0}} : (\boldsymbol{\phi}_j|_{eq} \otimes \boldsymbol{\phi}_i|_{eq}) \right] \tag{6.76}$$

$$= -(\mathbf{K}|_{eq})^{-1} \left[\frac{\partial^2 \mathbf{f}(\mathbf{u})}{\partial \mathbf{u} \partial \mathbf{u}}\bigg|_{\mathbf{u}=\mathbf{0}} : (\boldsymbol{\phi}_i|_{eq} \otimes \boldsymbol{\phi}_j|_{eq}) \right] = \frac{\partial \boldsymbol{\phi}_j}{\partial \eta_i}\bigg|_{eq}^{s} \, . \tag{6.77}$$

While the SMDs have been shown to be symmetric, such a claim cannot be made for the modal derivatives as given by (6.68).

(S)MDs in a Reduction Basis

A linear basis ($\boldsymbol{\Psi}$) consisting of VMs augmented with the corresponding MDs could be used to reduce the nonlinear system (Idelsohn and Cardona 1985a, b).

$$\boldsymbol{\Psi} = \left[\boldsymbol{\phi}_1|_{eq} \; \boldsymbol{\phi}_2|_{eq} \; \dots \; \boldsymbol{\phi}_m|_{eq} \; \dots \; \boldsymbol{\theta}_{ij}|_{eq} \dots \right], \tag{6.78}$$

where $\boldsymbol{\theta}_{ij} = \frac{\partial \boldsymbol{\phi}_i}{\partial \eta_j}\Big|_{eq}$ or $\frac{\partial \boldsymbol{\phi}_i}{\partial \eta_j}\Big|_{eq}^{s}$. From the chosen normalization in (6.72), any VM $\boldsymbol{\phi}_i$ is $\mathbf{M}$-orthogonal to all corresponding MDs $\frac{\partial \boldsymbol{\phi}_i}{\partial \eta_j}\Big|_{eq}$ $\forall j \in \{1, \dots, m\}$. However, this is not sufficient for $\boldsymbol{\Psi}$ to possess a full column rank, which is essential to have a non-singular reduced system. In general, one should perform an orthogonalization on $\boldsymbol{\Psi}$ before using it as a reduction basis.

Using MDs, one could expect a maximum basis size of m^2 (if all MDs are linearly independent). Since SMDs are symmetric, a basis $\boldsymbol{\Psi} \in \mathbb{R}^{n \times M}$ can be obtained using

(S)MDs, where $M = m + \frac{m(m+1)}{2}$ would be its maximum size. Indeed, in both cases inclusion of MDs in the basis is expected to increase the reduced number of unknowns quadratically with the number of VMs (m) in the reduction basis.

6.6.2 Wilson Vectors

Another popular way of constructing the reduction basis $\mathbf{V}$ is given by the Krylov vectors. This technique is also known under different denominations as Wilson vectors (Ming-Wu 1982), Arnoldi sequence (Arnoldi 1951) and load dependent vectors (Idelsohn and Cardona 1985). Unlike vibration modes, it consists of finding a proper reduction basis by using the information of the applied load distribution. Consider the linear problem

$$\mathbf{M}\ddot{\mathbf{u}} + \mathbf{K}\mathbf{u} = \mathbf{g}, \tag{6.79}$$

where damping has been neglected. We first start by computing the linear solution

$$\tilde{\mathbf{u}}_0 = \mathbf{K}^{-1}\mathbf{g}, \tag{6.80}$$

which is then mass normalized, i.e.,

$$\mathbf{u}_0 = \frac{1}{\sqrt{\tilde{\mathbf{u}}_0^T \mathbf{M} \tilde{\mathbf{u}}_0}} \tilde{\mathbf{u}}_0 \tag{6.81}$$

The static solution $\mathbf{u}_0$ is then used as acceleration distribution to obtain an inertial force $\mathbf{M}\mathbf{u}_0$ as pseudo-load to obtain a new vector $\mathbf{u}_1$ as

$$\mathbf{u}_1 = \mathbf{K}^{-1}\mathbf{M}\mathbf{u}_0. \tag{6.82}$$

The procedure can be made recursive, making sure that the newly computed vector is orthogonalized with respect to the previously calculated ones, and properly orthogonalized. A generic vector $\boldsymbol{u}_i$ of the sequence is first formed by computing $\hat{\mathbf{u}}_i$ as

$$\mathbf{K}\hat{\mathbf{u}}_i = \mathbf{M}\mathbf{u}_{i-1}, \tag{6.83}$$

which is subsequently made orthogonal to the already computed vectors $\mathbf{u}_j,\ j = 1, \ldots, i-1$ as

$$\tilde{\mathbf{u}}_i = \hat{\mathbf{u}}_i - \sum_{j=1}^{i-1} c_j \mathbf{u}_j, \tag{6.84}$$

where

$$c_j = \mathbf{u}_j^T \mathbf{M} \hat{\mathbf{u}}_i \ \text{ for } \ j = 1, \ldots, i-1. \tag{6.85}$$

Finally, the mode is mass normalized as

$$\mathbf{u}_i = \frac{1}{\sqrt{\tilde{\mathbf{u}}_i^T \mathbf{M} \tilde{\mathbf{u}}_i}} \hat{\mathbf{u}}_i. \tag{6.86}$$

The so obtained vectors represent a good basis for the linearized problem (6.79). With respect to the vibration modes, they do not require the solution of an eigenvalue problem, but only the solution of a sequence of linear problems sharing the same matrix of coefficients, which can be then factorized once for all. Note also that the sequence is initiated by the static solution $\mathbf{u}_0$, which is directly related to the shape of the applied load. In contrast to vibration modes, Wilson vectors do account for the particular load shape applied. As such, they tend to show faster convergence with respect to vibration modes, but would need to be recomputed in case another load shape is applied to the system.

Since the so obtained basis relates to the linearized problem, it is likely to perform poorly in the case of geometric nonlinearities. To overcome this limitation, modal derivatives of Wilson vectors can be computed in a way analogous to the one described in the previous section. In particular, say $\mathbf{V}_W = [\mathbf{u}_1, \ldots, \mathbf{u}_m]$ the basis of linear Wilson vectors, then one can compute the modal derivatives

$$\boldsymbol{\theta}_{W_{ij}} = \mathbf{K}^{-1} \frac{\partial \mathbf{K}}{\partial q_i} \mathbf{u}_j \tag{6.87}$$

and then form a reduction basis $\mathbf{V}$ as

$$\mathbf{V} = [\mathbf{V}_W \; \boldsymbol{\theta}_{W_{11}} \; \boldsymbol{\theta}_{W_{12}}, \ldots, \boldsymbol{\theta}_{W_{ij}}, \ldots, \boldsymbol{\theta}_{W_{mm}}], \quad i, j = 1, \ldots, m. \tag{6.88}$$

In order to avoid numerical issues, $\mathbf{V}$ can then be further orthogonalized.

6.7 Data-Driven Methods

All the techniques so far described can be classified as *model-driven* reduction techniques—that is, the ROM is constructed by ingredients that are inherent to the model. Another philosophy for deriving ROMS is based on data coming for the simulation the HFM for some significant cases. These trajectories will then furnish data which can be processed to obtain the reduction basis. This topic is vast, and here, we just briefly introduce the main concept. It is important to realize that as opposed to model- driven methods, which never require the solution of the HFM, the data-driven methods tend to require significant computational and storing resources. As such, their use is justifiable only when the obtained ROM—which tends to be highly efficient—is extensively used so to amortize the offline costs associated to its construction.

6.7.1 Proper Orthogonal Decomposition

The Proper Orthogonal Decomposition (POD) is a data-driven method that outputs a reduced order basis that best represents a given set of snapshots. The POD method is applied in various fields. In our context, the snapshots represent the structural configuration at sampled time instances.

Let $\mathbf{u}_i \in \mathbb{R}^n, i \in \{1, \dots, n_t\}$ be snapshot vectors obtained from the full solution of system (6.41), which are collected in the matrix $\mathbf{U} := \left[\mathbf{u}_1, \mathbf{u}_2, \dots, \mathbf{u}_{n_t}\right] \in \mathbb{R}^{n \times n_t}$. A lower dimensional POD basis $\mathbf{V} = [\mathbf{v}_1\ \mathbf{v}_2 \dots \mathbf{v}_m] \in \mathbb{R}^{n \times m}$ containing $m \ll n_t$ orthogonal vectors which *best* spans the vectors in this ensemble can be obtained by the solution to the following minimization problem:

$$\min_{\mathbf{v}_i \in \mathbb{R}^n} \sum_{j=1}^{n_s} \left\| \mathbf{u}_j - \sum_{i=1}^{m} (\mathbf{u}_j^T \mathbf{v}_i) \mathbf{v}_i \right\|_2^2 .$$

This is a least squares problem and the vectors in $\mathbf{V}$ are the left singular vectors of $\mathbf{U}$, obtained by the Singular Value Decomposition (SVD) of $\mathbf{U}$ of zero mean as

$$\mathbf{U} = \mathbf{L}\mathbf{S}\mathbf{R}^T, \tag{6.89}$$

where is $\mathbf{U}$ is factorized into unitary matrices $\mathbf{L} = [\mathbf{l}_1, \mathbf{l}_2, \dots, \mathbf{l}_n] \in \mathbb{R}^{n \times n}$ (containing the left singular vectors) and $\mathbf{R} \in \mathbb{R}^{n_t \times n_t}$ (containing the right singular vectors); and the diagonal (rectangular) matrix $\mathbf{S} \in \mathbb{R}^{n \times n_t}$ (containing corresponding singular values on the diagonal). These singular values represent the relative importance of corresponding vectors of $\mathbf{L}$ in forming the basis $\mathbf{V}$. If the singular values (and the corresponding singular vectors) are arranged in a descending order $S_{11} \geq S_{22} \geq \dots \geq 0$, it can be shown that

$$\sum_{j=1}^{n_s} \left\| \mathbf{u}_j - \sum_{i=1}^{m} (\mathbf{u}_j^T \mathbf{l}_i) \mathbf{l}_i \right\|_2^2 = \sum_{i=m+1}^{r} S_{ii}^2 .$$

Thus, the left singular vectors $\mathbf{l}_i$ corresponding to the highest singular values are the most relevant for constructing a reduction basis using POD. for our case, each (normalized) left vector $\mathbf{l}_i$ is scaled by the corresponding singular value σ_i, and associated to the time history contained in $\mathbf{r}_i^T$. Since an SVD can be performed over any solution snapshots of any general nonlinear problem, POD is seen as a versatile method. It should be noted, however, that such a reduction basis is optimal only for capturing the solution used to collect the snapshots. The procedure needs to be repeated (or a database of full solutions need to be created in the first place) to take other types of loading or boundary conditions into account.

Once a reduction basis is obtained, the same ensemble of full solution vectors can be used as training vectors for hyper-reduction to accelerate the computation of nonlinearity during the online stage. This is discussed in the following sections.

6.8 Hyper-reduction

With the term *hyper-reduction* we indicate techniques that aim at providing a computationally affordable computation of the reduced nonlinear terms $\tilde{\mathbf{f}}(\mathbf{q})$. In a FE context, $\tilde{\mathbf{f}}(\mathbf{q})$ requires assembly over the whole FE mesh, as

$$\tilde{\mathbf{f}}(\mathbf{q}) = \sum_{e=1}^{n_e} \mathbf{V}_e^T \mathbf{f}_e(\mathbf{V}_e \mathbf{q}), \tag{6.90}$$

and, as such, is computationally intensive, as it scales with the mesh size and not with the size of the reduction basis. As already seen, one possibility is to write $\tilde{\mathbf{f}}(\mathbf{q})$ directly in terms of $\mathbf{q}$ as

$$\tilde{\mathbf{f}}(\mathbf{q}) = \sum_{i=1}^{m} \sum_{j=1}^{m} B_r(i,j) q_i, q_j + \sum_{i=1}^{m} \sum_{j=1}^{m} \sum_{k=1}^{m} A_r(i,j,k) q_i, q_j, q_k, \tag{6.91}$$

where the coefficients $B_r(i,j)$ and $A_r(i,j,k)$ are identified by means of nonintrusive methods, or computed directly in case of polynomial nonlinearities. In more general cases, alternatives to this approach are available. In the following sections, we briefly present the Discrete Empirical Interpolation (DEIM) and some useful variants for the FE context, and the Energy Conserving Sampling and Weighting (ECSW) method. They differ in their philosophy, but they both aim at constructing $\tilde{\mathbf{f}}(\mathbf{q})$ cheaply and independently of the actual form of $\mathbf{f}$.

6.8.1 *Discrete Empirical Interpolation and Variants*

The main idea behind DEIM is to provide and approximation for the nonlinear forces $\mathbf{f}(\mathbf{u})$ as a combination of few vectors forming a base $\mathbf{F} = [\mathbf{f}_1, \ldots, \mathbf{f}_m] \in \mathbb{R}^{n \times m}$ spanning a subspace of dimension $m < n$, as

$$\mathbf{f} \approx \mathbf{F}\mathbf{c}, \tag{6.92}$$

where $\mathbf{c} \in \mathbb{R}^{m \times 1}$ the vector of unknown amplitudes of such force vectors. The system of Eqs. (6.92) cannot be solved, as it is overdetermined. In order to solve it, one option is to choose m distinct rows of $\mathbf{f}$. This can be done by premultiplying by the transpose of a boolean matrix $\mathbf{P} \in \mathbb{R}^{n \times m}$ as

$$\mathbf{P}^T \mathbf{f} = \mathbf{P}^T \mathbf{F}\mathbf{c}. \tag{6.93}$$

If $\mathbf{P}^T \mathbf{F}$ is non-singular, we can determine the amplitudes $\mathbf{c}$ by solving (6.93) as

$$\mathbf{c} = (\mathbf{P}^T\mathbf{F})^{-1}\mathbf{P}^T\mathbf{f}, \tag{6.94}$$

and therefore the approximate nonlinear force vector (6.92) becomes

$$\mathbf{f} \approx \mathbf{F}(\mathbf{P}^T\mathbf{F})^{-1}\mathbf{P}^T\mathbf{f}. \tag{6.95}$$

This is to say, the term $\mathbf{P}^T\mathbf{f}$ specifies that the nonlinear vector $\mathbf{f}$ needs to be calculated only at the entries specified by $\mathbf{P}^T$. The precomputed term $\mathbf{F}(\mathbf{P}^T\mathbf{F})^{-1}$ provides approximated values at the other entries not considered by $\mathbf{P}$. The projection basis $\mathbf{F}$ is computed from a POD of an appropriate set of snapshots of the nonlinear forces $\mathbf{F} = [\mathbf{f}(\mathbf{u}_1), \ldots, \mathbf{f}(\mathbf{u}_{n_s})]$. The selection points are determined from the greedy Algorithm 1. Details are given in Chaturantabut and Sorensen (2010).

Algorithm 1 DEIM points selection

Input: $\{\mathbf{u}_i\}_{i=1}^m \subset \mathbb{R}^n$ linearly independent
Output: $\boldsymbol{\wp} = [\wp_1, \ldots, \wp_m]^T \in \mathbb{R}^m$

$[|\rho|\ \wp_1] = \max |\mathbf{u}_1|$
$\mathbf{U} = [\mathbf{v}_1]$, $\mathbf{P} = [\mathbf{e}_{\phi 1}]$, $\boldsymbol{\wp} = [\wp_1]$
for $i = 2$ to m **do**
  Solve $(\mathbf{P}^T\mathbf{U})\mathbf{c} = \mathbf{P}^T\mathbf{u}_i$ for $\mathbf{c}$
  $\mathbf{r} = \mathbf{u}_i - \mathbf{U}\mathbf{c}$
  $[|\rho|\ \wp_i] = \max |\mathbf{r}|$
  $\mathbf{U} \leftarrow [\mathbf{U}\ \mathbf{u}_i]$, $\mathbf{P} \leftarrow [\mathbf{P}\ \mathbf{e}_{\wp_i}]$, $\boldsymbol{\wp} \leftarrow \begin{bmatrix} \boldsymbol{\wp} \\ \wp_i \end{bmatrix}$
end for

Inefficiency of DEIM Applied to FEM

The DEIM algorithm requires the evaluation of the nonlinear vector only at a few selected locations. This approach is particularly efficient when $\mathbf{f}(\mathbf{y})$ is scalar valued. By this, we mean that each component of the nonlinear force vector depends on the corresponding entry of the displacement vector: $f_i = f_i(y_i)$. In this case, the cost of the DEIM evaluation of $\mathbf{f}$ is the cost of the evaluation of m scalar-valued functions $f_{\wp_i}(\boldsymbol{\Phi}_{\wp_i}\mathbf{q})$. If this holds, the selection matrix $\mathbf{P}$ can be brought inside the function evaluation, as

$$\mathbf{P}^T\mathbf{f}(\boldsymbol{\Phi}\mathbf{q}) = \mathbf{f}(\mathbf{P}^T\boldsymbol{\Phi}\mathbf{q}). \tag{6.96}$$

Unfortunately, this does not hold when $\mathbf{f}$ comes from a FEM discretization. In a FE framework, each component f_i represents the generalized force component at one specific node. This entry depends on the generalized displacements of all the nodes belonging to the neighboring elements, i.e., a given degree of freedom is supported by a set of elements. As a consequence, $\mathbf{P}$ cannot be brought inside the function

$$\mathbf{P}^T\mathbf{f}(\boldsymbol{\Phi}\mathbf{q}) \neq \mathbf{f}(\mathbf{P}^T\boldsymbol{\Phi}\mathbf{q}). \tag{6.97}$$

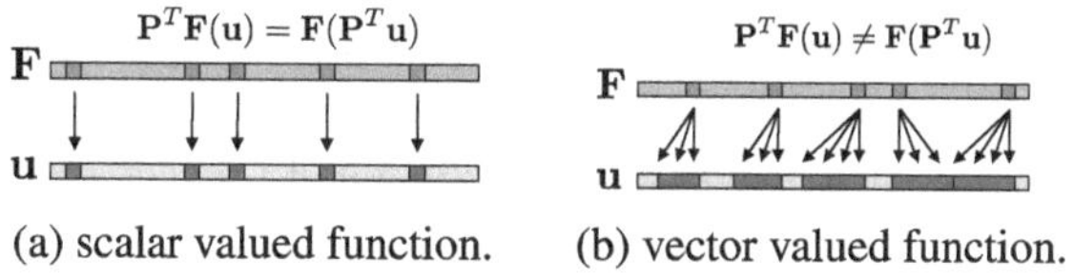

Fig. 6.7 Scalar-valued function (left) and vector valued function (right). In the first case, each component F_i of $\mathbf{F}$ depends on only one component u_j of $\mathbf{u}$. In the second case, F_i depends on several components of $\mathbf{u}$. This latter case depicts the FE discretization

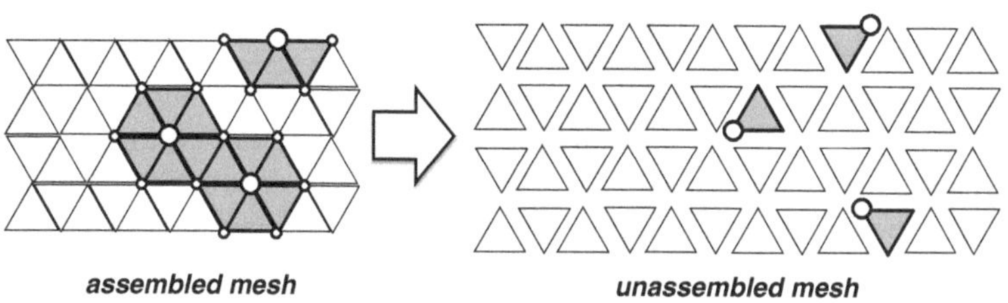

Fig. 6.8 Sparsity comparison for the evaluation of $\mathbf{f}$ at the DEIM points. In the case of assembled meshes (left), one evaluation implies an element function call for all the neighboring elements. If the mesh is treated as unassembled (left), only one element per DEIM point needs to be queried

This issue is illustrated in Fig. 6.7b and is responsible for the diminished inefficiency of DEIM when applied on FE schemes. In general, a generic component $f_{\wp_i}$ depends on a subset of the displacements $\bar{\mathbf{u}}_{\wp_i} \in \mathbb{R}^p$, where p depends on the type of element and the mesh connectivity. This implies that the evaluation of a single component f_i of the nonlinear force vector $\mathbf{f}$ requires a call to the element functions relative to all the neighboring elements.

Alternative DEIM Formulation

A possible workaround of the inefficiency discussed in the previous section is to apply DEIM on an unassembled force vector, i.e., on the stacking of all the elemental forces without performing element connection. In this way, the evaluation of a single entry of such vector will require just one element call. The concept is illustrated in Fig. 6.8. This modification is called Unassembled DEIM (UDEIM).
Unassembled DEIM

Essentially, one has to form a collection of snapshots $\mathbb{F}_u = \left[\mathbf{f}_u(\mathbf{y}_1), \ldots, \mathbf{f}_u(\mathbf{y}_{n_s})\right]$ of the unassembled forces obtained form a full model simulation. This can be formally written as

$$\mathbf{f}_u^T = \left[\mathbf{f}_{e_1}^T(\mathbf{y}_{e_1}), \ldots, \mathbf{f}_{e_{ne}}^T(\mathbf{y}_{e_{ne}})\right], \tag{6.98}$$

where $\mathbf{f}_{e_i}^T(\mathbf{y}_{e_i}) \in \mathbb{R}^{p\times 1}$ is the element force vector relative to the element i, $i = 1, \ldots, n_e$ and p is the number of elemental degrees of freedom. It follows from (6.98) that $\mathbf{f}_u^T \in \mathbb{R}^{n_u \times 1}$, where $n_u = pn_e$ is the size of the unassembled force snapshots. The vector $\mathbf{f}_u$ is projected onto a basis $\mathbf{U}_u$ obtained by a SVD of the unassembled force snapshots. The DEIM algorithm is then applied to $\mathbf{U}_u$ and the obtained indices will point to elemental degrees of freedom, instead of mesh ones. Once the amplitudes $\mathbf{c}_u$ have been obtained, an assembled approximation of $\mathbf{f}$ can be formed by

$$\mathbf{f} \approx \mathbf{U}_a(\mathbf{P}_u^T\mathbf{U}_u)^{-1}(\mathbf{P}_u^T\mathbf{f}_u), \tag{6.99}$$

where $\mathbf{U}_a$ is obtained by a FE assembly of $\mathbf{U}_u$ the element contribution. The details of the UDEIM algorithm are shown in Algorithm 2.

As indicated, each snapshot $\mathbf{f}_{u_i}$ contains $n_u > n$ components. Since the number of operations to obtain an SVD decomposition of a $r \times q$ matrix is proportional to r^2q, the computational cost to calculate the reduction basis for the nonlinear forces in the unassembled case can be significantly higher than the one related to the assembled forces snapshots of the original DEIM. To overcome this drawback, we propose the use of an element surrogate quantity, as discussed next.

Algorithm 2 UDEIM

Input: unassembled snapshots $\mathbb{F}_u = [\mathbf{f}_u^1, \ldots, \mathbf{f}_u^{n_s}] \in \mathbb{R}^{n_u \times n_s}$
1: $\text{SVD}(\mathbb{F}_u)$: $\mathbb{F}_u = \tilde{\mathbf{U}}_u\mathbf{\Sigma}_u\mathbf{V}_u^T$
2: select first m columns of $\tilde{\mathbf{U}}_u$: $\mathbf{U}_u = \tilde{\mathbf{U}}_u(:, 1 : m)$, with $m < n_s$
3: `DEIM`$(\mathbf{U}_u) \rightarrow \mathbf{P}_u$ (Algorithm 1)
4: ecompute $(\mathbf{P}_u^T\mathbf{U}_u)^{-1}$
5: assemble $\mathbf{U}_a \in \mathbb{R}^{n \times m}$: $\mathbf{U}_a :=$ `assemble`$(\mathbf{U}_u)$
6: $\mathbf{f} \approx \mathbf{U}_a(\mathbf{P}_u^T\mathbf{U}_u)^{-1}(\mathbf{P}_u^T\mathbf{f}_u)$

Surrogate Unassembled DEIM

The unassembled basis $\mathbf{U}_u$ contains the most relevant force modes where each element contribution is left unassembled. A DEIM selection on $\mathbf{U}_u$ (UDEIM) as outlined in Algorithm 1 then picks the most relevant elemental forces to form the interpolation. This way to proceed improves the sparsity of the force–displacement dependency but increases the effort of performing the SVD of the snapshots.

The main idea is that each elemental contribution can be represented by a surrogate quantity rather than the complete elemental force vector. If this holds, the size of the unassembled snapshot vectors can be reduced while preserving the advantage of operating on elements directly.

Let us assume that the contribution of element i to the nonlinear behavior can be represented by the sum $\mathbf{s}_i^e$ of all its generalized force components f_i^e, as

$$\mathbf{s}_i^e = \sum_{j=1}^{p} \mathbf{f}_j^e. \tag{6.100}$$

In case the generalized element forces do not share the same units (as in the shell element case here considered), the surrogate can be extended as

$$\mathbf{s}_i^e = \begin{bmatrix} \sum_{j=1}^{p_f} \mathbf{f}_j^e \\ \\ \sum_{j=1}^{p_m} \mathbf{m}_j^e \end{bmatrix}, \tag{6.101}$$

where m_i^e are the nodal elemental moments, and p_f and p_m are the number of element forces and moments, respectively. Then, a collection of surrogates snapshots $\mathbb{S} = \left[\mathbf{S}_1, \ldots, \mathbf{S}_{n_s}\right]$ can be built, where each vector $\mathbf{S}_i$ is given by

$$\mathbf{S}_i^T = \left[\mathbf{s}_1^e \ldots, \mathbf{s}_{n_e}^e\right], \tag{6.102}$$

and $\mathbb{S} \in \mathbb{R}^{s \times n_s}$, s being the size of the surrogate snapshots, and $s \ll n_u$. If the surrogate snapshots $\mathbf{S}_i$ are supposed to feature the same overall contribution as the corresponding unassembled snapshots $\mathbf{S}_u$, then we can assert that a SVD decomposition of $\mathbb{S}$ and $\mathbb{F}_u$ should yield very similar right singular vectors $\mathbf{V}$, at least for the dominant left singular vectors $\mathbf{U}_u$. This request can be written as

$$\begin{aligned} \text{SVD}(\mathbb{F}_u) : \ \mathbb{F}_u &= \mathbf{U}_u \boldsymbol{\Sigma}_u \mathbf{V}^T \\ \text{SVD}(\mathbb{S}) : \quad \mathbb{S} &= \mathbf{U}_{\mathbb{S}} \boldsymbol{\Sigma}_{\mathbb{S}} \mathbf{V}^T. \end{aligned} \tag{6.103}$$

It is useful to remind here that the right vectors $\mathbf{V}$ represent the time history of the corresponding left singular vectors $\mathbf{U}_u$. By claiming the decomposition (6.103) holds, we imply that the unassembled singular vectors $\mathbf{U}_u$ and the surrogates vectors $\mathbf{U}_{\mathbb{S}}$ share the same time behavior. Our numerical results indeed show that this indeed holds for the most significant modes. Since we are interested in only the spatial basis $\mathbf{U}_u$, the expensive decomposition SVD($\mathbb{F}_u$) can be avoided and the basis $\mathbf{U}_u$ can be formed by using the vectors $\mathbf{V}$ obtained by the less computationally expensive SVD($\mathbb{S}$), as

$$\begin{aligned} \mathbb{F}_u \mathbf{V} = \mathbf{U}_u \boldsymbol{\Sigma}_u \mathbf{V}^T \mathbf{V} = \mathbf{U}_u \boldsymbol{\Sigma}_u &= \bar{\mathbf{U}}_u, \\ \mathbf{U}_u &= \text{orth}(\bar{\mathbf{U}}_u), \end{aligned} \tag{6.104}$$

where $\mathbf{V}$ is calculated from the inexpensive SVD($\mathbb{S}$) and $\mathbf{U}_u = \text{orth}(\bar{\mathbf{U}}_u)$ is a Gram–Schmidt orthogonalisation such that $\mathbf{U}_u^T \mathbf{U}_u = \mathbf{I}$. In other words, if the surrogates singular vectors feature the same time dependency $\mathbf{V}$ as the unassembled singular vectors, we are able to compute $\mathbf{U}_u$ without performing the expensive $\mathbb{F}_u = \mathbf{U}_u \boldsymbol{\Sigma}_u \mathbf{V}^T$. Once $\mathbf{U}_u$ is found, DEIM can be applied in a UDEIM fashion, as discussed in the previous section. We call this variant of the procedure *Surrogate Unassembled* DEIM (SUDEIM). The details of the algorithm are outlined in Algorithm 3.

Algorithm 3 SUDEIM

Input: unassembled snapshots $\mathbb{F}_u = [\mathbf{F}_u^1, \ldots, \mathbf{F}_u^{n_s}] \in \mathbb{R}^{n_u \times n_s}$,
1: surrogate snapshots $\mathbb{S} = [\mathbf{S}^1, \ldots, \mathbf{S}^{n_s}] \in \mathbb{R}^{s \times n_s}$, $s < n$
2: SVD($\mathbb{S}$) : $\mathbb{S} = \tilde{\mathbf{U}}_{\mathbb{S}} \boldsymbol{\Sigma}_{\mathbb{S}} \mathbf{V}^T$
3: $\bar{\mathbf{U}}_u \approx \mathbb{F}_u \mathbf{V}$ and $\tilde{\mathbf{U}}_u = \text{orth}(\bar{\mathbf{U}}_u)$
4: select first m columns of $\tilde{\mathbf{U}}_u$: $\mathbf{U}_u = \tilde{\mathbf{U}}_u(:, 1 : m)$, with $m < n_s$
5: DEIM($\mathbf{U}_u$) $\rightarrow \mathbf{P}_u$ (Algorithm 1)
6: compute $(\mathbf{P}_u^T \mathbf{U}_u)^{-1}$
7: assemble $\mathbf{U} \in \mathbb{R}^{n \times m}$: $\mathbf{U} := \texttt{assemble}(\mathbf{U}_u)$
8: $\mathbf{F} \approx \mathbf{U}(\mathbf{P}_u^T \mathbf{U}_u)^{-1}(\mathbf{P}_u^T \mathbf{F}_u)$

Surrogate DEIM

It follows from the discussion above that the concept of the surrogate can be applied also to the assembled basis for the nonlinear forces **U**, as outlined in Algorithm 4. The computational saving lies in this case only in the cheaper SVD decomposition of the surrogate, while the sparsity of the force–displacement mapping is not improved. This variant is called *Surrogate* DEIM (SDEIM) (see Algorithm 4). Numerical results comparing these different variants of DEIM are presented and discussed in Tiso and Rixen (2013).

Algorithm 4 SDEIM

Input: snapshots $\mathbb{F} = [\mathbf{F}^1, \ldots, \mathbf{F}^{n_s}] \in \mathbb{R}^{n \times n_s}$,
1: surrogate snapshots $\mathbb{S} = [\mathbf{S}^1, \ldots, \mathbf{S}^{n_s}] \in \mathbb{R}^{s \times n_s}$, $s < n$
2: SVD($\mathbb{S}$) : $\mathbb{S} = \tilde{\mathbf{U}}_{\mathbb{S}} \boldsymbol{\Sigma}_{\mathbb{S}} \mathbf{V}^T$
3: $\bar{\mathbf{U}} \approx \mathbb{F}\mathbf{V}$ and $\tilde{\mathbf{U}} = \text{orth}(\bar{\mathbf{U}})$
4: select first m columns of $\tilde{\mathbf{U}}$: $\mathbf{U} = \tilde{\mathbf{U}}(:, 1 : m)$, with $m < n_s$
5: DEIM($\mathbf{U}$) $\rightarrow \mathbf{P}$ (Algorithm 1)
6: compute $(\mathbf{P}^T \mathbf{U})^{-1}$
7: $\mathbf{F} \approx \mathbf{U}(\mathbf{P}^T \mathbf{U})^{-1}(\mathbf{P}^T \mathbf{F})$

6.8.2 Energy Conserving Sampling and Weighting

The ECSW method (Farhat et al. 2014, 2015) is based on a different principle as compared to DEIM. Essentially, ECSW aims to identify a small set of elements E of the structure ($|E| \ll n_e$) to cheaply approximate $\tilde{\mathbf{f}}(\mathbf{q})$ as (cf. (6.45))

$$\tilde{\mathbf{f}}(\mathbf{q}) = \sum_{e=1}^{n_e} \mathbf{V}_e^T \mathbf{f}_e(\mathbf{V}_e \mathbf{q}) \approx \sum_{e \in E} \xi_e \mathbf{V}_e^T \mathbf{f}_e(\mathbf{V}_e \mathbf{q}), \tag{6.105}$$

where $\xi_e \in \mathbb{R}^+$ is a positive weight assigned to each element $e \in E$, which is empirically chosen to ensure a good approximation of the summation in (6.45). Clearly, if $|E| \ll n_e$, then the evaluation of the approximation in (6.105) is cheap when compared to the cost associated to to (6.45). In doing so, ECSW approximates the virtual work done by the internal forces on the set of vectors in the basis $\mathbf{V}$. As a consequence, the ECSW preserves the structure and the stability properties of the underlying full model (cf. Farhat et al. 2015).

The elements and weights are determined to approximate virtual work over the chosen training sets, which generally come from full solution run(s). If there are n_t training displacement vectors in the set with $\mathbf{u}^{(i)}$ representing the ith vector, then corresponding reduced unknowns $\mathbf{q}^{(i)}$ can be calculated using least squares as

$$\mathbf{q}^{(i)} = (\mathbf{V}^T\mathbf{V})^{-1}\mathbf{V}^T\mathbf{u}^{(i)}.$$

The element level contribution of projected internal force for each of the training vectors can be assembled in a matrix $\mathbf{G}$ as

$$\begin{gathered}\mathbf{G} = \begin{bmatrix} \mathbf{g}_{11} & \cdots & \mathbf{g}_{1n_e} \\ \vdots & \ddots & \vdots \\ \mathbf{g}_{n_t1} & \cdots & \mathbf{g}_{n_tn_e} \end{bmatrix} \in \mathbb{R}^{mn_t \times n_e}, \qquad \mathbf{b} = \begin{bmatrix} \mathbf{b}_1 \\ \vdots \\ \mathbf{b}_{n_t} \end{bmatrix} \in \mathbb{R}^{mn_t}, \\ \mathbf{g}_{ie} = \mathbf{V}_e^T\mathbf{f}_e(\mathbf{V}_e\mathbf{q}^{(i)}), \qquad \mathbf{b}_i = \tilde{\mathbf{f}}\left(\mathbf{q}^{(i)}\right) = \textstyle\sum_{e=1}^{n_e}\mathbf{g}_{ie}, \\ \forall i \in \{1, \ldots, n_t\},\ e \in \{1, \ldots, n_e\}. \end{gathered} \tag{6.106}$$

The set of elements and weights is then obtained by a sparse solution to the following Non-negative Least Squares (NNLS) problem

$$\boldsymbol{\xi} = \arg \min_{\tilde{\boldsymbol{\xi}} \in \mathbb{R}^{n_e}, \tilde{\boldsymbol{\xi}} \geq \mathbf{0}} \|\mathbf{G}\tilde{\boldsymbol{\xi}} - \mathbf{b}\|_2, \tag{6.107}$$

A sparse solution of (6.107) (i.e., a solution vector with lots of zeros) returns a sparse vector $\boldsymbol{\xi}$, the nonzero entries of which form the reduced mesh E used in (6.105) as

$$E = \{e : \xi_e > 0\}.$$

An optimally sparse solution of (6.107) is NP-hard to obtain. However, a greedy approach-based algorithm (Peharz and Pernkopf 2012), which finds a suboptimal solution, has been found to deliver an effective reduced mesh E (Farhat et al. 2014). For the sake of completeness, the procedure is listed here in Algorithm 5, where $\boldsymbol{\zeta}_E$ and $\mathbf{G}_E$ denote, respectively, the restriction of $\boldsymbol{\zeta} \in \mathbb{R}^{n_e}$ and columnwise restriction of $\mathbf{G}$ to the elements in the *active* subset E. The set Z is the disjoint *inactive* subset which contains the zero entry indices of $\boldsymbol{\xi}$ and $\boldsymbol{\zeta}$.

Algorithm 5 Sparse NNLS for ECSW

Input: $\mathbf{G}, \mathbf{b}, \tau$
Output: $\boldsymbol{\xi} \in \mathbb{R}^{n_e}$ sparse, $E \subset \{1, \ldots, n_e\}$

$E \leftarrow \emptyset$, $Z \leftarrow \{1, \ldots, n_e\}$, $\boldsymbol{\xi} \leftarrow \mathbf{0} \in \mathbb{R}^{n_e}$
while $\|\mathbf{G}\boldsymbol{\xi} - \mathbf{b}\|_2 > \tau\|\mathbf{b}\|_2$ **do**
  $\boldsymbol{\mu} \leftarrow \mathbf{G}^T(\mathbf{b} - \mathbf{G}\boldsymbol{\xi})$
  $[|\nu|, e] = \max(\boldsymbol{\mu})$ ▷ max: returns maximum value in a vector followed by its location (index)
  $E \leftarrow E \cup \{e\}$, $Z \leftarrow Z \backslash \{e\}$
  while true do
    $\boldsymbol{\zeta}_E \leftarrow \mathbf{G}_E^{\dagger}\mathbf{b}$ ▷ † represents pseudoinverse
    $\boldsymbol{\zeta}_Z \leftarrow \mathbf{0}$
    if $\boldsymbol{\zeta}_E > \mathbf{0}$ **then**
      $\boldsymbol{\xi} \leftarrow \boldsymbol{\zeta}$
      break
    end if
    $\eta = \min_{k \in E} \xi_k / (\xi_k - \zeta_k)$
    $\boldsymbol{\xi} \leftarrow \boldsymbol{\xi} + \eta(\boldsymbol{\zeta} - \boldsymbol{\xi})$
    $Z \leftarrow \{i | \xi_i = 0\}$
    $E \leftarrow \{1, \ldots, n_e\} \backslash Z$
  end while
end while

6.9 Hurty/Craig–Bampton Substructuring with ROMs and Interface Reduction

The material reviewed here comes primarily from Kuether et al. (2016) and Kuether et al. (2017). Several nonlinear substructuring methods have been developed to predict the dynamics of an assembly based on the dynamics of its subcomponents. For a review of the relevant literature, see Kuether et al. (2017).

This section focuses on geometric nonlinearities that are distributed throughout all the elements in the subcomponent FEA model. Recently, Wenneker (2013) used a basis defined by the Craig–Bampton and Rubin approach and augmented these with modal derivative vectors in order to account for the effects of geometric nonlinearity. The number of modal derivatives required scales quadratically with the number of component modes used, resulting in a rather large order reduced system. Perez (2012) was the first to explore the use of CB modes and a reduced set of constraint modes in conjunction with an indirect reduced-order modeling approach. He presented a thorough analysis of a complicated multi-bay frame reducing the linear model from 96,000 to 232 DOF that were a combination of fixed-interface modes and constraint modes reduced using proper orthogonal decomposition. Unfortunately, this model was still 2.6 times larger than an 89 mode ROM that he had created for the assembled system, so nonlinear modal substructuring was not pursued further.

The proposed nonlinear substructuring approach relies on a nonintrusive, reduced order modeling strategy that can generate a ROM from any geometrically nonlinear FEA model as described in Sect. 6.5. This ROM formulation is valid for beams, shells and solid structures whether flat, curved or interconnected and the components may be meshed with a variety of 1D, 2D, or 3D elements. However, the focus of this work and of most of the studies that use this type of ROM has been on flat and/or curved shell type structures where nonlinearity comes about as bending deformations induce in-plane stretching.

Starting with the linearized subcomponent modes as a basis, a low order set of nonlinear equations in parametric form are formulated in terms of the modal coordinates. For linear elastic finite element models with quadratic strain–displacement relations, the nonlinear modal restoring forces are a quadratic and cubic polynomial function of modal displacements. In Kuether et al. (2017), a Craig–Bampton nonlinear reduced order model (CB-NLROM) was used for each subcomponent using the fixed-interface and constraint modes, the nonlinearity was described by nonlinear stiffness parameters obtained using the Implicit Condensation and Expansion (ICE) approach discussed in Sect. 6.5.

One important application for this nonlinear modal substructuring techniques is as an efficient means of evaluating the effect of boundary conditions on geometric nonlinearity. Accurate modeling of in-plane forces becomes especially important during the analysis of coupled fluid–thermal–structural interactions as the in-plane resistance to thermal expansions can affect the onset of buckling, causing very different response behaviors to oscillating pressure loads. That stiffness is provided by a complicated airframe structure adjacent to the panel of interest, (a structure that may not even be fully designed yet when panel loads are considered), and it is usually simplified. For example, these boundaries are approximated as either pinned or fixed, or with discrete springs to approximately model the compliance of the adjacent airframe. The CB-NLROM procedure can be used to evaluate the effect of boundary conditions, i.e., on a structural modification problem where a geometrically nonlinear beam is modified by adding stiffness with an in-plane spring element.

Starting with the equations of motion for a conservative, N-DOF system discretized by the finite element method,

$$\mathbf{M}\ddot{\mathbf{x}} + \mathbf{K}\mathbf{x} + \mathbf{f}_{NL}(\mathbf{x}) = \mathbf{f}(t), \tag{6.108}$$

the $N \times 1$ nonlinear restoring force vector, $\mathbf{f}_{NL}(\mathbf{x})$, captures the interior forces introduced by the geometric nonlinearity and depends on the strain model used to model the large deformations. For the reduction procedure that follows, it is assumed that the FEA model is linear elastic with a quadratic strain–displacement relationship. The $N \times N$ linear mass and stiffness matrices, $\mathbf{M}$ and $\mathbf{K}$, respectively, are generated using linear finite element analysis, and the external loads are applied through the $N \times 1$ vector $\mathbf{f}(t)$.

As elaborated previously, a Hurty/Craig–Bampton model can be written as follows. Below, the nonlinear forces due to geometric nonlinearity are also included.

$$\begin{bmatrix} \mathbf{M}_{ii} & \mathbf{M}_{ib} \\ \mathbf{M}_{bi} & \mathbf{M}_{bb} \end{bmatrix} \begin{Bmatrix} \ddot{\mathbf{x}}_i \\ \ddot{\mathbf{x}}_b \end{Bmatrix} + \begin{bmatrix} \mathbf{K}_{ii} & \mathbf{K}_{ib} \\ \mathbf{K}_{bi} & \mathbf{K}_{bb} \end{bmatrix} \begin{Bmatrix} \mathbf{x}_i \\ \mathbf{x}_b \end{Bmatrix} + \begin{Bmatrix} \mathbf{f}_{NL,i}(\mathbf{x}) \\ \mathbf{f}_{NL,b}(\mathbf{x}) \end{Bmatrix} = \begin{Bmatrix} \mathbf{0} \\ \mathbf{f}(t) \end{Bmatrix}. \tag{6.109}$$

To reduce this partitioned system of equations, the Hurty/Craig–Bampton transformation matrix is used, which is repeated here for convenience.

$$\mathbf{x} = \begin{Bmatrix} \mathbf{x}_i \\ \mathbf{x}_b \end{Bmatrix} = \begin{bmatrix} \boldsymbol{\Phi}_{ik} & \hat{\boldsymbol{\Psi}}_{ib} \\ \mathbf{0} & \boldsymbol{I}_{bb} \end{bmatrix} \begin{Bmatrix} \mathbf{q}_k \\ \mathbf{x}_b \end{Bmatrix} = \mathbf{T}_{CB}\mathbf{q}. \tag{6.110}$$

The CB-NLROM of a subcomponent is defined by the low order set of equations,

$$\hat{\mathbf{M}}_{CB}\ddot{\mathbf{q}} + \hat{\mathbf{K}}_{CB}\mathbf{q} + \mathbf{T}_{CB}^{\mathrm{T}} \begin{Bmatrix} \mathbf{f}_{NL,i}\left(\mathbf{T}_{CB}^{\mathrm{T}}\mathbf{q}\right) \\ \mathbf{f}_{NL,b}\left(\mathbf{T}_{CB}^{\mathrm{T}}\mathbf{q}\right) \end{Bmatrix} = \mathbf{T}_{CB}^{\mathrm{T}} \begin{Bmatrix} \mathbf{0} \\ \mathbf{f}(t) \end{Bmatrix}, \tag{6.111}$$

where

$$\hat{\mathbf{M}}_{CB} = \mathbf{T}_{CB}^{\mathrm{T}} \begin{bmatrix} \mathbf{M}_{ii} & \mathbf{M}_{ib} \\ \mathbf{M}_{bi} & \mathbf{M}_{bb} \end{bmatrix} \mathbf{T}_{CB} = \begin{bmatrix} \mathbf{I}_{kk} & \hat{\mathbf{M}}_{kb} \\ \hat{\mathbf{M}}_{bk} & \hat{\mathbf{M}}_{bb} \end{bmatrix}, \tag{6.112}$$

$$\hat{\mathbf{K}}_{CB} = \mathbf{T}_{CB}^{\mathrm{T}} \begin{bmatrix} \mathbf{K}_{ii} & \mathbf{K}_{ib} \\ \mathbf{K}_{bi} & \mathbf{K}_{bb} \end{bmatrix} \mathbf{T}_{CB} = \begin{bmatrix} \boldsymbol{\Lambda}_{kk} & \mathbf{0}_{kb} \\ \mathbf{0}_{bk} & \hat{\mathbf{K}}_{bb} \end{bmatrix}. \tag{6.113}$$

The linear portion is exactly that of the CB reduced order model for any linear subsystem. Most commercial finite element packages are not able to directly extract the functional form of the nonlinear restoring force, $\mathbf{f}_{NL}(\mathbf{T}_{CB}^{T}\mathbf{q})$, in terms of the modal coordinates. However, because the geometric nonlinearity is known to be expressible in terms of quadratic and cubic polynomials for the common large displacement strain models, the same functional form will hold for the fixed-interface modes of the structure. As a result, the nonlinear term in Eq. (6.111) becomes

$$\mathbf{T}_{CB}^{\mathrm{T}} \begin{Bmatrix} \mathbf{f}_{NL,i}\left(\mathbf{T}_{CB}^{\mathrm{T}}\mathbf{q}\right) \\ \mathbf{f}_{NL,b}\left(\mathbf{T}_{CB}^{\mathrm{T}}\mathbf{q}\right) \end{Bmatrix} = \boldsymbol{\theta}(\mathbf{q}) \tag{6.114}$$

and each row of the nonlinear function $\theta(\mathbf{q})$ defines the nonlinear portion of the rth modal equation as

$$\theta_r(\mathbf{q}) = \sum_{i=1}^{m}\sum_{j=1}^{m} B_r(i,j)q_i, q_j + \sum_{i=1}^{m}\sum_{j=1}^{m}\sum_{k=1}^{m} A_r(i,j,k)q_i, q_j, q_k \tag{6.115}$$

The nonlinear stiffness coefficients can then be determined using the Implicit Condensation method as explained in Sect. 6.5. The resulting nonlinear model can also be expressed in the Nash form, which is convenient for what follows.

$$\mathbf{N}_1(\mathbf{q}) = \frac{\partial \boldsymbol{\beta}}{\partial \mathbf{q}} = \left[\frac{\partial \boldsymbol{\beta}}{\partial q_1}\ \frac{\partial \boldsymbol{\beta}}{\partial q_2} \cdots \frac{\partial \boldsymbol{\beta}}{\partial q_m}\right], \tag{6.116}$$

$$\mathbf{N}_2(\mathbf{q}) = \frac{\partial \alpha}{\partial q} = \left[\frac{\partial \alpha}{\partial q_1} \; \frac{\partial \alpha}{\partial q_2} \cdots \frac{\partial \alpha}{\partial q_m} \right]. \tag{6.117}$$

The CB-NLROM in Eq. (6.111) is then rewritten in matrix form with the nonlinear stiffness matrices in the equations above as

$$\hat{\mathbf{M}}_{CB}\ddot{\mathbf{q}} + \hat{\mathbf{K}}_{CB}\mathbf{q} + \frac{1}{2}\mathbf{N}_1(\mathbf{q})\,\mathbf{q} + \frac{1}{3}\mathbf{N}_2(\mathbf{q})\,\mathbf{q} = \mathbf{T}_{CB}^{\mathrm{T}} \begin{Bmatrix} \mathbf{0} \\ \mathbf{f}(t) \end{Bmatrix} \tag{6.118}$$

These ROMs can then be coupled using the methods outlined earlier, and their behavior depends on the adequacy of the HCB basis for the modes of interest as discussed earlier.

As an alternative to Implicit Condensation, one can also use modal derivatives with an enforced displacements procedure to create a ROM for a substructure. This is detailed in Chap. 3 of Wu (2018). Substructuring can also be extended to flexible multibody systems, as detailed in Chaps. 4 and 5 of Wu (2018). The content of Chap. 4 in Wu (2018) was published in Wu and Tiso (2016).

References

Arnoldi, W. E. (1951). The principle of minimized iterations in the solution of the matrix eigenvalue problem. *Quarterly of Applied Mathematics*, *9*, 17–29.

Chaturantabut, S., & Sorensen, D. (2010). Nonlinear model reduction via discrete empirical interpolation. *SIAM Journal on Scientific Computing*, *32*(5), 2737–2764. https://doi.org/10.1137/090766498.

Farhat, C., Avery, P., Chapman, T., & Cortial, J. (2014). Dimensional reduction of nonlinear finite element dynamic models with finite rotations and energy-based mesh sampling and weighting for computational efficiency. *International Journal for Numerical Methods in Engineering*, *98*(9), 625–662. https://doi.org/10.1002/nme.4668.

Farhat, C., Chapman, T., & Avery, P. (2015). Structure-preserving, stability, and accuracy properties of the energy-conserving sampling and weighting method for the hyper reduction of nonlinear finite element dynamic models. *International Journal for Numerical Methods in Engineering*, *102*, 1077–1110. https://doi.org/10.1002/nme.4820.

G'eradin, M., & Rixen, D. (1997). *Mechanical Vibrations: Theory and Application to Structural Dynamics* (2nd Ed.). Hoboken: Wiley. ISBN: 0-471-97524-9

Gordon, R.W., & Hollkamp, J. J. (2011). Reduced-order models for acoustic response prediction. AFRL-RB-WP-TR-2011-3040

Hill, T. L., Cammarano, A., Neild, S. N., & Wagg, D. (2014). An analytical method for the optimisation of weakly nonlinear systems. In *Proceedings of the 9th International Conference on Structural Dynamics-EURODYN*. Porto, Portugal.

Hill, T. L., Cammarano, A., Neild, S. N., & Wagg, D. (2015). Interpreting the forced response of a two-degree-of-freedom nonlinear oscillator using backbone curves. *Journal of Sound and Vibration*, *349*, 276–288.

Hollkamp J. J., G.R.W., Spottswood, S.M. (2005). Nonlinear modal models for sonic fatigue response prediction: a comparison of methods. *Journal of Sound and Vibration*, *284*, 1145–1163.

Idelsohn, S. R., & Cardona, A. (1985) A load-dependent basis for reduced nonlinear structural dynamics. *Computers & Structures*, *20*(1), 203 – 210. https://doi.org/10.1016/0045-7949(85)90069-0. Special Issue: Advances and Trends in Structures and Dynamics.

Idelsohn, S. R., & Cardona, A. (1985a). A load-dependent basis for reduced nonlinear structural dynamics. *Computers & Structures*, *20*, 203–210. https://doi.org/10.1016/0045-7949(85)90069-0.

Idelsohn, S. R., & Cardona, A. (1985b). A reduction method for nonlinear structural dynamic analysis. *Computer Methods in Applied Mechanics and Engineering*, *49*(3), 253–279. https://doi.org/10.1016/0045-7825(85)90125-2.

Kerschen, G., Peeters, M., Golinval, J. C., & Vakakis, A. F. (2009) Nonlinear normal modes. part i. a useful framework for the structural dynamicist. *Mechanical Systems and Signal Processing23*(1), 170–194. Copyright 2008, The Institution of Engineering and Technology 10330862 0888-3270 nonlinear normal modes NNMs large-scale engineering structures time-frequency analysis structural dynamics nonlinear vibrating structures wavelet transform modal analysis.

Kuether, R. J., Allen, M. S., & Hollkamp, J. J. (2016). Modal substructuring for geometrically nonlinear finite element models. *AIAA Journal*, *54*(2), 691–702. https://doi.org/10.2514/1.J054036.

Kuether, R. J., Allen, M. S., & Hollkamp, J. J. (2017). Modal substructuring of geometrically nonlinear finite element models with interface reduction. *AIAA Journal* (accepted).

Kuether, R. J., Renson, L., Detroux, T., Grappasonni, C., Kerschen, G., & Allen, M. S. (2015) Nonlinear normal modes, modal interactions and isolated resonance curves. *Journal of Sound and Vibration*, *351*(299-310). https://doi.org/10.1016/j.jsv.2015.04.035.

Mignolet M. P., Przekop, A., Rizzi, S. A., & Spottswood, S. M. (2013). A review of indirect/non-intrusive reduced order modeling of nonlinear geometric structures. *Journal of Sound and Vibration*, *332*(10), 2437–2460.

Muravyov, A. A., & Rizzi, S. A. (2003). Determination of nonlinear stiffness with application to random vibration of geometrically nonlinear structures. *Computers &Structures*, *81*(15), 1513–1523.

Nelson, R. B. (1976). Simplified calculation of eigenvector derivatives. *AIAA Journal*, *14*(9), 1201–1205. http://arc.aiaa.org/doi/abs/10.2514/3.7211?journalCode=aiaaj

Peharz, R., & Pernkopf, F. (2012). Sparse nonnegative matrix factorization with ℓ^0-constraints. *Neurocomputing*, *80*, 38–46.

Perez, R. A. (2012). Multiscale reduced order models for the geometrically nonlinear response of complex structures.

Rosenberg, R. M. (1960). Normal modes of nonlinear dual-mode systems. *Journal of Applied Mechanics*, *27*, 263–268.

Siddhi, V. (2005). *A generalized approach for calculation of the eigenvector ensitivity for various eigenvector normalizations*. Master's thesis, University of Missouri - Columbia. https://mospace.umsystem.edu/xmlui/bitstream/handle/10355/4244/research.pdf?sequence=3

Tiso, P., & Rixen, D. J. (2013). Discrete empirical interpolation method for finite element structural dynamics. In G. Kerschen, D. Adams, & A. Carrella (Eds.), *Topics in nonlinear dynamics* (Vol. 1, pp. 203–212). New York: Springer.

Touzé, C., Vidrascu, M., & Chapelle, D. (2014). Direct finite element computation of non-linear modal coupling coefficients for reduced-order shell models. *Computational Mechanics*, *54*(2), 567–580. https://doi.org/10.1007/s00466-014-1006-4.

Vakakis, A. F. (1997). Non-linear normal modes (nnms) and their applications in vibration theory: an overview. *Mechanical Systems and Signal Processing11*(1), 3–22. Compilation and indexing terms, Copyright 2009 Elsevier Inc. Compendex 1997163536841 08883270 Nonlinear normal mode Shock isolation designs Vibrational energy.

Weeger, O., Wever, U., & Simeon, B. (2016). On the use of modal derivatives for nonlinear model order reduction. *International Journal for Numerical Methods in Engineering*, pp. n/a–n/a. https://doi.org/10.1002/nme.5267.

Wenneker, F. (2013). Component mode synthesis for geometrically nonlinear structures.

Wilson, E. L., Yuan, M. W. & Dickens, J. M. (1982). Dynamic analysis by direct superposition of Ritz vectors. *Earthquake Engineering & Structural Dynamics10*(6), 813–821. https://doi.org/10.1002/eqe.4290100606.

Witteveen, W., & Pichler, F. (2014). Efficient model order reduction for the nonlinear dynamics of jointed structures by the use of trial vector derivatives. In *IMAC-XXXII: International Modal Analysis Conference, Orlando, FL*. Bethel, CT: Society for Experimental Mechanics.

Woinowski-Krieger, S. (1950). The effect of axial force on the vibration of hinged bars. *Journal of Applied Mechanics, 17*.

Wu, L. (2018). *Model order reduction and substructuring methods for nonlinear structural dynamics*. Dissertation, Delft University of Technology.

Wu, L., & Tiso, P. (2016). Nonlinear model order reduction for flexible multibody dynamics: a modal derivatives approach. *Multibody System Dynamics*, *36*(4), 405–425. https://doi.org/10.1007/s11044-015-9476-5.

Chapter 7
Weakly Nonlinear Systems: Modeling and Experimental Methods

Abstract The prior chapter presented rigorous theory and methods for nonlinear systems, which is necessary in general because many nonlinear systems exhibit strong modal coupling due to the nonlinearity; this is commonly the case for the geometrically nonlinear structures that were the focus of that chapter. However, one of the most common sources of nonlinearity in built-up structures is the joints, and in many cases, these introduce only a weak stiffness nonlinearity together with a significant damping nonlinearity. In this case, and in many others that are relevant to industry, one can obtain good estimates of the response of the structure using a weakly nonlinear model in which the linear modes of the structure are presumed to be preserved and coupling between modes is neglected. This chapter provides a brief introduction to these concepts.—*Chapter Authors: Randall Mayes and Matt Allen*

7.1 Modal Models for Weakly Nonlinear Substructures and Application to Bolted Interfaces

Much of the uncertainty in finite element models for built-up structures comes from the interfaces (e.g., bolted joints, riveted connections, press-fits). They often cause the structure to behave nonlinearly, making experiments more challenging, and to model them accurately requires a nonlinear model that can be dramatically more expensive. See Brake (2018), Bhushan (2013), Ferri (1995) for a review of these issues. This work focuses on methods by which joints are represented in a simplified way, so that the structure's response can be simulated quickly and the model updated to correlate with experimental measurements.

In most industrial applications, the bolts are designed to retain integrity, so most of the joints nonlinearity is attributed to micro-slip friction in which the joint remains intact but there is partial slipping for some material near the edges of the contact (Groper 1985; Gaul and Lenz 1997). Commercial finite element codes typically include the ability to simulate the static and dynamic response of flexible structures with Coulomb friction between contacts. For example, Abaqus (2014) includes the capability to model friction using Penalty and Lagrange methods. However, once the contact interfaces are meshed with sufficient resolution to capture micro-slip and the

M. S. Allen et al., *Substructuring in Engineering Dynamics*,
CISM International Centre for Mechanical Sciences 594,
https://doi.org/10.1007/978-3-030-25532-9_7

penalty stiffness and stabilization are set to acceptable levels, the simulations become very expensive for static response (Jewell et al. 2017), and completely prohibitive when a dynamic response is desired that spans many cycles of oscillation. To seek to address this, harmonic balance methods have been developed that computes the frequency response directly (Petrov and Ewins 2003; Bograd et al. 2011). In order to model micro-slip behavior accurately, such models typically require hundreds of friction elements per interface (Di Maio et al. 2016). The resulting simulations are still quite expensive and may be impractical if the model contains many joints. Furthermore, when updating a model to correlate with measurements, one must typically run many simulations while varying the parameters for the joint, a daunting prospect with current models.

A recent work by Festjens et al. (2013) presented an alternative to traditional time and frequency domain methods. They computed the quasi-static response of the structure to a loading representative of a certain mode, and then used the results of that simulation to compute the effective natural frequency and damping for the structure when it vibrates in that mode. Their predictions were validated by comparing them to the transient response computed by integrating a full-order nonlinear finite element model, which included the preload in the fastener and the resulting Coulomb friction between the contacting surfaces. Their methodology was found to provide good estimates of the effective damping and natural frequency, however the costs are still quite high if many iterations on the model must be run, i.e., for model updating applications.

One way of reducing the computational burden associated with joints is to replace the joint region with a simpler element that captures its effective stiffness and damping. Then, instead of enforcing frictional contact between individual nodes or element faces, those degrees of freedom can instead be tied (e.g., by multi-point constraints) to two ends of a single nonlinear element whose constitutive formulation is designed to reproduce micro-slip behavior internally. Such a modeling approach is the premise of Segalman's "whole-joint" model (Segalman 2006). The nonlinear element may be a hysteretic model such an Iwan (1966), Segalman (2005) model or a Bouc–Wen (Ismail et al. 2009) or LuGre model. This is appealing because joints are often only one small part of a complicated, assembled structure, and various experimental studies have shown that the presence of one or more bolted joints induces only a slight softening in the natural frequencies, accompanied with a large increase in the damping as the structure is excited to higher response amplitudes (Di Maio et al. 2016; Smallwood et al. 2000; Hartwigsen et al. 2004; Abad et al. 2012; Dossogne et al. 2017). Unfortunately, no methods exist in literature that can predict the parameters of these hysteresis elements based on joint geometry, surface roughness, material, etc. So, one must use a model updating approach to determine the parameters. For example, an experiment may be performed in which the transient-free response is measured and then the instantaneous natural frequency and damping for each mode are extracted from those measurements (Song et al. 2004; Deaner et al. 2015; Roettgen et al. 2017; Lacayo et al. 2017). Then, model updating can then be used to find the joint parameters that bring the model into an agreement with the measurements as was done in Lacayo and Allen (2018).

7.2 Test Methods for Identifying Nonlinear Parameters in Pseudo-Modal Models

The nonlinear modal models described in the previous section are shown schematically in Fig. 7.1, where the oscillator representing each mode contains nonlinear elements in parallel to the linear spring and damper for each mode in order to produce the desired deviations from linear behavior. This method assumes that stiffness is almost constant, but damping can be quite nonlinear. Modes are assumed not to interact and mode shapes are assumed to not change. The Complex Mode Indicator function plot in Fig. 7.2 shows the difference between a traditional linear modal model response taken from low input force modal data and response of the system from a higher level input. One can see that the frequencies appear to drop slightly but the damping changes drastically at the higher amplitude response. With this approach applied to such data, a modal filter can be utilized to effectively separate each modal response (Mayes et al. 2016; Roettgen et al. 2017).

Modal Filter

A modal filter is a spatial filter that uses the mode shapes to pull out the response of a single mode and rejects the responses of all other modes. By using such a filter, one can analyze the response associated with that single shape and determine how nonlinear it is or assume a functional form for the frequency and damping and fit parameters to identify the model. It should be emphasized that if the response of other modes seeps through the filter, the estimates of stiffness or damping as a function of amplitude can be polluted. As a result, the modal filter must be very effective in order to reliably quantify the nonlinearity associated with the mode shape of interest.

The standard modal filter is a vector of constants, here denoted as $\boldsymbol{\Psi}$, multiplied by a vector of actual responses, $\mathbf{x}$, which produces a single modal response q for mode i, as

$$\boldsymbol{\Psi}^T\mathbf{x} = q_i. \tag{7.1}$$

The traditional full modal filter may be derived from the standard modal substitution as

$$\mathbf{x} = \boldsymbol{\Psi}\mathbf{q}. \tag{7.2}$$

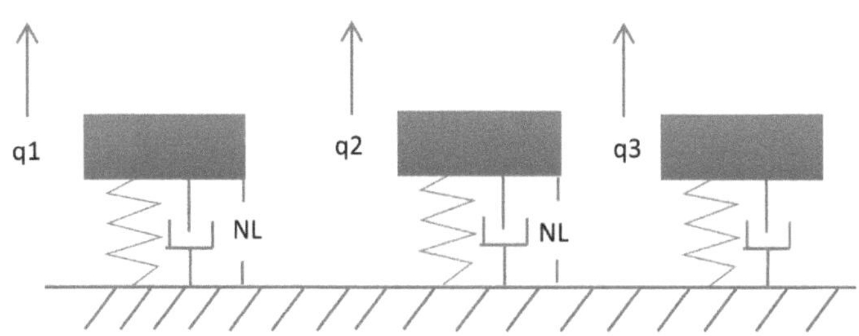

Fig. 7.1 Nonlinear pseudo-modal model with 3 DOF

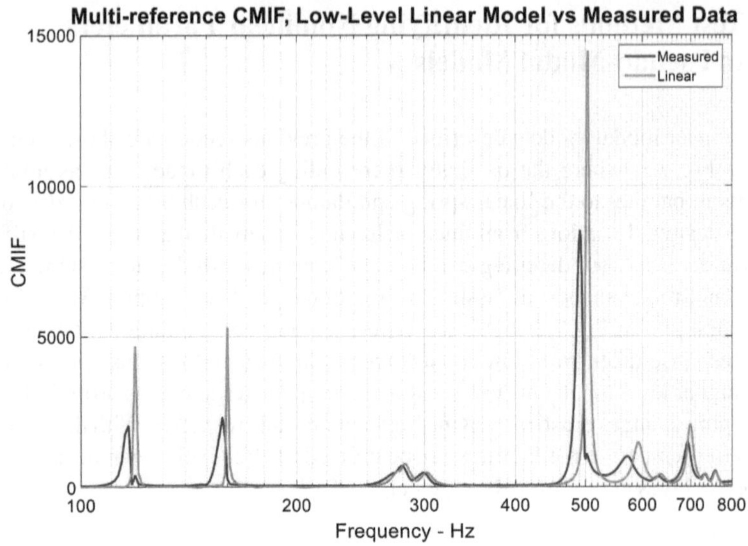

Fig. 7.2 CMIF of linear model versus measured data

Assuming that one has more responses than modes and the responses are appropriately place one can use

$$\mathbf{q} = \boldsymbol{\Psi}^{+}\mathbf{x}, \tag{7.3}$$

where each row of the pseudoinverse of $\boldsymbol{\Psi}$ can be used as a modal filter for the associated mode. A variation on this is to use only a single mode of phi, but it is not as effective as the full modal filter. The SMAC modal filter (Mayes and Johansen 1998) assumes that one has found the best linear estimate of frequency and damping for a mode in a standard modal test. In this case, one can analytically write the standard real modes FRF using the associated frequency, damping and drive point mode shape, and use the measured FRFs from the test in the following equation:

$$\boldsymbol{\Psi}^{T}\mathbf{H}_{x} = \mathbf{H}_{qi}, \tag{7.4}$$

where the only unknown is the vector $\boldsymbol{\Psi}$. This is solved in the frequency domain and transposed so that the frequency lines are associated with the rows with

$$\boldsymbol{\Psi} = \mathbf{H}_{x}^{T+}\mathbf{H}_{qi}^{T}. \tag{7.5}$$

In our experience, neither of the approaches above always performs better than the other, however, usually the SMAC modal filter is more robust. A typical set of filtered data for an attempt to extract a modal response in the mode above 550 Hz for all three methods is shown in Fig. 7.3.

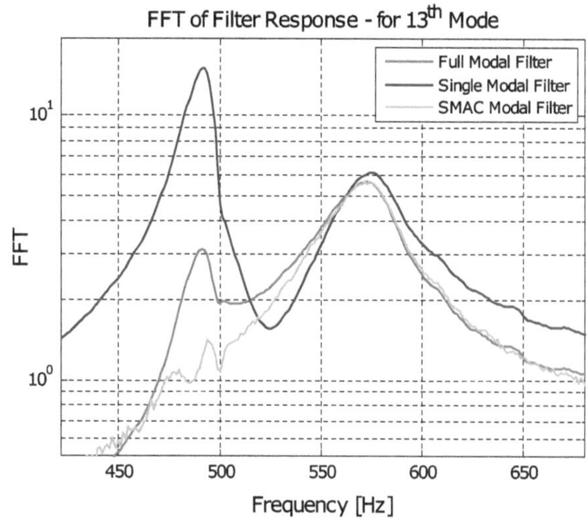

Fig. 7.3 Full, single and SMAC modal filters applied to obtain the mode above 550 Hz

Test Methods

The most attractive response for this approach is a modal filtered response from an impact. Immediately after the impact, there are no external forces acting, so any changes in the frequency or damping as a function of amplitude can be observed in a straightforward analysis of the free decay time response. A version of the Restoring Force Surface method can be used to fit the parameters in the assumed nonlinear form. Below is the equation that is utilized from this method for an assumed form of quadratic polynomials to fit the frequency and damping as a function of amplitude. In this case, the entire right-hand side is known from the modal filtered data and the frequency and damping extracted from a low level modal test. The functions of q on the left-hand side are formed from the modal filtered q. This leaves only the quadratic and cubic constants of the damping and stiffness polynomial to be extracted from a least squares problem. The rows can be either time response, or response in the frequency domain.

$$\begin{bmatrix} |\dot{q}|\,\dot{q} & \dot{q}^3 & |q|\,q & q^3 \end{bmatrix} \begin{bmatrix} c_1 \\ c_2 \\ k_1 \\ k_2 \end{bmatrix} = f - \ddot{q} - c_0\dot{q} - k_0 q.$$

Impact Methods

As stated above, the free decay from an impact is very attractive from an analysis point of view. This makes and impact test desirable. The first one does the low- level modal test with impact to get the linear estimates of frequency and damping for the right-hand side above. Then one simply does a high-level impact and extracts the

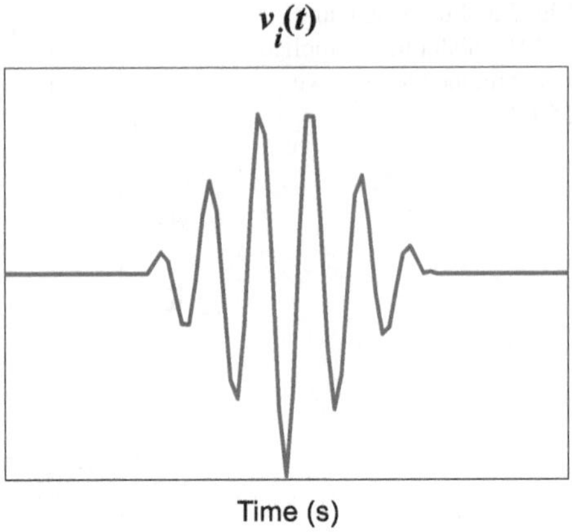

Fig. 7.4 Windowed sinusoid time domain

modal response q with a modal filter to obtain the response for other amplitudes. This is very easy to implement if the modal test has already been set up.

A disadvantage of the impact method has been observed in that since all modes active in the drive point are excited at once, the drive point accelerometer can overload at some impact levels, which provides the limit to how much the modes can be excited. Also, because many modes are excited, the modal filter is taxed to the greatest extent. A tailored shaker input can overcome these disadvantages and will be discussed in the next section.

Shaker Windowed Sinusoid

A way to overcome the disadvantages of the impact method is to use a shaker with an input tailored to the local frequency band of a single mode. This has two benefits. One is that there are many modes that are not excited at all, so the modal filtering is easier in the sense that many fewer modes must be filtered out. The second is that all the energy of the shaker and amplifier can be poured into a single mode to drive it to higher amplitudes so that the nonlinear parameters are calibrated for the higher amplitudes. A windowed sinusoid signal has been developed (Pacini et al. 2017), which allows easy tailoring of the input to excite the mode of interest. The sinusoid frequency is set to that of the resonant frequency of the mode. Then a triangular window is multiplied by the sinusoid signal as seen in Fig. 7.4. A narrow triangular window provides a broad input around the resonant frequency in the frequency domain. A wide triangular window provides a narrow input around the resonant frequency in the frequency domain. The frequency domain plot is shown in Fig. 7.5 (Other windows are available as well.).

The signal is fed into the amplifier and the force is measured. A frequency response function between the force and input voltage can be calculated. If the force drops out too much (a common problem in shaker testing), then the FRF can be used to calculate a voltage input that will fill in the dropout of the force at resonance with a force correction. The corrected force is plotted in red in Fig. 7.6 with the desired force in green and the initial force from the windowed sinusoid in blue. A time domain

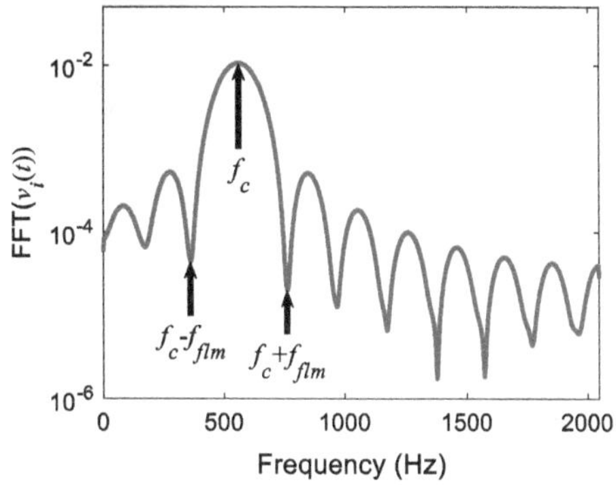

Fig. 7.5 Windowed sinusoid frequency domain

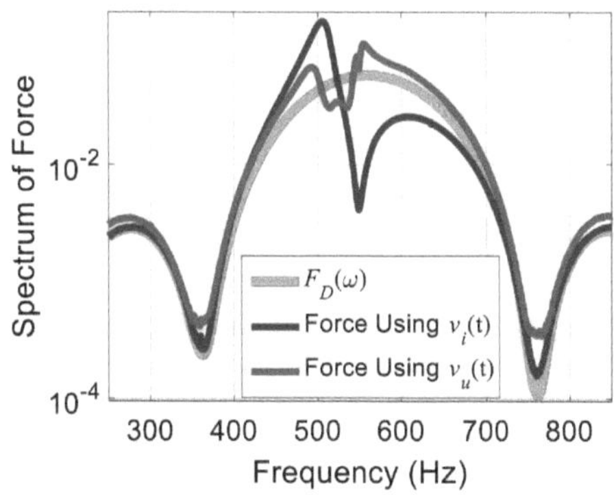

Fig. 7.6 Correcting the force to emulate the desired windowed sinusoid

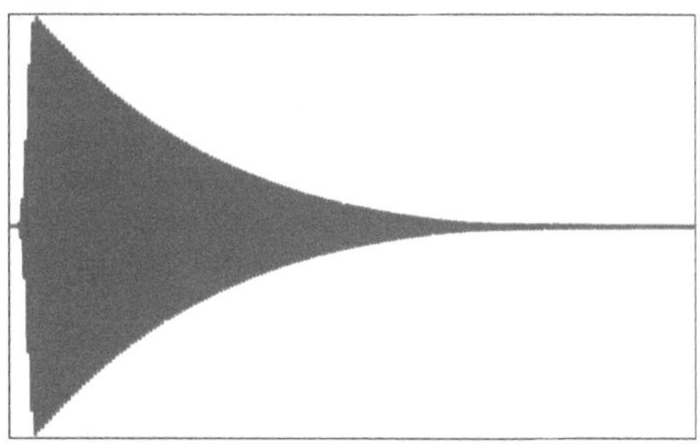

Fig. 7.7 Time response after modal filter applied

modal filtered response is shown from a shaker input in Fig. 7.7. Figure 7.8 shows hardware which was tested for nonlinearities of the type described here.

After the high-level burst signal is applied, the data are reduced to one q through the modal filter. The frequency domain RFS nonlinear parameter fitting process described earlier can then be applied. Experience has shown that if the nonlinearity excited by this method becomes severe enough to create impact and rattle, the method is not adequate for such cases. In other cases, it has worked well.

Fig. 7.8 Hardware demonstrating weak nonlinearity due to bolted interfaces

References

Abad, J., Franco, J. M., Celorrio, R., & Lezáun, L. (2012). Design of experiments and energy dissipation analysis for a contact mechanics 3D model of frictional bolted lap joints. *Advances in Engineering Software*, *45*(1), 42–53. https://doi.org/10.1016/j.advengsoft.2011.09.021.

Abaqus analysis user's guide, User Manual, Simulia (2014).

Bhushan, B. (2013). *Concepts and applications of tribology* (2nd ed.). Hoboken: Wiley.

Bograd, S., Reuss, P., Schmidt, A., Gaul, L., & Mayer, M. (2011). Modeling the dynamics of mechanical joints. *Mechanical Systems and Signal Processing*, *25*(8), 2801–2826. https://doi.org/10.1016/j.ymssp.2011.01.010.

Brake, M. R. (2018). *The mechanics of jointed structures: Recent research and open challenges for developing predictive models for structural dynamics* (1st ed.). Berlin: Springer.

Deaner, B. J., Allen, M. S., Starr, M. J., Segalman, D. J., & Sumali, H. (2015). Application of viscous and Iwan modal damping models to experimental measurements from bolted structures. *Journal of Vibration and Acoustics*, *137*(2), 021,012. https://doi.org/10.1115/1.4029074.

Di Maio, D., Schwingshackl, C., & Sever, I. A. (2016). Development of a test planning methodology for performing experimental model validation of bolted flanges. *Nonlinear Dynamics*, *83*, 983–1002. https://doi.org/10.1007/s11071-015-2382-9.

Dossogne, T., Jerome, T. W., Lancereau, D. P. T., Smith, S. A., Brake, M. R. W., Pacini, B., et al. (2017). Experimental assessment of the influence of interface geometries on structural dynamic response. In M. Allen, R. Mayes, & D. Rixen (Eds.), *Dynamics of coupled structures* (Vol. 4, pp. 255–261)., Conference Proceedings of the Society of Experimental Mechanics Series Berlin: Springer. https://doi.org/10.1007/978-3-319-54930-9_22.

Ferri, A. A. (1995). Friction damping and isolation systems. *Journal of Mechanical Design*, *117*, 196–206. https://doi.org/10.1115/1.2836456.

Festjens, H., Chevallier, G., & Dion, J. L. (2013). A numerical tool for the design of assembled structures under dynamic loads. *International Journal of Mechanical Sciences*, *75*, 170–177. https://doi.org/10.1016j.ijmecsci.2013.06.013.

Gaul, L., & Lenz, J. (1997). Nonlinear dynamics of structures assembled by bolted joints. *Acta Mechanica*, *125*(1), 169–181. https://doi.org/10.1007/BF01177306.

Groper, M. (1985). Microslip and macroslip in bolted joints. *Experimental Mechanics*, *25*(2), 171–174. https://doi.org/10.1007/BF02328808.

Hartwigsen, C. J., Song, Y., McFarland, D. M., Bergman, L. A., & Vakakis, A. F. (2004). Experimental study of non-linear effects in a typical shear lap joint configuration. *Journal of Sound and Vibration*, *277*(1–2), 327–351. https://doi.org/10.1016/j.jsv.2003.09.018.

Ismail, M., Ikhouane, F., & Rodellar, J. (2009). The hysteresis Bouc-Wen model, a survey. *Archives of Computational Methods in Engineering*, *16*(2), 161–188. https://doi.org/10.1007/s11831-009-9031-8.

Iwan, W. D. (1966). A distributed-element model for hysteresis and its steady-state dynamic response. *Journal of Applied Mechanics*, *33*(4), 893–900. https://doi.org/10.1115/1.3625199.

Jewell, E., Allen, M. S., Lacayo, R. (2017). Predicting damping of a cantilever beam with a bolted joint using quasi-static modal analysis. In *Proceedings of the ASME 2017 International Design Engineering Technical Conference IDETC 2017*, Cleveland, OH, USA.

Lacayo, R. M., & Allen, M. S. (2018) Updating structural models containing nonlinear iwan joints using quasi-static modal analysis. *Mechanical systems and signal processing*. Submitted 2017.

Lacayo, R. M., Deaner, B. J., & Allen, M. S. (2017). A numerical study on the limitations of modal Iwan models for impulsive excitations. *Journal of Sound and Vibration*, *390*, 118–140. https://doi.org/10.1016/j.jsv.2016.11.038.

Mayes, R. L., Johansen, D. D. (1998). A modal parameter extraction algorithm using best-fit reciprocal vectors. Requires guessing initial frequencies, mode vectors and damping ratios and using optimization to improve the result. Sounds pretty weak, although it allows one to play around with the result. Check future papers. Interesting links to papers on "Modal Filters".

Mayes, R. L., Pacini, B. R., & Roettgen, D. R. (2016). A modal model to simulate typical structural dynamic nonlinearity. *Dynamics of coupled structures* (pp. 57–76). Berlin: Springer.

Pacini, B. R., Mayes, R. L., Owens, B. C., & Schultz, R. (2017). Nonlinear finite element model updating, part i: Experimental techniques and nonlinear modal model parameter extraction. *Dynamics of coupled structures* (pp. 263–274). Berlin: Springer.

Petrov, E. P., & Ewins, D. J. (2003). Analytical formulation of friction interface elements for analysis of nonlinear multi-harmonic vibrations of bladed disks. *Journal of Turbomachinery*, *125*(2), 364–371. https://doi.org/10.1115/1.1539868.

Roettgen, D. R., & Allen, M. S. (2017). Nonlinear characterization of a bolted, industrial structure using a modal framework. *Mechanical systems and signal processing* (Vol. 84, Part B, pp. 152–170). https://doi.org/10.1016/j.ymssp.2015.11.010.

Roettgen, D. R., Allen, M. S., Kammer, D. C., & Mayes, R. L. (2017). Substructuring of a nonlinear beam using a modal iwan framework: Part i: Nonlinear modal model identification. *Dynamics of coupled structures* (pp. 165–178). Berlin: Springer.

Segalman, D. J. (2005). A four-parameter Iwan model for lap-type joints. *Journal of Applied Mechanics*, *72*(5), 752–760. https://doi.org/10.1115/1.1989354.

Segalman, D. J. (2006). Modelling joint friction in structural dynamics. *Structural Control and Health Monitoring*, *13*(1), 430–453. https://doi.org/10.1002/stc.119.

Smallwood, D. O., Gregory, D. L., Coleman, R. G. (2000). Damping investigations of a simplified frictional shear joint. In *71st Shock and Vibration Symposium*, Arlington, Virginia.

Song, Y., Hartwigsen, C., McFarland, D., Vakakis, A., & Bergman, L. (2004). Simulation of dynamics of beam structures with bolted joints using adjusted Iwan beam elements. *Journal of Sound and Vibration*, *273*, 249–276. https://doi.org/10.1016/S0022-460X(03)00499-1.

